Hydrological and Limnological Aspects of Lake Monitoring

Water Quality Measurements Series

Series Editor
Philippe Quevauviller,
European Commission, Brussels, Belgium

Forthcoming Titles in the Water Quality Measurements Series

**Detection Methods for Algae, Protozoa and Helminths
in Fresh and Drinking Water**
Edited by André Van der Beken, Giuliano Ziglio and
Franca Palumbo

Hydrological and Limnological Aspects of Lake Monitoring

EDITED BY

PERTTI HEINONEN
Finnish Environment Institute, Helsinki, Finland

GIULIANO ZIGLIO
Universita di Trento, Italy

ANDRÉ VAN DER BEKEN
Techware, Brussels, Belgium

JOHN WILEY & SONS, LTD
Chichester • New York • Weinheim • Brisbane • Singapore • Toronto

Other Wiley Editorial Offices

John Wiley & Sons, Inc., 605 Third Avenue,
New York, NY 10158-0012, USA

WILEY-VCH Verlag GmbH, Pappelallee 3,
D-69469 Weinheim, Germany

Jacaranda Wiley Ltd, 33 Park Road, Milton,
Queensland 4064, Australia

John Wiley & Sons (Asia) Pte Ltd, 2 Clementi Loop #02-01,
Jin Xing Distripark, Singapore 129809

John Wiley & Sons (Canada) Ltd, 22 Worcester Road,
Rexdale, Ontario M9W 1L1, Canada

British Library Cataloguing in Publication Data

A catalogue record for this book is available from the British Library

ISBN 0-471-89988-7

Typeset in Great Britain by Dobbie Typesetting Ltd, Tavistock, Devon
Printed and bound in Great Britain by Antony Rowe Ltd, Chippenham, Wiltshire
This book is printed on acid-free paper responsibly manufactured from sustainable forestation
for which at least two trees are planted for each one used for paper production.

Contents

List of Contributors

Jon Lasse Bratli, Norwegian Pollution Control Authority, PO Box 8100, N-0032 Oslo, Norway.

Ana-Cristina Cardoso, Environment Institute, European Commission Joint Research Centre, I-21020 Ispra, Italy. Email *ana-cristina.cardosoa@ei.jrc.it*

Pertti Eloranta, Department of Limnology and Environmental Protection, University of Helsinki, PO Box 27, FIN-00014 Helsinki, Finland. Email *pertti.eloranta@helsinki.fi*

Tom Frisk, Häme Regional Environment Centre, PO Box 297, FIN-33101 Tampere, Finland. Email *tom.frisk@vyh.fi*

Pertti Heinonen, Finnish Environment Institute, PO Box 140, FIN-00251 Helsinki, Finland. Email *pertti.heinonen@vyh.fi*

Sirpa Herve, Central Finland Regional Environment Centre, PO Box 110, FIN-40101 Jyväskylä, Finland. Email *sirpa.herve@vyh.fi*

Veli Hyvärinen, Finnish Environment Institute, PO Box 140, FIN-00251 Helsinki, Finland. Email *veli.hyvarinen@vyh.fi*

Kari Kallio, Finnish Environment Institute, PO Box 140, FIN-00251 Helsinki, Finland. Email *kari.y.kallio@vyh.fi*

Esa Koskenniemi, West Finland Regional Environment Centre, PO Box 262, FIN-65101 Vaasa, Finland. Email *esa.koskenniemi@vyh.fi*

Jussi Kukkonen, Department of Biology, University of Joensuu, PO Box 111, FIN-80101 Joensuu, Finland. Email *jussi.kukkonen@joensuu.fi*

Esko Kuusisto, Finnish Environment Institute, PO Box 140, FIN-00251 Helsinki, Finland. Email *esko.kuusisto@vyh.fi*

Tim Lack, European Topic Centre on Inland Waters (ETC/IW), Medmenham Laboratory, PO Box 16, Buckinghamshire SL7 2HD, United Kingdom. Email *lack@wrcplc.co.uk*

Kirsti Lahti, Finnish Environment Institute, PO Box 140, FIN-00251 Helsinki, Finland. Email *kirsti.lahti@vyh.fi*

Hannu Lehtonen, Department of Limnology and Environment Protection, University of Helsinki, PO Box 27, FIN-00014 Helsinki, Finland. Email *hannu.lehtonen@helsinki.fi*

Ahti Lepistö, Finnish Environment Institute, PO Box 140, FIN-00251 Helsinki, Finland. Email *ahti.lepisto@vyh.fi*

Jaakko Mannio, Finnish Environment Institute, PO Box 140, FIN-00251 Helsinki, Finland. Email *jaakko.mannio@vyh.fi*

Pentti Minkkinen, Chemical Technology, Lappeenranta University of Technology, PO Box 20, FIN-53851 Lappeenranta, Finland. Email *pentti.minkkinen@lut.fi*

Seppo Niemelä, Department of Applied Chemistry and Microbiology, University of Helsinki, PO Box 56, FIN-00014 Helsinki, Finland. Email *seppo.niemela@helsinki.fi*

Jorma S. Niemi, Finnish Environment Institute, PO Box 140, FIN-00251 Helsinki, Finland. Email *jorma.niemi@vyh.fi*

R. Maarit Niemi, Finnish Environment Institute, PO Box 140, FIN-00251 Helsinki, Finland. Email *maarit.niemi@vyh.fi*

Steve Nixon, European Topic Centre on Inland Waters (ITC/IW), Medmenham Laboratory, PO Box 16, Buckinghamshire SL7 2HD, United Kingdom. Email *snixon@etc.iw.eionet.eu.int*

Jaakko Paasivirta, Department of Chemistry, University of Jyväskylä, PO Box 35, FIN-40351 Jyväskylä, Finland. Email *jpta@cc.jyu.fi*

Guido Premazzi, Environment Institute, European Commission Joint Research Centre, I-21020 Ispra, Italy. Email *guido.premazzi@ei.jrc.it*

Philippe Quevauviller, European Commission, DG XII-C5 (MO73 3/09), Rue de la Loi 200, B-1049 Brussels, Belgium. Email *philippe.quevauviller@cec.eu.int*

Heikki Simola, Karelian Institute Section of Ecology, University of Joensuu, PO Box 111, FIN-80101 Joensuu, Finland. Email *heikki.simola@joensuu.fi*

Kaarina Sivonen, Department of Applied Chemistry and Microbiology, University of Helsinki, PO Box 56, FIN-00014 Helsinki, Finland. Email *kaarina.sivonen@helsinki.fi*

Jouko Soveri, Finnish Environment Institute, PO Box 140, FIN-00251 Helsinki, Finland. Email *jouko.soveri@vyh.fi*

Heikki Toivonen, Finnish Environment Institute, PO Box 140, FIN-00251 Helsinki, Finland. Email *heikki.toivonen@vyh.fi*

André Van der Beken, TECHWARE, c/o CIBE/BIWN, 70 Rue of Laines/ Wolstraat 70, B-1000 Brussels, Belgium. Email *coordinator@techware.org*

Bertel Vehviläinen, Finnish Environment Institute, PO Box 140, FIN-00251 Helsinki, Finland. Email *bertel.vehvilainen@vyh.fi*

Matti Verta, Finnish Environment Institute, PO Box 140, FIN-00251 Helsinki, Finland. Email *matti.verta@vyh.fi*

Torgny Wiederholm, Department of Environmental Assessment, Swedish University of Agricultural Sciences, PO Box 7050, S-750 07 Uppsala, Sweden. Email *torgny.wiederholm@ma.slu.se*

Eva Willén, Department of Environmental Assessment, Swedish University of Agricultural Sciences, PO Box 7050, S-750 07 Uppsala, Sweden. Email *eva.willen@ma.slu.se*

Giuliano Ziglio, Dipartimento di Ingegneria, Civile ed Ambientale, University of Trento, Via Mesiano di Povo, 77, I-38100 Trento, Italy. Email *giuliano.ziglio @ing.unitn.it*

Series Preface

PHILIPPE QUEVAUVILLER
Series Editor

Water is a fundamental constituent of life and is essential to a wide range of economic activities. It is also a limited resource, as we are frequently reminded by the tragic effects of drought in certain parts of the world. Even in areas with high precipitation, and in major river basins, the over-use and mismanagement of water have created severe constraints on availability. Such problems are widespread and will be made more acute by the accelerating demand on freshwater arising from trends in economic development.

Despite the fact that water-resource management is essentially a local, river-basin based activity, there are a number of areas of action where it is advisable to pool efforts for the purpose of understanding relevant phenomena (e.g. pollutions, geochemical studies, etc.), thus developing technical solutions and/ or defining management procedures. One of the keys for successful co-operations, aimed at studying hydrology, water monitoring, biological activities, etc., is to achieve and ensure good water quality measurements.

Quality measurements are essential to demonstrate the comparability of data obtained worldwide and they form the basis for correct decisions related to management of water resources, monitoring issues, biological quality, etc. Besides the necessary quality control tools developed for various types of physical, chemical and biological measurements, there is a strong need for education and training related to water quality measurements. This need has been recognized by the European Comission, where the latter has funded a series of training courses on this topic, covering aspects such as monitoring and measurements of lake recipients, measurements of heavy metals and organic compounds in drinking and surface water, use of biotic indexes, and methods to analyse algae, protozoa and helminths.

This book series will ensure a wide coverage of issues related to water quality measurements, in particular the topics of the above mentioned courses. In addition, other aspects related to quality control tools (e.g. certified reference materials for the quality control of water analysis) and wastewater monitoring will also be considered.

This present book on 'Hydrological and Limnological Aspects of Lake Monitoring' is the first one of the series. It has been written by experts in lake sciences and offers the reader an extensive overview of all of the important aspects that need to be tackled when starting lake monitoring studies.

The Series Editor – Philippe Quevauviller

Philippe Quevauviller began his research activities in 1983 at the University of Bordeaux I, France, where he gained a PhD degree in oceanography. Between 1984 and 1987, he was Associate Researcher at the Portuguese Environment Ministry where he performed a multidisciplinary study (sedimentology and geochemistry) of the coastal environment of Galé and of the Sado Estuary, Portugal. In 1988, he became Associate Researcher at the Dutch Ministry for Public Works, where he investigated the organotin contamination levels of Dutch coastal environments. Since 1989, he has been managing various Research and Technological Development (RTD) projects in the frame of the Standards, Measurements and Testing Programme (formerly BCR) of the European Commission in Brussels. In 1999, he obtained an HDR (Diplôme d'Habilitation à Diriger des Recherches) in chemistry at the University of Pau, France from a study of the quality assurance of chemical species' determinations in environmental elements.

Philippe Quevauviller has published (as author and co-author) more than 180 scientific publications, 50 reports and 2 books for the European Commission, and has acted as editor and co-editor for 19 special issues of scientific journals and 6 books.

TECHWARE (TECHnology for WAter REsources)
c/o CIBE-BIWM
Wolstraat 70, rue aux Laines, B-1000 Brussels, Belgium
Tel: +32-2-518 88 93
Fax: +32-2-502 67 35
e-mail: coordinator@techware.org
website: http://keywater.euro.net

This book was prepared within the framework of a task force of TECHWARE
(TECHnology for WAter REsources) – a non-profit association of universities
and enterprises focusing on training – which has organised a series of short
courses on water quality measurements, funded by the European Commission.

Preface

PERTTI HEINONEN, GIULIANO ZIGLIO AND
ANDRÉ VAN DER BEKEN

The specific aim of the particular course which this book refers to, is to attempt a simulation of the broader view of the lake ecosystem, in order to encourage those people who are responsible for management and decision-making aspects of measurements to move from a separate to an integrated approach, and even better, to a comprehensive one for this kind of activity.

A lake, as an individual ecosystem, is intrinsically complex and can have different responses with regard to the quality of its water and, more generally, the status of the system, showing different fragility to human pressure factors. The lake ecosystems, with respect to hydrological and limnological aspects, could be described and studied as separate sub-components, such as the recipient, the dynamic processes developing in the ecosystem, the external pressure factors, the management activity, and the data/information collection/ evaluation processes. Each of these sub-components have their own specific complexity; furthermore, within each of them, specific interactions and interdependencies can be envisaged.

As regards the recipient, the following are of significance: water–sediment interaction, micro/macro biota growth and their relationship in the food chain with chemical and other physical determinants, dynamic energy transfer in the water mass, etc. The complexity of the sub-component processes can be related to the following aspects: water balance (including ground water contribution), dynamics of mixing transport and advection of water and their dissolved/ suspended compounds, the physical and chemical processes regulating deposition sorption and bioavailability to the living biota of nutrients and xenobiotic compounds, etc.

All these and other aspects included in this book should be considered when planning 'measurement' activities. The level of complexity to be dealt with has to be evaluated carefully, because it greatly influences the cost and requires even more adequate competencies. Measurement activities concerning hydrological and limnological aspects are strongly diversified with respect to determinants, time and space scales, organizations, etc., when considering the purposes involved, as follows:

(a) to describe and observe phenomena on a selected space scale and complexity level;
(b) to explicate surveillance and control, according to certain administrative procedures;
(c) to monitor certain variables/parameters, in an attempt to relate lake status to pressure factors, and evaluate the efficacy of measures taken at regional, national and international levels;
(d) to perform research in the field, as well as in the laboratory, having specific objectives;
(e) to calibrate and validate models, with the aim of using these in scenario/vision activities for pre-evaluating the status response to the simulated measures to be taken.

The most important problems currently impacting lake resources are as follows:

(a) nutrient loads (with eutrophication and possible health effects);
(b) water acidification (with impacts on biodiversity and metal cycling);
(c) the presence of endocrine disrupters (persistent organic compounds and bioaccumulatable compounds with potential environmental and health consequences);
(d) the bioaccumulation of mercury in the food chain.

However, various resources and tools are available to manage and tackle these threats; we can refer here to basic environmental knowledge, new and more advanced technologies for lake quality monitoring, co-ordination at national and European level in monitoring strategies and data handling, control regulation and surveillance activities.

This book has solicited particular contributions from and has concentrated attention on the Nordic countries because of the specifc interest and sensibility developed by these countries on lake quality monitoring and management. Since the principal aims are to have a broader comprehensive approach and to describe methodologies, these facts do not adversely affect the book.

For a comprehensive approach, the following recommendations and perspectives should be pointed out:

(a) to increase the use of statistical tools in design and state interpretation, including quality assurance and quality control programmes within and between laboratories;
(b) to search for coherence between different indicators when applied in monitoring or surveillance activities;
(c) to produce more information than data;
(d) to manage more dynamic information than static information.

The editors are perfectly aware of the intrinsic difficulties within this approach and have tried to offer a contribution which gives an overall view of the lake system, avoiding, whenever possible, the use of specialized language. In this way we hope to allow a better understanding of this complex reality and to stimulate contacts and dialogues among people in charge of 'measurement' activities and those involved in research or other activities concerning lake ecosystems. We gratefully acknowledge the various authors in their efforts to achieve these objectives.

Part One
Abiotic Processes in Lakes

Chapter 1.1
Hydrology of Lakes

ESKO KUUSISTO AND VELI HYVÄRINEN

Hydrological and Limnological Aspects of Lake Monitoring
Edited by Pertti Heinonen, Giuliano Ziglio and André Van der Beken
©2000 John Wiley & Sons, Ltd, ISBN 0 471 89988 7

1.1.1   INTRODUCTION

It has been estimated that there are 12 million lakes on the Earth; their total area is 2.7 million km^2, while the total volume is 166 000 km^3. The 10 largest in area account for 33% and the 10 largest in volume for as much as 90% of the corresponding world totals.

There are almost one and a half million lakes in Europe, if small water bodies with an area down to 0.001 km^2 are included. Of these, at least 500 000 natural lakes are larger than 0.01 km^2 (Kristensen and Hansen, 1994). Many of them appeared 10 000–15 000 years ago, having being formed or reshaped by the last glaciation period, the Weichsel. Most of them are located in northern Europe, the Nordic Countries and the Karelo–Kola part of the Russian Federation, where lakes cover 5–10% of the surface area. Lakes are also common in Iceland, Ireland and in north-western parts of the UK. In central Europe, there are, in addition to high-altitude small lakes, some larger lakes on the margin of the Alps (e.g. Lake Geneva, Lake Garda, Lake Maggiore and Lake Constance), in the Dinarian Alps (the ancient Lake Ohrid and Lake Prespan) and on the Hungarian plain (Lake Balaton and Lake Neusiedler). In countries little affected by the glaciation period, such as Portugal, Spain, France, Belgium, southern England, central Germany, the Czech Republic, the Slovak Republic and the central European part of the Russian Federation, natural lakes are few.

The total area of European lakes and reservoirs is about 300 000 km^2, of which reservoirs make up almost one third. More than 10 000 major reservoirs have been constructed in Europe. The Volga basin alone has a reservoir area of 38 000 km^2 (Mordukhai-Boltovskoi, 1979). The total volume of European lakes is 3300 km^3, and that of reservoirs is 800 km^3; these figures obviously exclude the Caspian Sea.

1.1.2   WHAT IS A LAKE?

A unique definition of a lake does not exist. Most textbooks on lakes start without actually attempting to define the object that they are going to deal with. From a geological point of view, any workable definition should take into account two distinct parts; namely the basin and the water body. From a hydrological point of view, a lake should be distinguished from a wide river section, i.e. how much must a river widen before one calls the place a lake?

The question of the narrowness is very typical, e.g. in some parts of Canada and in the Finnish Lake District. A lake may consist of several basins, separated by short straits. The mean water level in the downstream basins is a few centimetres lower than in the upstream basins. Do we have in this case a single lake or several lakes?

The minimum size of a water body which can be called a lake should also be considered. However, let this be a matter to be discussed eternally. In any case, a water body smaller than a lake should be called a pond. Welch (1952) defined a pond as 'a very small, very shallow body of standing water, in which quiet water and extensive occupancy by higher aquatic plants are common characteristics'. It sounds obvious that eutrophication can change a lake into a pond.

As a summary, a water body should fill the following requirements in order to be regarded as a lake:

(a) it should fill or partially fill a basin or several connected basins;
(b) it should have essentially the same water level in all parts, with the exception of relatively short occasions caused by wind, thick ice cover, large inflows, etc.;
(c) even if the water body may be located in the immediate vicinity of the sea coast, it does not have a regular intrusion of sea water;
(d) the water body should have so small an inflow-to-volume ratio that a considerable portion of suspended sediment is captured;
(e) the area of the water body should exceed a specified value, e.g. 1 ha, at mean water level.

1.1.3   LAKE CLASSIFICATIONS

Classifications of lakes have been suggested in order to improve the possibilities of estimating quantitatively the dynamic, thermal and biological processes taking place in such bodies. A detailed genetic classification has already been presented by Hutchinson (1957). Morphometric classifications include a variety of parameters and the subdivision of the lake basins according to their shape. The earliest thermal classification may be that made by Forel (1901), who distinguished between polar, sub-polar, temperate and tropical lakes. A more detailed thermal classification was presented by Keller (1974).

A water-balance classification of lakes can be based on three simple criteria (Szesztay, 1974):

(a) the inflow factor, which can be calculated by using the following expression:

$$IF = IN/(IN + P) \tag{1}$$

where IN = inflow into the lake and P = lake precipitation

(b) The outflow factor, which can be calculated by using the following
    expression:

$$OF = OU/(OU + E) \qquad (2)$$

where OU = outflow from the lake and E = lake evaporation
(c) The magnitude of the mean annual flux of incoming or outgoing waters.

A closed lake obviously has an outflow factor of 0%, while the outflow
factor of a throughflow lake is almost 100%. Of the large lakes in the world,
Lake Victoria might have the lowest inflow factor, i.e. below 20%. In the large
Finnish lakes, inflow factors are typically 70–90%, with outflow factors being
slightly higher than this.

1.1.4   THERMAL CONDITIONS

The temperature of water bodies is essentially determined by the radiation
balance, i.e. the flux of latent heat and convective heat supply. Higher water
temperatures tend to increase evaporation and outgoing radiation, both of
which, in turn, have a cooling effect. Moreover, changed runoff dynamics,
together with changes in the transport of suspended sediment, can also
influence the water temperature.

The annual thermal cycle of a moderately deep, temperate lake is
characterized by the following important dates:

(a) the break-up date, i.e. when a rapid increase of temperature begins as a
    result of mixing of the water mass due to wind;
(b) the date of the maximum mean temperature of the water mass, i.e. when
    the amount of heat energy stored in the water mass reaches its maximum;
(c) the date of the beginning of the autumn homothermy;
(d) the date of the water-density maximum, i.e. when the whole water mass has
    a temperature of 4.0 °C.
(e) the freezing date, i.e. when the effect of wind mixing ceases and winter
    conditions begin;
(f) the date of the minimum mean temperature of the water mass, i.e. when the
    amount of heat energy reaches its minimum.

In large Finnish lakes, the date of the maximum mean temperature usually
occurs in August, although the deepest layers continue to warm up until the
end of September. The autumn homothermy starts in early October and lasts
3–6 weeks; thereafter, it usually takes 2–3 weeks before the lake freezes. The
mean temperature of the water mass by this freezing time is around 1.0 °C.

1.1.5   ICE CONDITIONS

Some morphometric parameters have a close connection with the ice conditions of lakes. For lakes of equal area, an early freezing-up is associated with the following characteristics:

• small mean and maximum depths;
• small effective fetch;
• large shore development;
• large insulosity;
• direction of major axis perpendicular to the direction of prevailing winds.

Of the fifty largest lakes in the world, about half are completely or partially covered by ice in winter. Numerous lakes in high latitudes or at high altitudes have an ice-cover season longer than the open-water period. Some lakes in high mountains, in the Arctic islands and in the Antarctic are always covered by ice.

There are 50–100 lakes in the world with freezing and breakup data series which are over a century long. In Finland, the longest series is from Lake Kallavesi, where the observations first started in the autumn of 1833. Since that time, both freezing and breakup dates have shifted by about two weeks 'towards a milder climate'.

1.1.6   CLIMATE CHANGE AND EUROPEAN LAKES

Aquatic ecosystems are excellent integrators of changes in climate and catchment conditions. The results of the complex interplay between climate, vegetation and soil will be summarized in streams and lakes, and further on in marine ecosystems. In lakes with minor catchment areas, the direct effects of changed climate and atmospheric deposition control the future development.

Water quality changes are largely determined by changes in leaching. In acid-sensitive lakes, the increased nitrogen leaching might disturb the positive development started as a result of reductions in sulfur emission.

In agriculturally loaded lakes, an increase in the primary production of phytoplankton is likely due to a longer growing season, higher nutrient loads and higher CO_2 concentrations. Changes in population and community structure of the aquatic biota are also likely to occur.

In areas with high lake densities, the position within the catchment area can strongly influence the response of individual lakes to climate shifts, thus leading to divergence in biogeochemical patterns of change across a region (Webster *et al.*, 1996).

The response of European lakes to climate change can be discussed by dividing the lakes into four categories.

1.1.6.1   Deep, temperate lakes

Typical representatives of this group are, e.g. Lakes Maggiore (Italy), Geneva (Switzerland/France), Ness (Scotland) and Constance (Germany/Switzerland/ Austria) with mean depths of 177, 153, 132 and 90 m, respectively. Due to these great depths and relatively mild winters, there is usually no ice cover.

Most of these lakes are warm monomictic (one turnover), while the deepest ones are warm oligomictic (irregular and seldom). Convective overturn occurs in winter or early spring. In some lakes, there is a correlation between annual air temperature and the temperature of the hypolimnion; in this case the lake can even be used as a filtered indicator of climatic change or variations (Livingstone, 1993).

The future climate change may suppress the turnover in deep monomictic lakes, thus giving them the classification of oligomictic. This implies the enhancement of anoxic bottom conditions and an increased risk of eutrophication. The oxygen conditions can also be expected to deteriorate due to increased bacterial activity in deep waters and surficial bottom sediment.

Although many of these lakes have a long residence time, the role of increased evaporation in concentrating nutrients in the water mass is relatively small. Changes in catchment conditions will be more important.

The maritime lakes in this group (particularly in the UK and Ireland) are strongly influenced by cyclones coming from the Atlantic. The increased power of these weather phenomena could affect the stratification, and consequently the biological conditions in these lakes. Moreover, a correlation between the average summer biomass of zooplankton and the position of the Gulf Stream has been found in UK lakes (George and Taylor, 1995).

1.1.6.2   Shallow, temperate lakes

Lake Balaton (596 km², 3 m) in Hungary, Lake Shkodra (368 km², 8 m) in Albania and Lake Müritz (114 km², 8 m) in Germany are typical examples of this group.

High water temperatures will result in intensified primary production and bacterial decomposition. The probability of harmful extreme events, e.g. mass production of algae, will increase. The impacts may extend to fish life; with changes in species composition and reduced fish catches being anticipated. The use of the expression 'thermal pollution' is well justified for these lakes.

In lakes with relatively long retention times, increased evaporation causes conservative solutes to concentrate. This effect may be enhanced by decreased annual inflows in southern Europe and east European lowlands.

For Lake Balaton, Szilagyi and Somlyody (1991) estimated that the increased dissociation of inorganic carbon will probably be a more important contributor to acidification and to ionic composition than acid deposition.

Doubling the atmospheric CO_2 concentration would result in a decrease of the pH value by 0.2, and would cause a significant increase in the salt content and hardness of the lake water.

1.1.6.3   Boreal lakes

Lake Ladoga (Russia) (17 670 km^2, 51 m), Lake Onega (Russia) (9670 km^2, 30 m) and Lake Vänern (Sweden) (5670 km^2, 27 m) are the largest lakes in this group, with these also being the three largest lakes in Europe. This group includes about 120 lakes with an area exceeding 100 km^2.

Most lakes of the boreal zone are dimictic with two overturns in a year. Shortening of the ice cover period is the most obvious consequence of climate change in these lakes (Huttula *et al.*, 1996). This could improve the oxygen conditions in winter and spring. A longer ice-free period might also result in increased turbidity due to erosion from exposed land surfaces. A longer and stronger summer stratification might have harmful effects on water quality in the hypolimnion. This will obviously depend on future wind speeds, which nobody is yet willing to predict.

A simulation of Lake Ladoga by Meyer *et al.* (1994) in a $2 \times CO_2$-climate scenario resulted in the following changes:

(a) No ice cover was formed. In this century, a complete ice cover has occurred in Lake Ladoga in about 80% of winters. The lack of ice cover may stimulate fog formation. The lake's biota, presently including a number of unique species, might also be affected by the absence of ice cover.
(b) The lake will remain dimictic. Intensive cooling during winter will lead to inverse stratification in March and April.
(c) Summer stratified conditions seem to be preserved as at present. Convective mixing can potentially have more significant consequences due to the longer duration of overturn periods.

1.1.6.4   Mountain and arctic lakes

These are mainly small water bodies in central European and Scandinavian mountains and in the Arctic.

The lakes in this group are generally considered to be particularly sensitive to environmental changes. They usually have small catchments with limited chemical and biological erosion, and their simple and labile ecosystems react quickly to environmental stress and changes. Many lakes are ultra-oligotrophic; the ice cover season is long.

A small mountain lake can be biologically very isolated. It closely reflects the statement of one of the fathers of limnology, S. T. Forbes, from the year 1887 (Hutchinson, 1957):

It forms a little world within itself – a microcosm within which all the elemental forces are at work and the play of life goes on in full, but on so small a scale as to bring it easily within the mental grasp.

Even if mountain lakes are connected by channels, physical and ecological constraints limit species migration between them. In a warming climate, there is no escape route; the only possibility for survival is adaptation.

In the Arctic, melting permafrost may fatally threaten lake ecosystems. In some cases, it may threaten the whole existence of the lake, i.e. ground thaw together with enhanced evaporation may cause the lake to disappear.

One risk in mountain lakes may be the future enhancement of UV-B radiation. As a function of altitude, this component increases by 20% per 1000 m (Blumthaler and Rehwald, 1992). Transparent water tends to maximize the UV-B dose to aquatic alpine organisms.

1.1.7   CONCLUSIONS

Knowledge of the hydrology of lakes is essential for their proper use and conservation. Water quality is closely linked to the water and energy budgets, mixing, stratification and other physical aspects of lakes. Without a morphometric description of a lake, the quantitative analysis of the thermal and biological processes is impossible. If a lake has an ice cover in winter, a completely different approach to the analysis of heat budget and dynamic processes is needed.

On the geological time scale, lakes are transitory features on the Earth's surface. Their life expectancies may vary from a short spell between two floods to millions of years. The variation of water levels or the rate of sedimentation are indicators of the future development of a lake. Particularly interesting are closed lakes – these are laboratories for the study of basic climatic, hydrological and geological phenomena.

Considerable progress has been made in the modelling of lake hydro-dynamics in recent decades. However, the complexity of water motions in lakes still hinders a detailed analysis, and simplified strategies have to be applied.

Long hydrological data series of lakes have gained in importance with increasing evidence of climate change, with the latter having essential effects on lakes in many regions of the world. In high latitudes, the shortening of the ice-cover period will be the most obvious consequence. In addition, increased water temperatures may result in intensified primary production and harmful changes in water quality.

REFERENCES

Blumthaler, M. and Rehwald, W., 1992. Solar UV-A and UV-B radiation fluxes at two alpine stations at different altitudes, *Theor. Appl. Climatol.*, **46**, 39–44.

Forel, F. A., 1901. *Handbook der Seenkunde*, Verlag von J. Engelhorn, Stuttgart.

George, D. and Taylor, A., 1995. UK lake plankton and the Gulf Stream, *Nature (London)*, **378**, 139.

Hutchinson, G. E., 1957. *A Treatise on Limnology*, Vol. I, Wiley, New York.

Huttula, T., Peltonen, A. and Kaipainen, H., 1996. Effects of Climatic Changes on Ice Conditions and Temperature Regime in Finnish Lakes, The Final Report of SILMU, 4/96, Helsinki, Finland, Academy of Finland, 167–172.

Keller, R., 1974. Physico-geography – object and problem of research, *IAHS Bull.*, **19**, 63–71.

Livingstone, D., 1993. Temporal structure in the deep-water temperature of four Swiss lakes: A short-term climatic change indicator?, *Verh. Int. Verein. Limnol.*, **25**, 75–81.

Kristensen, P. and Hansen, H. O. (Eds), 1994. European Rivers and Lakes: Assessment of their Environmental State, EEA Environmental Monographs 1, European Environment Agency, Silkeborg, Denmark.

Meyer, G., Masliev, I. and Somlyody, L., 1994. Impact of Climate Change on Global Sensitivity of Lake Stratification, IIASA Report, WP-94-28.

Mordukhai-Boltovskoi, D., 1979. The River Volga and its Life, Dr W. Junk, The Hague.

Szesztay, K., 1974. Water balance and water level fluctuations of lakes, *IAHS Bull.* **19**, 73–84.

Szilagyi, F. and Somlyody, L., 1991. Potential impact of climatic changes on water quality in lakes, *IAHS Publ.*, **206**, 79–86.

Webster, K., Kratz, T., Bowser, C. and Magnuson, J., 1996. The influence of landscape position on lake chemical responses to drought in northern Wisconsin, *Limnol. Oceanogr.* **41**, 977–984.

Welch, P. S., 1952. *Limnology*, McGraw-Hill, New York.

Chapter 1.2

Hydrological Forecasting and Real-Time Monitoring: The Watershed Simulation and Forecasting System (WSFS)

BERTEL VEHVILÄINEN

Hydrological and Limnological Aspects of Lake Monitoring
Edited by Pertti Heinonen, Giuliano Ziglio and André Van der Beken
©2000 John Wiley & Sons, Ltd, ISBN 0 471 89988 7

1.2.1 INTRODUCTION

Real-time information from runoff, discharges and water levels are needed in water resources management and water quality monitoring. A hydrological model is a tool for expanding and improving the availability of such data. In this chapter, we present a system based on hydrological models for giving extensive coverage of hydrological real-time data and forecasts over Finland.

A real-time monitoring and forecasting system based on hydrological watershed models has been widely used in Finland since 1990. The main operating part of the watershed simulation and forecasting system (WSFS) consists of 20 watershed models, which simulate the hydrological cycle by using standard meteorological data. The operation of a watershed model consists of meteorological and hydrological data collection, a basic simulation run, the updating of the model accuracy according to observations, model runs with different regulatory rules for regulated lakes, a forecasting run with weather forecasts and weather statistics and the delivery of the forecast to the users (and to the Internet). Owing to the large number of forecasts made, the entire operating system has been developed into a fully automatic form. Forecast and simulation results are presented as graphs of discharges, water levels, water equivalents of snow, areal precipitation, soil evaporation, lake evaporation and daily temperatures. If needed, forecasts can cover up to a maximum period of six months.

Maps are available of water level, water equivalent of snow, daily snowmelt, runoff, soil moisture deficit and soil evaporation over Finland. The address of the home page for the WSFS is:

http://www.vyh.fi/tila/vesi/ennuste/index.html

The forecasts for different lakes and rivers can be chosen by 'clicking' the watershed of interest on the map of Finland.

1.2.2 GENERAL DESCRIPTION OF THE SYSTEM

The independent systems to which the WSFS is connected are as follows:

- the hydrological data register (HYTREK);
- the operative watershed management system (VKTJ);
- the automatic real-time water level and discharge station net (PROCOL);
- the weather stations of the Finnish Meteorological Institute (FMI);
- the European Centre of Medium-Range Weather Forecasts (ECMWF) via the FMI, which provides weather forecasts.

The WSFS reads the watershed data from the registers, runs forecasts and distributes results to the Regional Environment Centres and to the Internet.

1.2.3 THE DATA SOURCES OF THE SYSTEM

The FMI sends (by e-mail) daily precipitation data from 170 stations and temperature data from 48 stations. A 10-day precipitation and temperature forecast from the ECMWF is delivered to the WSFS via the FMI. The watershed models also need potential evaporation observations, utilizing data from 20 stations which are reporting with a 1-month delay, or simulating potential evaporation by a temperature-dependent model (Table 1.2.1).

Hydrological data, water levels and discharges, are gathered from different sources. For real-time forecasting, the most important source is the PROCOL system (Puupponen, 1988), which delivers water level and discharge data in real-time to the registers and models. The other source is the VKTJ system in which water level and discharge data are stored manually. Such real-time hydrological data are crucial for an accurate forecasting system; the watershed models are updated according to this information.

Most of the hydrological data can be obtained from HYTREK with a 1–2 month delay. These data can be used to update further sub-basins, which thus

Table 1.2.1 Meteorological and hydrological data used in the watershed models.

Observation	Institute	Number	Delivery
Precipitation			
Synoptic stations	FMI	48	Daily
Precipitation stations	FMI	170	Daily
Temperature			
Synoptic stations	FMI	48	Daily
Potential evaporation			
Class-A pan	FEI	22	Monthly
Water level			
Hytrek	FEI	272	Monthly
Procol	FEI	50	Daily
VKTJ	FEI	51	Daily
VKTJ	FEI	15	Weekly/monthly
Discharge			
Hytrek	FEI	175	Monthly
PROCOL	FEI	50	Daily
VKTJ	FEI	36	Daily
VKTJ	FEI	11	Weekly/monthly
Snow line			
Hytrek	FEI	117	Biweekly

increases the accuracy of the watershed models. Snow-line measurements are available from HYTREK (with some delay) and are used to check the accuracy of areal snow simulations of the watershed models (see Table 1.2.1).

1.2.4 WATERSHED-MODEL IMPLEMENTATION

The basic component of a watershed model is a conceptual hydrological runoff model (Bergström, 1976; Vehviläinen, 1994) which simulates runoff by using precipitation, potential evaporation, and temperature observations as input data (Figure 1.2.1). The main parts of the hydrological model are precipitation,

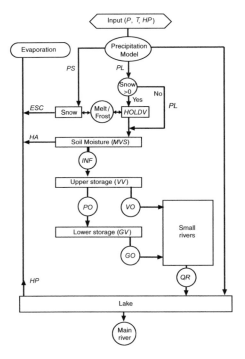

Figure 1.2.1 The structure of the rainfall-runoff model: ESC (mm d^{-1}), evaporation from snow cover; GO (mm d^{-1}), outflow from ground-water storage; GV (mm), ground-water storage; HA (mm d^{-1}), actual soil evaporation; $HOLDV$ (mm), liquid-water storage of snowpack; HP (mm d^{-1}), potential evaporation; INF (mm d^{-1}), inflow into sub-surface storage; MVS (mm), soil-moisture storage; P (mm d^{-1}), precipitation; PL (mm d^{-1}), liquid precipitation; PO (mm d^{-1}), inflow into ground-water storage; PS (mm d^{-1}), solid precipitation; QR (mm d^{-1}), runoff; T (°C), temperature; VO (mm d^{-1}), outflow from sub-surface storage; VV (mm), sub-surface (upper) storage.

snow, soil moisture, sub-surface, and ground water models. This hydrological model is calibrated more or less specifically for all of the sub-basins in the watershed depending on the available data.

Watershed model implementation begins by dividing the watershed into sub-basins according to the classification of Finnish river basins presented by Ekholm (1993). The aim is to divide the watershed into small homogeneous sub-basins according to elevation, land use, snow distribution and lakes. The number of sub-basins within a watershed model is typically 30–100; for each of these, the hydrological runoff model is calibrated. The area of a sub-basin ranges from 50 to 500 km^2. Regulated, large unregulated, observed, and otherwise important lakes are described by the lake model. This allows the correct simulation of water levels and outflow in a lake and improves the simulation of areal runoff and discharges. Finally, the basic hydrological runoff and lake models are connected with river models to form the watershed model.

The optimization criteria in the calibration are the sum of the square of the differences between the observed and simulated water equivalents of snow, discharge, and water level. All available observations are used in the calibration and thus up to 100 different calibration criteria can be available in a watershed-model calibration.

1.2.5 OPERATIONAL USE OF WATERSHED MODELS

The WSFS has an automatic model-updating system which has been developed in the Finnish Environment Institute. This updating system guarantees that the watershed models are in the best possible state before the forecast is evaluated according to observations, and also makes the updating possible, a task which is impossible to do manually due to the large amount of simulated observation points (275) and sub-basins (1194). Model updating is carried out against the water level and discharge data gathered from different registers. When new watershed data become available, the updating procedure corrects the model simulation by changing the areal values of temperature, precipitation and potential evaporation so that the observed and simulated discharges, water levels and water equivalents of snow are equal.

Short-term forecasts are the relevant forecasts for watersheds with short response times and low lake percentages, where the time interval between snowmelt or rainfall event and flood is only a few days. These watersheds with short response times need real-time data from discharges and water levels and continuous updating to maintain the quality of simulations and forecasts. The latter must be made daily in flood periods. The 10-days temperature and precipitation forecasts from the ECMWF are the main meteorological input for the forecasting period, while after this, statistical values are then used.

In long-term forecasting, the statistical precipitation, temperature and potential evaporation data are more important than the 10-days weather forecasts. The hydrological forecast is based on mean (50%), 5 or 10%, and 90 or 95% precipitation sums for periods of 1, 2, 3, and 6 months. Especially at the beginning of winter, long-term forecasts are sensitive to temperature; thus the 25, 50 and 75% probability values for temperature are used for the first month.

Forecasts are usually made once or twice a week, even for the largest watersheds with long response times. This is done partly to test the entire system from the data collection, for the delivery of results for possible problems, and to correct these in time before the forecasts are generally needed. For watersheds with short response times, twice a week is too seldom a forecasting frequency during floods; thus forecasting runs are started by the system whenever rainfall, discharge or water levels exceed a given limit.

Watershed forecasts are used for the supervision of water levels, discharges, snow, soil moisture and runoff formation. In flood situations, watershed models are used to plan the regulation of lakes and reservoirs to minimize flood damage. The forecast of possible overtopping of river embankments helps the Regional Environment Centres to take necessary precautions in advance. The ability of watershed models to simulate water equivalents of snow is valuable when estimating flood potentials during snowmelt periods in real-time.

In more slowly responding watersheds with abundant lakes, the forecasts are used for long-term planning of regulation. The precipitation between the forecast day and the future flood peak event strongly affects the final results. Statistical precipitation, temperature and potential evaporation series must be used to provide the needed information.

The computer network in the FEI, and especially the Internet, gives excellent possibilities for delivery of watershed forecasts to the users. Point forecasts for water level and discharge for over 50 sites are delivered to the Internet; this is the most effective delivery system used with watershed models. Forecasts are available for nearly all possible users and the system is very reliable. In addition, the quality of the forecast pictures available on the Internet are better than with other delivery systems. The Regional Environment Centres can then inform and supervise all local authorities and organisations which need this information in their work. In the case of flood danger, the Regional Environment Centres and FEI then inform the press, radio and television.

1.2.6   MAP-BASED USER INTERFACE

A map-based user interface developed for the WSFS makes it possible to examine on a map the hydrological variables simulated by watershed models in different sub-basins, with 3000 of these altogether covering 40% of the area of Finland. First, the watershed of interest of the map-based user interface is chosen. From the chosen watershed with the first level sub-basin division, one can then go to the second and even to the third level sub-division. In each level, all of the data are available, i.e. snow water equivalent, soil moisture, discharges, storages, lake level, inflow and groundwater storage.

Via an 'output' icon, it is possible to store any simulated daily data into a file for further use. This possibility is intended especially for users who need discharge and runoff data for areas and rivers with no observations. The map-based user interface is a source of simulated discharge values, when it is impossible or too expensive to make direct observations. The time range for the simulated data is three months backwards from the day of the model run. Longer series are also available by request. The simulated data are also used for real-time watershed monitoring and water resources management. The quality of simulated data is maintained by continuous updating of the watershed models against the observed water level and discharge values.

1.2.7 CONCLUSIONS

The main use of lake inflow forecasts is the management of regulated lakes. Watershed models could also be very effective tools in general water resources planning; however, they are seldom used for this purpose at present. The problems arising with low-flow periods, e.g. water supply during droughts, have also been solved by using watershed models in a few cases.

In the case of observation break-ups in water levels and discharges, the simulated data from watershed models can be used to fill the gaps in registers. Furthermore, the comparison between model simulations and observation data quickly reveals most of the observation and recording errors; thus watershed-model simulations can be used as a first quality control process for the data in registers.

Contrary to what was previously believed, pollution due to agriculture and forestry has proved to be much more important than point-source pollution. The evaluation of rural pollution from agriculture and forestry needs runoff and discharge data from relatively small areas. Watershed models which simulate discharges for small sub-basins (50–500 km^2) will prove to be very valuable information sources in this context. One of the major discharge information sources for this system will be the WSFS.

Large watershed models can be used to evaluate the effects of climate change on water resources, especially for snow cover, discharge and water level changes.

REFERENCES

Bergström, S., 1976. Development and Application of a Conceptual Runoff Model for Scandinavian Catchments, SMHI, Nr R117, Norrköping, Sweden.

Ekholm, M., 1993. Suomen Vesistöalueet (Drainage Basins in Finland), Publications of the Water and Environment Administration – Series A126, Helsinki, Finland.

Puupponen, M., 1988. Real-time Hydrological Data Collection at the Finnish National Board of Waters and the Environment, Nordisk Hydrologisk Konferens, Rovaniemi, Finland, 1988; NHP Report 22, Vol. 2.

Vehviläinen, B., 1994. The Watershed Simulation and Forecasting System in the National Board of Waters and Environment, Publications of the Water and Environment Research Institute, No. 17, National Board of Waters and the Environment, Helsinki, Finland.

Chapter 1.3
Fluxes of Water and Materials into Lakes via Groundwater

JOUKO SOVERI[†]

[†]Deceased

Hydrological and Limnological Aspects of Lake Monitoring
Edited by Pertti Heinonen, Giuliano Ziglio and André Van der Beken
©2000 John Wiley & Sons, Ltd, ISBN 0 471 89988 7

1.3.1 INTRODUCTION

Groundwater input to lakes and seas has received increasing attention during the last ten years, as better techniques have been developed to quantify this flow and assess its hydrological and ecological importance. This source term occurs wherever a continental aquifer is hydraulically connected to the coastal areas of lakes and seas. The magnitude of groundwater discharge to coastal waters varies based on upgradient forcing due to precipitation, surface-water runoff, evaporation, infiltration rates and aquifer permeability. These main components of the water budget are fairly accurately estimated. However, few attempts have been made to quantify and qualify the groundwater component of flow into the lakes.

Groundwater that flows directly into lakes contributes not only to the lakes's water volume but also to its chemical balance. Groundwater and lakes often have different chemical characteristics. In some regions, groundwater has been demonstrated to have detrimental effects on coastal waters, while in other regions coastal productivity may depend on seepage-derived nutrients. Temporal variations in groundwater flow may also have seasonal controls on surface water biological productivity in regions where this input provides important nutrients (Cable, 1998).

1.3.2 INTERACTION AND HYDROLOGICAL CYCLE

There are two main aspects of this interaction. First, the flow of groundwater to support lakes and river flow and, secondly, the flow from rivers and lakes to groundwater. The former is a common occurrence in temperate regions, whereas the latter occurs widely in arid regions. Figure 1.3.1 shows a simplified conceptual model that illustrates the subject area of interactions between surface water and groundwater storages. Whenever there is a flow of water between lakes and aquifers, in either direction, there is a relationship between the quality and quantity of water in the two systems.

Groundwater flow is defined as flow within the saturated zone. In catchments with more than one aquifer, the base flow component may be subdivided according to the contributing formation. The proportion of direct runoff or base flow may vary substantially from one basin to another and from month to month because of the effect of different soil types, geology, land use, topography, stream patterns and changes in precipitation, evaporation and temperature.

In temperate regions, groundwater recharge is derived mainly from precipitation less evaporation, where evaporation is defined as including transpiration and interception losses from vegetation. In arid regions, however,

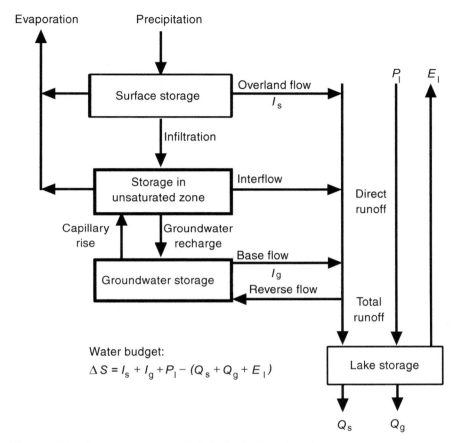

Figure 1.3.1 A conceptual model of the hydrological cycle.

where annual potential evaporation exceeds precipitation, groundwater recharge is frequently derived from temporary rivers or lakes that are in flood.

The storage, flow and quality characteristics of surface water and groundwater are frequently dissimilar. For this reason, the interaction is important in water resource development since advantage may be taken of the differing characteristics to increase yields or improve the quality of water supplies. Changes in one part of the hydrological cycle may induce beneficial or detrimental changes in another part of the cycle (Wright, 1980).

1.3.3 GROUNDWATER RECHARGE AND PROCESSES

Recharge is the term used to describe the water that enters the groundwater reservoir. It can flow downwards, upwards or sideways, and may only indirectly originate from precipitation. One of the most important factors in the determination of groundwater resources is the estimation of natural recharge to aquifers. Groundwater recharge occurs when the residual precipitation has penetrated to the groundwater table. The recharge due to snowmelt is of greatest significance in the northern and central parts of the Nordic Countries, where the main recharge event occurs as a result of the snowmelt in spring.

When groundwater is recharged mainly from rainfall (as in temperate environments), the changes in groundwater level depend on the characteristics and state of the aquifer and overlying soil. Important factors are the degree of saturation, effective porosity and permeability of the soil, and the distance that infiltrating water has to travel to reach the groundwater table. These factors all contribute to the delay in the reaction of groundwater table to rainfall. In Finland and in the other Nordic Countries, this delay is usually short, as the soil layer is generally rather thin and predominantly Quaternary. This causes the groundwater table to follow closely the annual rhythm of seasonal changes. In most geological formations of Finland, groundwater recharge is strongly seasonal, taking place mainly in the spring and autumn months.

The rise in groundwater level due to melting water is proportional to the effective porosity of the soil and the ratio of the volume of interconnected pore space available for fluid transmission to the bulk volume of the soil or rock. The effective porosity varies between till and glaciofluvial soil on average from about 5 to 30%.

The hydrostatic pressure at the surface of the groundwater corresponds to the prevailing atmospheric pressure. The groundwater surface loosely follows the topography of the soil surface. The relationship between the soil surface and the groundwater table varies in different soil types.

The groundwater table fluctuates in response to external factors in both confined and unconfined aquifers. Variations in level may be either relatively rapid, as in the case of seasonal variation, or long-term with a duration of several years. The short-term variations may be caused by changes in atmospheric pressure, earth tremors, precipitation or seepage of surface water. In addition, human interference, such as the regulation of watercourses, drainage, and earthworks, may cause rapid, either short-term or permanent, changes in level. Groundwater level fluctuations in a given locality tend to occur in a regular manner so that these may be used as an index of the regime characteristics (Soveri, 1985).

In temperate and humid environments, groundwater flow will tend to be towards lakes and rivers, where it becomes the base flow component of the surface water. Groundwater can penetrate into a lake by many different ways, e.g. as juveline water generated as a result of processes of degasification of the earth's mantle, or with river runoff as base flow or as groundwater flow, recharging in watershed areas and discharging directly into the lake bypassing the river network. The different pathways of fluid movements, which may be considered as 'groundwater' in the coastal zone, have been described by Buddemeier (1996) (see Figure 1.3.2).

When modelling water and solute fluxes of the total amount of water of underground origin present in seas or lakes, it must be also taken into account that about one third of river discharge is of groundwater origin (Zektser and Dzhamalov, 1988). During prolonged periods of dry weather, a high proportion of river flow tends to be derived from groundwater seepage. Thus, the quality of groundwater frequently tends to dominate the quality of dry-weather river flows. Groundwater is generally of good quality, but if it is polluted then there is the risk of surface waters becoming polluted, especially during low-flow conditions when there is a minimum of dilution of base flow. A relatively common example of river pollution by groundwater is that caused by the discharge of mine drainage to watercourses. This type of pollution may occur when mine water is pumped or when there is a natural overflow from a disused mine.

The geochemical cycle on land can be characterized by four major reservoirs, namely the atmosphere, the biosphere, soil and rock, where water is the primary agent for moving the chemical materials between them (Figure 1.3.3).

Inputs of materials to soil and vegetation originate partly as precipitation from the atmosphere, and partly from soil, as a result of weathering. Outputs of materials occur mainly by baseflow transport of particulate and dissolved material, but also to the atmosphere through gas emission from soil and

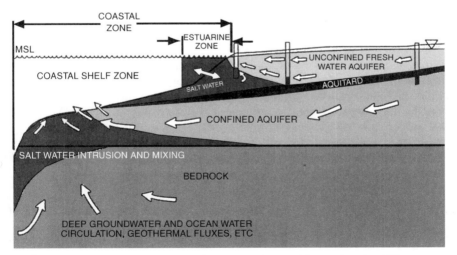

Figure 1.3.2 Cross-sectional views of the coastal zone, illustrating the different types and pathways of groundwater (Redrawn from Buddemeier, 1996).

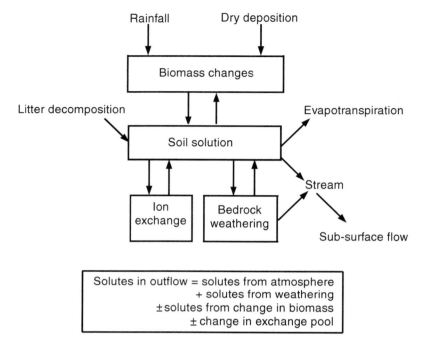

Figure 1.3.3   Schematic cycle of some chemical materials at the continental surface.

vegetation (CO_2, NH_3, H_2S, etc.), aerosol emission from vegetation and from particulate emission (dust, forest fires, etc.).

In most cases, a given element will be present in this input–output cycle in various different forms, e.g. gases, ions, molecules, dissolved and particulate organic material and inorganic particulate matter. If a given element is not enriched by soil weathering or stored in the ecosystem, then the total output should be equal to the total input. When rock weathering is low, the groundwater output is greatly influenced by atmospheric inputs (Maybeck, 1983; Soveri, 1985).

1.3.4   SOME CASE STUDIES AND METHODOLOGIES

Groundwater flows directly into lakes have been examined in some detail in recent years in several countries, e.g. the USA, at the Great Lakes Environmental Research Laboratory (GLERL), in co-operation with Poland and Russia.The overall challenge of this project is to study the groundwater flux to Lake Michigan and to two Polish lakes. Geological, hydrogeological, geophysical, isotopic and hydrodynamical techniques will be applied in this work. This study will require two methods for a comprehensive analysis. First, the available data will be applied to a groundwater model in order to determine the regional groundwater contribution to the lake. Secondly, detailed investigations will be conducted at specific sites. Hydrogeological, isotopic and geophysical data will be collected in the field and analyzed for anomalies in the lake water and sediment resulting from groundwater discharge. In addition, measurements of groundwater gradients, fluxes and water chemistry will be obtained by using a variety of instruments, including piezometers, seepage meters and ground penetrating radar. Results obtained from the second approach will be used to correct regional estimates and improve the methodology (GLERL, 1996).

In Japan in 1991, Lake Sai, one of the so-called Fuji Five Lakes, was flooded. There were several typhoons that year, with precipitation amounting to 450 mm during a two-month period. The maximum water level of Lake Sai was recorded at 8.73 m above normal and there was no perennial river flowing into the lake. After hydrological studies had been carried out, the fact was evidenced that the lake's water level had increased in part due to the release of groundwater. Isotope analyses (oxygen, deuterium and tritium) and water quality analyses of samples taken from Lake Sai and observation wells around the lake were conducted. Irregularities in the lake water quality were observed in the zone where it was estimated that groundwater flowed into the lake and led to posit the existence of springs at the bottom of the lake, spewing forth groundwater generated on Mount Fuji. Submarine-camera investigations confirmed the existence of lake bottom springs located at those points where irregularities in water quality had been observed in the first part of the study (Marui *et al.*, 1995).

A new project 'Groundwater Inflow from Coastal Aquifers to the Baltic Sea-Monitoring Proposal', supported by the Nordic Council of Ministers, started in 1998. The main aim of this project is to evaluate the national inflow and groundwater load using the existing data, on which basis the load calculations (quality and quantity) will be made. Conceptual hydrogeological models will be developed for selected pilot areas in order to form a computational monitoring method for groundwater inflow to the Baltic Sea. The results will then be used for an assessment of the effects of human impact on the quality of groundwater discharged to the Baltic Sea.

1.3.5 CONCLUSIONS

Various improvements have been made in recent years in the methods for assessing the interaction between lakes and groundwater. These improvements have been in several fields, such as instrumentation, the use of tracers and models. A particularly useful aid in understanding the process has been the use of tracers. For example, environmental isotope techniques now provide hydrologists, biologists and environmental scientists with a method for studying the actual mass transfer of water and to focus on the importance of groundwater seepage out of submerged aquifers in the supply of nutrients and contaminants to lakes. In many situations, the use of mathematical models are advisable, and indeed often essential to investigate the interaction between surface water and groundwater. A complete knowledge of the hydrogeology and the geometry of a given problem area is rarely, if ever, available. Thus, estimates have to be made of the suitable parameter values to be used in the models.

REFERENCES

Buddemeier, R. W., 1996. Groundwater flux to the ocean: definitions, data, applications, uncertainties, Proceedings of the International Symposium, Groundwater Discharge in the Coastal Zone, Moscow, Russia, 6–10 July, 1996; LOICZ Reports and Studies No. 8, LOICZ, 1790 AB Den-Burg Texel, The Netherlands, 16–21.

Cable, J. E., 1998. Assessing sources of sub-surface fluid inputs to marine waters using 222Rn/224Ra, Ocean Sciences Meeting, San Diego, USA, 9–13 February, 1998; *EOS, Trans. Am. Geophys. Un.*, **79**(1).

GLERL, 1996. Large Lakes Groundwater Budget, Information sheet from Great Lakes Environmental Research Laboratory, Ann Arbor, MI, USA.

Maybeck, M., 1983. Dissolved loads of rivers and surface water quantity/quality relationships. Atmospheric inputs and river transport of dissolved substances, IAHS Publication No.141, Galliard Ltd, Great Yarmouth, UK.

Marui, A., Yasuhara, M., Kono, T., Sato, Y., Kakiuchi, M., Hiyma, T., Suzuki, Y. and Kitagava, M., 1995. Water quality and lake bottom springs of Lake Sai on Mount Fuji's northern slope, *J. Jpn Ass. Hydrol. Sci.*, **25**(1).

Soveri, J., 1985. Influence of Meltwater on the Amount and Composition of Ground Water in Quaternary Deposits in Finland, Publications of the Water Research Institute, Helsinki, Finland, Vol. 63.

Wright, C. E. (Ed.), 1980. Surface water and groundwater interaction, Studies and Reports in Hydrology, No 29, Unesco, Paris.

Zekster, I. S and Dzhamalov, R. G., 1988. Role of Groundwater in the Hydrological Cycle and in Continental Water Balance, IHP III Project 2.3, Unesco, Paris.

Chapter 1.4

Nitrogen Leaching from Forested Soils into Watercourses

AHTI LEPISTÖ

Hydrological and Limnological Aspects of Lake Monitoring
Edited by Pertti Heinonen, Giuliano Ziglio and André Van der Beken
©2000 John Wiley & Sons, Ltd, ISBN 0 471 89988 7

1.4.1 INTRODUCTION

Nitrogen is a limiting element for terrestrial and marine production. Human activities such as combustion and agriculture, as well as growing populations, have severely impacted the global nitrogen cycle over the past decades. Today, human conversion of N_2 to more reactive N species, the global 'fertilization experiment', is promoting a wide range of environmental problems; from the effects on atmospheric chemistry, to acidification of freshwater systems, and to marine eutrophication. Eutrophication has been one of the main water problems in Europe for several decades. Mainly appearing first in lakes receiving domestic or industrial effluents, it is now affecting many coastal areas and lakes in rural areas without identified point sources of pollution.

Since the contribution from forests to the total nitrogen load to the watercourses and the sea is significant, it is important to quantify the load of different fractions of nitrogen and to analyze the magnitude of spatial variation and the reasons behind this variation. Spatial variability may be explained by natural catchment characteristics, but may also be a response to climatic factors and anthropogenic factors such as N deposition and forest management. The effects of such factors have been indicated by upward trends in nitrogen loads and concentrations. The atmospheric N input to forests in Europe and North America has increased dramatically during recent decades, due to the emission of NO_x from combustion processes and of NH_3 from agricultural activities. Forest ecosystems may accumulate considerable amounts of N in biomass and soil organic matter, but there is increasing concern that forest ecosystems may be overloaded or saturated with N from atmospheric deposition, thus leading to increased nitrate leaching.

Both nitrogen and phosphorus leaching may considerably affect ground and surface water quality and contribute to the eutrophication of watercourses and marine areas. Regarding large biogeochemical cycles, the N cycle has, quantitatively, been subject to much more severe changes. Possible factors affecting the leaching of nitrates are natural fluxes of N, hydrology, forest management activities, N deposition and climate change.

1.4.2 NUTRIENT CYCLING AND NITROGEN RETENTION

The quantitative assessment of the nitrogen cycle in forested ecosystems is difficult, because of the many gaseous, aqueous and particulate forms of nitrogen, and the large number of complex pathways involved. Nitrogen exists in inorganic and organic forms in both the solution and the solid phases of soils. Inorganic N in soil solutions includes nitrate (NO_3^-), nitrite (NO_2^-) and ammonium (NH_4^+), in addition to the dissolved gaseous forms (ammonia (NH_3), nitrous oxide (N_2O), other N oxides, etc.).

The nitrogen transformations which operate in soils are shown in Figure 1.4.1. More than 90% of the N in soils is present in organic forms not available to plants. For this N to become available, it must first be transformed (ammonification) by micro-organisms to NH_4^+, which is available for plant uptake and/or for oxidation to NO_3^- (nitrification) by nitrifying micro-organisms (see Figure 1.4.1). The process of denitrification (anaerobic NO_3^- reduction to N_2 or N_2O) in forest soils has received increasing attention in recent years, since this process might balance some of the N inputs. Most observations still indicate very low denitrification rates in soils of undisturbed forests ($< 1 kg N ha^{-1} a^{-1}$) (Gundersen and Bashkin, 1994).

Forests constitute large and highly heterogenous pools of nitrogen, and the retention of nitrogen is usually high in forest ecosystems. The forest floor often acts as a major site of nutrient retention (Vogt *et al.*, 1986). Stand age can be a critical factor regulating the losses of nitrogen by leaching. During early regrowth, the large accumulation of biomass facilitates nitrogen retention and minimizes drainage losses. With time, the rate of nitrogen accumulation peaks and older stands require less net nitrogen inputs and show greater leaching losses.

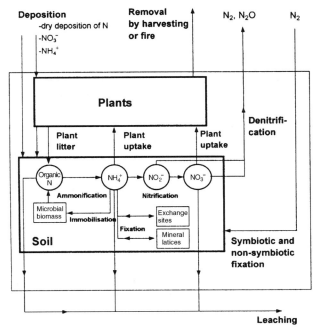

Figure 1.4.1 The nitrogen cycle with N transformations and fluxes.

1.4.3   RUNOFF GENERATION AND CHARACTERISTICS
OF BOREAL FOREST SOILS

In humid regions, chemical flux and cycling are intimately linked to the hydrological cycle. Small catchments provide a natural framework for various types of research, e.g. studies concerning nutrient leaching and cycling, mass balances, effects of landuse change and hydrological processes.

A fundamental premise of many hydrochemical studies is that hydrological processes, i.e. the source, pathway and residence time of water, in a catchment exert a strong control on the water chemistry (e.g. Hooper and Shoemaker, 1986). One of the key areas in any attempt to understand variability in catchment outputs is runoff generation. How is water routed through the catchment and how is runoff generated?

Hewlett and Hibbert (1967) put forward the variable-source-area concept of runoff generation, as a basis for understanding the catchment response to storm events. This dynamic framework for storm runoff generation was based on the notions that infiltration is seldom a limiting factor in forested environments and that subsurface stormflow (Kirkby, 1978), rather than overland flow *per se*, is capable of making a significant contribution to the flood hydrograph (Bonell, 1993). Modeling and field experiments from hydrometric and environmental tracer studies support the idea of a continuum in both spatial and temporal occurrence of infiltration-excess overland flow, saturation overland flow and subsurface stormflow, within individual catchments under different conditions of rainfall, antecedent soil moisture and intensity of land use impacts (Bonell, 1993).

In forested catchments in Boreal zones, a large proportion of runoff is generated by water recharge from rather shallow soils. The dominance of till soils is also typical of these forested areas, where saturated hydraulic conductivity is often high at the soil surface and decreases rapidly with depth. During conditions of low flow, water drains slowly from the catchment, because the most conductive superficial soil layers are unsaturated. Fresh inputs of rainfall to the soil increase the saturated depth in the soil profile and reduce the proportion of air-filled pores in the unsaturated zone. The resulting increase in transmissivity leads to an increase in runoff from the catchment (Bishop, 1991).

Analyses of the chemical or isotopic composition of streamwater during runoff events provide information on the integrated result of the various processes which contribute to streamflow generation. Hydrograph separation, i.e. the dividing of runoff into 'old' and 'new' components by using isotopes, has been carried out since the early 1970s, predominantly in mountainous areas in central Europe and in Canada, and later in Scandinavia. During the past two to three decades, it has been pointed out in many isotope studies that pre-event water is the dominating component ($>50\%$) of runoff, even during the high flow period. These separation analysis results, together with artificial tracers, provide information about the relative age of the water and the residence time for biogeochemical processes to occur in the soil, and are valuable for development and verification of more realistic distributed catchment models. Reviews of the nature of environmental isotopes and their use in hydrograph separations have been provided by, e.g. Buttle (1994) and Herrmann (1997), whereas links between runoff formation processes and nutrient leaching have recently been discussed by Lepistö *et al.* (1997).

1.4.4   RELATIONSHIPS BETWEEN RUNOFF AND CONCENTRATIONS

Part of the variance in streamwater concentration is usually a function of the streamflow. This comes about as a result of two different kinds of physical phenomena. One is dilution, i.e. a solute may be delivered to the stream at a reasonably constant rate, whereas the flow changes over time. The result of this situation is a decrease in concentration with increasing flow. The other process is washoff, i.e. a solute, sediment, or a constituent attached to the sediment can be delivered to the stream primarily from (surface-saturation) overland flow, or from streambank erosion. In these cases, concentrations as well as fluxes tend to increase with increasing flow (Helsel and Hirsch, 1992).

The relationships found in regression analyses are often non-linear and the patterns may be very different for different substances as well as for different catchments. When nitrogen export from a catchment is estimated, calculations are often based on linear interpolation of concentrations. However, the estimated load may be very different from the 'true' load, since concentrations rarely change linearly over time. There is considerable short-term variation in nitrogen concentrations in forest streams, which is influenced by water flow paths and transit times and by seasonal variability of biological processes (e.g. Turner and Macpherson, 1990; Andersson and Lepistö, 1998). Incorporation of knowledge about such links into load estimates would significantly improve the accuracy and precision of these export estimates. The proportion of direct surface runoff, or near-surface runoff, with short transit times may be especially high in the initial phase of the increasing flow, when the total runoff volume is still low (Turner and Macpherson, 1990). Increased concentrations in these situations may be explained by the near-surface flow paths and flushing of accumulated nitrates.

1.4.5   FOREST MANAGEMENT, HARVESTING AND DRAINAGE

Some areas of Europe, particularly in the Nordic and western regions, have witnessed the intensification of wood-producing techniques and practices over the last few decades. This has led to some extent, to the evolvement of the mechanization of harvesting of trees, the draining of 'wet' forests and the use of fast-growing tree species for clear-felling at regular intervals (Stanners and Bourdeau, 1995).

Forest harvesting results in direct removal of nitrogen in forest biomass, alters mineralization and nitrification rates and increases water and N fluxes from catchments. The degree of these impacts depends on the intensity of harvesting. Transpiration, interception, and hence evapotranspiration, are generally reduced by forest harvesting, which produces more soil water available for the remaining plants and/or increased water movement to streams or groundwater (Swank and Johnson, 1994). Increased nitrification in the forest floor after cutting may significantly increase amounts of nitrate N available for leaching. Of all forest management activities, harvesting is perhaps the most disruptive to element cycles. The paired small-catchment method has proved to be a useful tool for studying the effects of harvesting and other forest management activities on element dynamics in forests (e.g. Mann *et al.*, 1988). The long-term impact of various forest management practices have been studied in eastern Finland, where clear-cutting led to almost 40 kg N ha^{-1} (specific load of land affected) cumulative export during the 12-year monitoring period in a peatland-dominated catchment, with no decrease on even going back to the pre-treatment level. In a nearby mineral-soil catchment with protective zones, the export was considerably lower (Ahtiainen and Huttunen, 1999).

Drainage is a prerequisite for the utilization of peatlands and wet mineral soils in forestry. Drainage of organic soils causes increased mineralization in the aerated soil profiles, thus leading to increased amounts of ammonium-N which are available for nutrient uptake, adsorption or immobilization. Ammonium-N may further nitrificate, but if the environmental conditions (e.g. pH) are not present, this nitrogen will remain in an available form leading to increased outflow (Sikora and Keeney, 1983). Supplementary drainage works (e.g. cleaning of ditches) may have the same effect as the original drainage, with increased erosion and leaching of substances having once been sedimented on to the ditches. The integrated impacts of atmospheric deposition and forestry activities, together with the catchment characteristics on spatial leaching of N fractions, were studied by using data obtained from the 20 forested catchments in Finland and Sweden (Lepistö *et al.*, 1995). The most important factors for explaining the spatial variability of organic N losses were clearly forestry activities, drainage and clear-cutting.

1.4.6 NITROGEN DEPOSITION

Recent measurements of elevated nitrate leaching from certain high-elevation forests suggest that these forests have reached saturation; cumulative nitrogen deposition inputs have exceeded the capacity of these systems to accumulate nitrogen (Aber *et al.*, 1989).

Inputs of nitrogen to a forested catchment (Figure 1.4.1) may occur through: (1) wet deposition of NO_3^- and NH_4^+, and organic nitrogen, (2) dry deposition, (3) N_2 fixation, or (4) as a result of forest management, through fertilizing. The atmospheric nitrogen load to terrestrial ecosystems in Europe and North America has increased dramatically during recent decades (Grennfelt and Hultberg, 1986; Galloway and Likens, 1981). The main sources are emission of NO_x from combustion processes and emission of NH_3 from agricultural activities. Inorganic N inputs in throughfall range between <1 and about $70\,kg\,ha^{-1}\,a^{-1}$, from recent European data obtained from 140 forest ecosystems (Dise *et al.*, 1998).

There is growing evidence that otherwise undisturbed catchments may show effects of increased N deposition (Dise and Wright, 1995). Increased nitrogen deposition can have two types of indirect effects at the ecosystem level, namely biomass accumulation, or nitrate leaching when nitrogen is no longer buffered by biomass uptake. The range in N fluxes can span from complete retention to nitrogen-saturated sites (Dise and Wright, 1995). Detailed process studies at forested sites in Europe showed that nitrate leaching was more related to an index for ecosystem 'N status' than to the N input (Gundersen *et al.*, 1998), and recent analyses using European databases indicate an empirical relationship between forest floor C/N ratio and nitrate leaching (Dise *et al.*, 1998).

In Boreal zones, both N deposition and leaching levels are generally low. Some signs of elevated leaching due to N deposition over the period from the 1970s to the 1990s were detected in two intensively studied, fertile catchments in southern Finland (Lepistö, 1996). The experimental increase of N deposition to mid-European levels ($40\,kg\,N\,ha^{-1}\,a^{-1}$) in the Gårdsjön catchment (in Sweden) caused a dramatic increase in the N loss during the five years following the addition (Moldan and Wright, 1998). In Norway, during the period 1993–1995, a study on N inputs and losses to streamwater was carried out in 19 sub-catchments of the mountainous-areas-dominated Bjerkreim river ($685\,km^2$) in southwestern Norway. From this study, about 70% of the total N fluxes was estimated to be of atmospheric origin (Kaste *et al.*, 1997).

Increased concentrations of oxidized N in freshwater systems, resulting from increased leakage from catchments, would promote acidification. The relative importance of nitrogen compounds to the acidity of surface waters is increasing, because the sulfate/nitrate ratio in precipitation is decreasing and the future reduction in NO_x emissions will be smaller than for SO_2. The model simulations have indicated that N emission controls are extremely important to enable the maximum recovery in response to S emission reductions (Forsius *et al.*, 1998). In watersheds such as the Bjerkreim river, nitrate may already contribute significantly (up to 40%) to acidity (Hessen *et al.*, 1997).

1.4.7 ECOSYSTEM EXPERIMENTS AND MANIPULATION

European concern over the cause and consequences of acidification of soils and surface waters, forest decline and nutrient enrichment of terrestrial and aquatic ecosystems led in the late 1980s to the establishment of the NITREX (Nitrogen Saturation Experiments) and EXMAN (Experimental Manipulation of Forest Ecosystems in Europe) projects. In the NITREX work, N was either added to or removed from ambient atmospheric deposition in order to simulate major changes in deposition, while in the EXMAN work ambient atmospheric deposition was experimentally altered in chemical composition and/or quality (Wright and Rasmussen, 1998). Gundersen *et al.* (1998), after analyzing the NITREX data for the impacts of four to six years of treatment in five coniferous forests, found that nitrate leaching responded within the first year of treatment at all sites, whereas reponses in vegetation and soil were delayed. Changes in nitrate leaching were small at the low-N-status sites and substantial at the high-N-status sites.

1.4.8 CLIMATE-CHANGE IMPACTS

The major N fluxes occur under 'normal' climate conditions in late autumn and early spring (boreal zones), and also in winter (temperate zones). Climatic shifts, due to a possibly increasing temperature, may entail feedbacks of the seasonal N flux, affecting both acidification and eutrophication responses. One major feedback that could be seen under warmer and drier conditions, is increased mineralization and mobilization of the huge stores of organic N in northern peat and forested soils (e.g. Hessen *et al.*, 1997). Typical northern boreal coniferous forest soils contain several tons of N per hectare, and there is a gradient with decreasing soil stores of N towards the south. Mobilization even of a very minor part of these stores could significantly alter both N concentrations and N transport.

1.4.9 BUFFER ZONES AS SOURCES OR SINKS OF NITRATE

The buffer zone is a chemically and hydrologically complex environment which, together with the streambed, can profoundly influence some aspects of stream chemistry. Because of wet soils adjacent to the streams, riparian buffers may occur between agricultural or urban activities in the uplands of the catchments and small streams. These riparian areas have been shown to be very valuable for the removal of non-point-source pollution from drainage water. Several investigators have measured $>90\%$ reductions in sediment and nitrate concentrations in agricultural water flowing through the riparian areas (Gilliam, 1994). The excellent review by Hill (1996) gives a detailed description of the current status of knowledge on the removal of nitrate. In addition, Cirmo and McDonnell (1997) have recently reviewed the research carried out concerning the links between hydrological flowpaths and the biogeochemical environment controlling N transport in near-stream zones.

Groundwater is frequently present at a shallow depth beneath the riparian area, and vegetation and soil processes may therefore modify the chemistry of groundwater before it enters the stream (e.g. Lowrance *et al.*, 1985). One might assume that atmospheric N deposition on saturated areas will contribute more or less instantly to the N leaching during flow events. The buffer zones may play an important role in the reduction of nutrient export, having a potential to capture nitrate leached from upper parts of the catchment, either by nutrient uptake or denitrification. Unfortunately, the major water and N fluxes to streams in boreal zones occur during dormant periods when no N uptake by vegetation occurs.

1.4.10 CONCLUSIONS

The driving forces of large-scale environmental changes, i.e. N deposition, land-use practices and expected climate change, operate simultaneously and predictions concerning their integrated effects are needed. Long-term monitoring in representative or experimental catchments is regarded as important when assessing environmental changes, together with intensifying process studies in hydrology, hydrochemistry, atmospheric deposition, biology and catchment soils, and with the use of integrated models. The precise and accurate description of ecosystems within small catchments may serve as ideal references for an evaluation and scaling of remote sensing methods.

One of the biggest limitations to a broader application of watershed studies is the uncertainty regarding how well the understanding of ecosystem processes at the small catchment scale can be transferred to significantly larger landscape units. There are large uncertainties about the relative importance of different processes at different spatial scales. For example, processes determined to be of key importance at the scale of 1 km^2 may be unimportant or overwhelmed by other processes that operate only at scales greater than, e.g. 100 km^2. Hypotheses developed at the small watershed scale must be tested at larger scales to determine the controlling processes at each scale. Recent advances in geographic information systems (GIS), remote sensing, and environmental modelling should be jointly exploited to test new hypotheses. Further extrapolation of the understanding of systems at the individual stand and site level to the large units, catchments, ecotypes, watersheds, landscapes and regions has a high priority in future work.

REFERENCES

Aber, J. O, Nadelhoffer, K. J., Steudler, P. and Melillo, J. M., 1989. Nitrogen saturation in northern forest ecosystems, *Bioscience*, **39**, 378–386.

Ahtiainen, M. and Huttunen, P., 1999. Long-term effects of forestry managements on water quality and loading in brooks, *Boreal Environ. Res.*, **4**, 101–114.

Andersson, L. and Lepistö, A., 1998. Links between runoff generation, climate and nitrate-N leaching from forested catchments, *Water, Air, Soil Pollut.*, **105**, 227–237.

Bishop, K. H., 1991. Episodic Increases in Stream Acidity, Catchment Flow Pathways and Hydrograph Separation, PhD Thesis, University of Cambridge.

Bonell, M., 1993. Progress in the understanding of runoff generation dynamics in forests, *J. Hydrol.*, **150**, 217–275.

Buttle, J. M., 1994. Isotope hydrograph separations and rapid delivery of pre-event water from drainage basins, *Progr. Phys. Geog.*, **18**, 16–41.

Cirmo, C. P. and McDonnell, J. J., 1997. Linking the hydrological and biogeochemical controls of nitrogen transport in near-stream zones of temperate-forested catchments: a review, *J. Hydrol.*, **199**, 88–120.

Dise, N. B. and Wright, R. F., 1995. Nitrogen leaching from European forests in relation to nitrogen deposition, *For. Ecol. Manage.*, **71**, 153–161.

Dise, N., Matzner, E. and Gundersen, P., 1998. Synthesis of nitrogen pools and fluxes from European forest ecosystems, *Water, Air, Soil Pollut.*, **105**, 143–154.

Forsius, M., Guardans, R., Jenkins, A., Lundin, L. and Nielsen, K. E. (Eds), 1998. Integrated Monitoring: Environmental Assessment through Model and Empirical Analysis – Final Results from an EU/LIFE Project, Finnish Environment No. 218, Finnish Environment Institute, Helsinki, Finland.

Galloway, J. N. and Likens, G., 1981. Acid precipitation: Importance of nitric acid, *Atmos. Environ.*, **15**, 1081–1085.

Gilliam, J. W., 1994. Riparian wetlands and water quality, *J. Environ. Qual.*, **23**, 896–900.

Grennfelt, P. and Hultberg, H., 1986. Effects of nitrogen deposition on the acidification of terrestrial and aquatic ecosystems, *Water, Air, Soil Pollut.*, **30**, 945–963.

Gundersen, P. and Bashkin, V. N., 1994. Nitrogen cycling, in Moldan, B. and Cerny, J. (Eds), Biogeochemistry of Small Catchments: A Tool for Environmental Research, John Wiley & Sons, Chichester, SCOPE Report 51, 255–283.

Gundersen, P., Emmett, B. A., Kjonaas, O. J., Koopmans, C. J. and Tietema, A., 1998. Impact of nitrogen deposition on nitrogen cycling in forests: a synthesis of NITREX data, *For. Ecol. Manage.*, **101**, 37–55.

Helsel, D. R. and Hirsch, R. M., 1992. Statistical Methods in Water Resources, *Studies in Environmental Science*, No. 49, Elsevier, Amsterdam.

Herrmann, A., 1997. Global review of isotope hydrological investigations, in Oberlin, G. and Desbos, E. (Eds), FRIEND (Flow Regimes from International Experimental and Network Data), Third Report: 1994–1997, Unesco, Paris, Ch. 6, 307–316.

Hessen, D. O., Henriksen, A., Hindar, A., Mulder, J., Torseth, K. and Vagstad, N., 1997. Human impacts on the nitrogen cycle: a global problem judged from a local perspective, *Ambio*, **26**, 321–325.

Hewlett, J. D. and Hibbert, A. R., 1967. Factors affecting the response of small watersheds to precipitation in humid areas, in Sopper, W. E. and Lull, H. W. (Eds), *Forest Hydrology*, Proceedings of the International Symposium on Forest Hydrology, Pennsylvania State University, Philadelphia, PA, USA, August 29–September 10, 1965, Pergamon, New York, 275–290.

Hill, A. R., 1996. Nitrate removal in stream riparian zones, *J. Environ. Qual.*, **25**, 743–755.

Hooper, R. P. and Shoemaker, C. A., 1986. A comparison of chemical and isotopic hydrograph separation, *Water Resour. Res.*, **22**, 1444–1454.

Kaste, Ø., Henriksen, A. and Hindar, A., 1997. Retention of atmospherically-derived nitrogen in subcatchments of the Bjerkheim river in southwestern Norway, *Ambio*, **26**, 296–303.

Kirkby, M. J. (Ed.), 1978. *Hillslope Hydrology*, John Wiley & Sons, Chichester.

Lepistö, A., Andersson, L., Arheimer, B. and Sundblad, K., 1995. Influence of catchment characteristics, forestry activities and deposition on nitrogen export from small forested catchments, *Water, Air, Soil Pollut.*, **84**, 81–102.

Lepistö, A., 1996. Hydrological Processes Contributing to Nitrogen Leaching from Forested Catchments in Nordic Conditions, *Monographs of the Boreal Environment Research* No. 1, Finnish Environment Institute, Helsinki.

Lepistö, A., Andersson, L., Herrmann, A. and Holko, L., 1997. Hydrological processes in forested catchments, in Oberlin, G. and Desbos, E. (Eds), FRIEND (Flow Regimes from International Experimental and Network Data), Third Report: 1994–1997, Unesco, Paris, Ch 6, 317–329.

Lowrance, R. R., Leonard, R. and Sheridan, J., 1985. Managing riparian ecosystems to control nonpoint pollution, *J. Soil Water Conserv.*, **40**, 87–91.

Mann, L. K., Johnson, D. W., West, D. C., Cole, D. W., Hornbeck, J. W., Martin, C. W., Riekirk, H., Smith, C. T., Swank, W. T., Tritton, L. M. and Van Lear, D. H., 1988. Effects of whole-tree and stem-only clearcutting on postharvest hydrologic losses, nutrient capital, and regrowth, *For. Sci.*, **34**, 412–428.

Moldan, F. and Wright, R. F., 1998. Changes in runoff chemistry after five years of N addition to a forested catchment at Gårdsjön, Sweden, *For. Ecol. Manage.*, **101**, 187–197.

Sikora, L. J. and Keeney, D. R., 1983. Further aspects of soil chemistry under anaerobic conditions, in Gore, A. J. P. (Ed.), *Ecosystems of the World, Vol. 4A, Mires: Swamp, Bog, Fen and Moor: General Studies*, Elsevier, Amsterdam, 247–256.

Stanners, D. and Bourdeau, P. (Eds), 1995. *Europe's Environment, The Dobris Assessment*, European Environment Agency, Copenhagen, Denmark, 464–478.

Swank, W. T. and Johnson, C. E., 1994. Small catchment research in the evaluation and development of forest management practices, in Moldan, B. and Cerný, J. (Eds), *Biogeochemistry of Small Catchments: A Tool for Environmental Research*, John Wiley & Sons, Chichester, SCOPE Report 51, 383–408.

Turner, J. V. and Macpherson. D. K., 1990. Mechanisms affecting streamflows and stream water quality: An approach via stable isotope, hydrogeochemical and time series analysis, *Water Resour. Res.*, **26**, 3005–3019.

Vogt, K. A., Grier, C. C. and Vogt, D. J., 1986. Production, turnover and nutrient dynamics of above- and belowground detritus of world forests, *Adv. Ecol. Res.*, **15**, 303–377.

Wright, R. F. and Rasmussen, L., 1998. Introduction to the NITREX and EXMAN projects, *For. Ecol. Manage.*, **101**, 1–7.

Chapter 1.5

Chemical Variables in Lake Monitoring

SIRPA HERVE

Hydrological and Limnological Aspects of Lake Monitoring
Edited by Pertti Heinonen, Giuliano Ziglio and André Van der Beken
©2000 John Wiley & Sons, Ltd, ISBN 0 471 89988 7

1.5.1   INTRODUCTION

The lake ecosystem can in theory be divided into two different compartments, which, however, closely affect each other, namely the biotope and the biocoenosis. When there are some quality changes in the biotope, the biocoenosis also changes either immediately or after a short delay, and vice versa. All of the living organisms, for instance in eutrophic lakes, induce very clear chemical variations and changes in the biotope, in oxygen concentrations or in the concentrations of different substances and compounds in the different layers of the lake water.

The biotope is the abiotic part of the ecosystem. It forms an environment in which a set of organisms is created, i.e. the biocoenosis, according to its characteristics. The biotope can be characterized by different hydrological (water level of lakes, discharges in rivers, etc.), physical (thermal and optical conditions) and chemical variables.

The biocoenosis, on the other hand, consists of all of the living organisms found in the lake ecosystem concerned, i.e. bacterioplankton, phytoplankton, zooplankton, bottom fauna, periphytic communities, macrophytes, fishes, etc. The biocoenosis is the living part of the ecosystem. Of course, a biocoenosis can best be characterized by different biological methods, phytoplankton counting, species diversity, etc. Many of these themes will also be discussed in this present book (see Part Two).

The most important chemical variables measured in a lake are as follows (e.g. Hutchinson, 1957; Wetzel, 1983):

(a) the oxygen status, including the organic compounds affecting the oxygen budget;
(b) the nutrient (phosphorus and nitrogen) status;
(c) the acidity status;
(d) the harmful substances, including heavy metals and a large number of organic compounds.

1.5.2   THERMAL PROPERTIES

Before discussing the chemical variables usually measured in lake research and monitoring, some important principles concerning the thermal properties of lakes need to be mentioned (see also Chapter 1.1). Reliable research and monitoring of lakes, especially the assessment of chemical and biological results, always require relevant information about the thermal variations that take place during one year in the lake. All of the examples cited in this chapter are from northern latitudes.

In northern countries, four different thermal situations can be found in a lake during the period of one year:

- summer stratification;
- autumn circulation;
- winter stratification (or inverse stratification);
- spring circulation.

The longest of these seasons in Finland is the winter stratification, which in southern Finland varies between 100 and 140 days. In the northernmost part of Finland, the winter season in a lake may even extend over 200 days. The lakes are then covered with ice and the winter stratification is quite solid. The ice and the snow that cover such lakes prevent light penetration to the water mass, and the dominant vital function is bacterial decomposition.

The shortest thermal season is the spring circulation after the breakup of ice. The spring circulation restores the poorer oxygen budget of the water very quickly. The autumn circulation is, on the other hand, remarkably long and significant with regard to the oxygen situation in the following winter. Of course, variations in the lengths of the different thermal conditions may be quite remarkable from year to year.

When water cools, its density increases, but it reaches its maximum value just before freezing, i.e. at $+4\,°C$. When the cooling of the water continues below $+4\,°C$, the volume starts to increase slowly, until it is suddenly extended by about one eleventh at the moment of freezing. This special feature guarantees that life can continue even in northern lakes that regularly receive an ice cover each year.

The density difference corresponding to water temperature is insignificant at around $+4\,°C$, but as the temperature increases, the corresponding density difference also increases.

During the summer stratification (Figure 1.5.1), different thermal layers can usually be found in every deep lake. These layers are traditionally termed as follows:

- epilimnion – the upper region of the lake, which is more or less uniformly warm and circulating;
- hypolimnion – the deep, cold, and relatively undisturbed region;
- thermocline – the region of rapid decrease in temperature which separates the epilimnion from the hypolimnion.

The total depth of the lake has a notable influence on the type of summer stratification in this body. The shape of the lake concerned and the prevailing wind directions are also important factors. During a windy and cold summer, the stratification type of the lake is totally different from a summer when calm and warm weather dominates.

In deep lakes, the epilimnion and hypolimnion can usually be outlined very easily in the summer with precise measurements of temperature. In typical Finnish shallow lakes, on the contrary, the hypolimnion is in many lakes

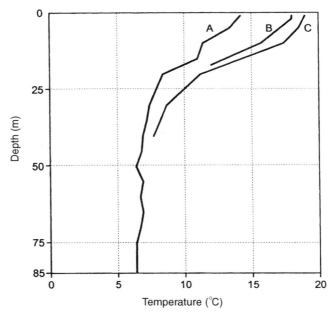

Figure 1.5.1 Thermal stratifications in summer for different types of lakes: A, deep lake with a large volume; B, shallow lake; C, relatively deep lake with a small volume.

missing, and the whole water volume consists of a totally circulating and productive epilimnion.

The maximum theoretical temperature during winter stratification is +4 °C. If the autumn is long and relatively windy, the whole water mass cools during the autumn circulation to clearly below +4 °C, i.e. down to +2 °C, and in the biggest and deepest lakes, even down to +1.0–+1.5 °C. The temperature just below the ice cover is +0.1 °C, and the difference between the epilimnion and hypolimnion temperatures is very small. This has a great effect on the oxygen budget of the lake.

On the contrary, if the winter comes very early, the freezing of the lake can already take place after a very short circulation time. The temperature below the ice is +0.1 °C, but in the deepest part of the lake it may even be +4 °C. Under this condition, the bacterial decomposition of organic matter, which uses oxygen, is much faster and may lead to a total deficit of oxygen, especially near the bottom of the lake.

1.5.3 OXYGEN

Oxygen is the most important gas that dissolves in water from the atmosphere and is the most dominant chemical factor regulating the life in waters. Its content in most natural waters determines the nature and quantity of the bacterial, zoobenthos and fish communities that are present.

Oxygen is dissolved in water from the atmosphere only when the water is open at the surface between air and water, and is dependent on the air pressure and the temperature of the water. The effect of air pressure is clearly less important, and could be disregarded in practical water studies. Water temperature has, on the other hand, a significant impact on the amounts of oxygen that are being dissolved.

Oxygen is dissolved in the water during the whole open water period with respect to the epilimnion. Oxygen is, however, dissolved in the entire water mass only during the circulation periods (spring and autumn). Of these, the autumn circulation is very significant. When the circulation takes a long time, the oxygen content in the whole water mass has time to increase. At the end of the autumn circulation, the oxygen contents are, therefore, usually near the theoretical maximum, according to the prevailing temperature, even in water areas with a high load.

The oxygen content of water is usually expressed in milligrams per litre. Table 1.5.1 presents certain water temperatures and the corresponding theoretical amounts of oxygen that can be dissolved at equilibrium, i.e. the saturation values.

The oxygen saturation values decrease when the temperature increases, but not linearly. In order to be able to compare better the oxygen results obtained at different temperatures, the results of the oxygen analyses are expressed, in addition to the concentration values, as so-called saturation values, which express the oxygen content as a percentage of the theoretical saturation value, according to the temperature of the water sample.

Further processes that increase the contents of dissolved oxygen in water are the assimilation of carbon dioxide by the macrophytes, and above all by the assimilation by small microscopical algae, the primary production in waters, i.e. photosynthesis. In this process, plants with chlorophyll utilize carbon dioxide and different nutrient salts in the water to produce organic substances with the help of light energy, according to the following simple reaction:

Table 1.5.1 Saturation oxygen contents of water at different temperatures.

Temperature (°C)	Oxygen content (mg/L^{-1})
0.0	14.2
5.0	12.4
10.0	10.9
15.0	9.8
20.0	8.8
25.0	8.1

$$6CO_2 + 6H_2O \longrightarrow C_6H_{12}O_6 + 6O_2$$

carbon dioxide + water $\longrightarrow$ organic substances + oxygen

The more carbon dioxide is bound, then the stronger the process of photosynthesis of plants with chlorophyll will be, and the more oxygen will be released in this reaction. The released oxygen is dissolved in the surrounding water. Especially in cases where pollution is caused by domestic wastewater with a high nutrient content, the algae production in the recipient watercourses increases considerably. In these cases, the amounts of oxygen can clearly exceed the saturation value during light periods.

In dark conditions, the dominant process is the respiration of algae (a reaction that is contrary to photosynthesis), in which case oxygen is consumed in the metabolic activity that produces carbon dioxide. The more eutrophic the lake is, then the larger the difference between the maximum and minimum contents of the dissolved oxygen measured over 24 h will become. When results are interpreted, it is therefore very important to know at what time of the day the oxygen sample has been taken. When the samples are taken in continuous monitoring, the exact same times of the day should be used to improve the comparability.

A water body also obtains oxygen through additional water with high oxygen content, and in Nordic countries above all, through melt waters in the spring. Especially in polluted waters, the importance of these melt waters in improving the general oxygen status may be considerable. The significance of this phenomenon is at its greatest in shallow waters, which usually have a very low oxygen content at the end of the winter.

The samples for oxygen determinations are always taken as a vertical series, starting from the uppermost layer in the epilimnion (usually 1 m) and finishing in the hypolimnion at a depth that is 1 m above the bottom sediment. The number of samples needed for the estimation of the oxygen budget of the lake depends totally on the thermal stratification of the lake. During circulation periods, the oxygen status can be measured using less samples than in the case where the difference between the epilimnetic and hypolimnetic temperatures is large. Volume-weighted values of the oxygen content are best suited for the comparison of oxygen situations between different lakes and for the same lake during different years. The oxygen status in the hypolimnion, is also, on the whole, a good monitoring variable.

The oxygen concentration is determined by a conventional and well tried titrimetric technique, i.e. the Winkler method. The analysis is started in the field after the vertical measurements of the temperature, and the results are expressed as $mg L^{-1}$, or as percentages of the saturation values.

In oligotrophic lakes (Figure 1.5.2), the difference in oxygen concentrations between the epilimnion and hypolimnion is relatively small, even in deep lakes. The oxygen curve in this case is known as *orthogradic*. In eutrophied lakes, the

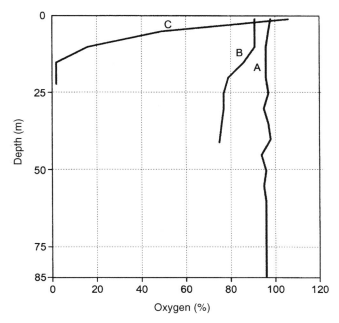

Figure 1.5.2 Oxygen concentrations in different types of lakes during summer stratification: A, oligotrophic lake; B, dystrophic lake; C, eutrophic lake.

situation is quite different. In the epilimnion, primary production is dominant, and therefore the oxygen content may also increase significantly. Saturation values of over 150% are not rare in a hypereutrophic lake. In the hypolimnion, on the contrary, the decomposition of organic matter produced in primary production is dominant, and the oxygen concentrations decrease. The steepest part of the oxygen curve is usually in the thermocline, where the change of the temperature that regulates the decomposition is the greatest. This form of the oxygen curve is termed *clinogradic*.

In a polluted or eutrophied lake, the oxygen situation changes during the winter-time. At the beginning of the winter, after autumn circulation, the oxygen concentrations may be quite normal in the whole water mass, even near the bottom. During winter, the intensity of oxygen consumption is highest near the bottom layer, where the temperature is the highest. The consumption of oxygen per time unit is at its greatest at the beginning of the winter, after the icing, and it decreases towards the end of the winter. During a long winter, the consumption may even lead to a total oxygen deficit.

In a totally polluted lake, the oxygen deficit is great even in the surface water, and a totally deoxygenated situation already starts in the thermocline. When there is no oxygen, reduced compounds, such as methane and hydrogen sulfide, start to develop in the hypolimnion. When these compounds come into contact with the water in the epilimnion during the following overturn, they deprive the water of oxygen and may use the whole oxygen reserve of the epilimnion for their own oxidation. Many deaths of fish that take place, above all in the spring, may result directly from this lack of oxygen, which arises in connection with the overturn and may last only a short time.

The oxygen concentration is a very good variable in monitoring the long-term trend of water quality in polluted or eutrophied lakes. The monitoring does require, however, very careful simultaneous registration of the weather conditions.

1.5.4 NUTRIENTS

Nutrients are substances which directly influence the primary production of a water body. Nutrients come to waters naturally, either from the ground or, on the other hand, as a result of human activities. In addition to wastewaters, the nutrient content in waters is increased by non-point sources of pollution, and above all as nutrients from agricultural areas. With regard to eutrophy, the most important nutrients are phosphorus and nitrogen. Their reciprocal effect can be examined by considering the following general photosynthesis formula:

$$1HPO_4^{2-} + 16NO_3^- + 106CO_2 + 122H_2O + 18H^+ \rightarrow$$
$$(CH_2O)_{106}(NH_3)_{16}H_3PO_4 + 138O_2 \tag{2}$$

In the organic matter that is created during the photosynthesis (e.g. algal biomass), the needed weight ratio of phosphorus and nitrogen is about 1:7, calculated as above. In practice, this means that 7 g of nitrogen is needed for 1 (bound) g of phosphorus in the photosynthesis reaction. If there is only, for example, 5 g of nitrogen, it is the nitrogen that is the so-called minimum factor regulating the production. Similarly, if there is, for example, 10 g of nitrogen, it is the amount of phosphorus that can reduce the growth.

In most Scandinavian lakes, phosphorus is usually the limiting factor for primary production. Finnish water bodies are by nature oligotrophic, i.e. they have naturally low nutrient contents. The contents vary from only a few micrograms in an oligotrophic lake to two to three hundred micrograms of phosphorus per litre in eutrophic lakes. Total phosphorus and phosphate-phosphorus are determined by conventional spectrophoto-metric methods.

Phosphorus can be found in waters either as inorganic salts, phosphates, or as organically bound species. When the algae mass starts to die, the phosphorus bound to this organic matter becomes mineralized. When the part of the decomposing mass of algae that has not been decomposed in the epilimnion reaches the sediment bottom, the phosphorus is immobilized as insoluble ferric phosphate. As long as the oxygen content is high enough ($>0.5\,\text{mg}\,\text{L}^{-1}$), the phosphorus stays in this bound form on the bottom of the lake. If all of the oxygen is consumed from the bottom, and above all of the water layer near the bottom, the iron is reduced from the ferric to the ferrous form. In this case, the bound phosphorus is again soluble in the water.

Another nutrient that is important with regard to the eutrophication of waters is nitrogen. It can mostly be found in waters as organic compounds bound to the humus, and in the water as soluble, inorganic salts, nitrates, ammonium and nitrite. Nitrogen is dissolved from the atmosphere in the same way as oxygen. The significance of this gaseous nitrogen for the nutrient budget is, however, small when compared to the nitrogen salts.

The form in which the actual soluble nitrogen salts occur depends on the oxygen status of the water body. When there is oxygen in the water, the nitrogen salts are mainly nitrates (NO_3). If the water in the hypolimnion looses its oxygen, nitrogen can also be found in the form of ammonia (NH_4). Between these two, often in the metalimnion, nitrite (NO_2) can be detected. The most important nitrogen compounds in eutrophication research are the NH_4 and NO_3 nitrogen species. The analyses of these compounds are carried out with spectrophotometry. The total organic nitrogen is also determined.

1.5.5   OTHER CHEMICAL VARIABLES

The natural water quality of water bodies primarily depends on the quality and composition of the bedrock and ground of the catchment. The bedrock that was revealed by the glacial period and the moraine that it moved and piled up has a very low solubility within the entire Fennoscandia, which is why the salt content of the inland waters is very low.

In northern latitudes, the swamping of land is another factor that influences the natural status of waters considerably. There is a large amount of humic substances dissolved from swamps in almost all of the northern water bodies, thus making the water brown to different degrees and causing a clearly acid reaction of the water.

The organic matter of lake waters is determined directly by analysing the total organic carbon (TOC) or, more frequently, indirectly by analysing the biochemical oxygen demand ($BOD_{5,7\,d}$), especially from rivers, or the chemical oxygen demand (COD), as the permanganate or dichromate demand. Some estimations of the concentrations of organic matter can also be drawn from water colour and, in the simplest manner, by Secchi depth measurements.

The acidification of lakes is a real risk in Finland, Sweden and Norway, countries where natural leaching from the ground and bedrock is small. Acid neutralizing capacity (ANC) measurements analysed by the Gran alkalinity method, have been a very useful tool in estimating the possible trends in the acidification phenomena of lakes with low buffering capacity. The base cation level ($Ca + Mg + Na + K$) is also a very suitable variable for estimating the lake water sensitivity to acidification.

Heavy metals are usually determined by atomic absorption spectroscopy. The concentrations of heavy metals in Finnish freshwaters are at a very low level. In Table 1.5.2, the mean concentrations of some heavy metals are presented. These heavy metals still come to waters primarily in the wastewaters of different industrial plants. Their discharges have, however, clearly decreased.

Many of the harmful substances in lakes or lake sediments are of organic origin. For reliable determination of these compounds specific methods are needed, which require complex and even sophisticated analytical procedures.

1.5.6   COMBINATION OF BIOLOGICAL AND CHEMICAL METHODS

Analysis of chlorophyll *a* has for a long time been regarded as a very good estimate for monitoring and assessing the eutrophication status of lakes. Chlorophyll samples are usually taken from the most productive layer of the epilimnion by vertical sampling, as a rule from the surface to a depth of 2 m. The Finnish standard method for analysing chlorophyll *a* is based on eluting with ethanol. The method is simple and quick, and is also very economical when compared to microscopic evaluation of the phytoplankton biomass. The

Table 1.5.2 Distribution of metal concentrations ($\mu g \, L^{-1}$) of the main rivers discharging into the Baltic (data taken from the Water Quality Register of the Finnish Environment Institute for the years 1994–95).

Metal	Distribution of concentrations (%)						n^a
	10	25	50	75	90	100	
As	0.18	0.27	0.55	0.77	1.0	2.9	583
Cd	0.03	0.03	0.03	0.04	0.1	3.9	632
Cr	0.37	0.59	1.0	2.25	3.98	21.0	587
Cu	0.52	0.905	1.7	3.12	4.9	40.0	632
Hg	0.01	0.01	0.01	0.01	0.02	0.08	185
Ni	0.44	0.76	1.805	3.6	5.76	37.8	594
Pb	0.09	0.16	0.35	0.79	1.4	7.6	626
Se	0.4	0.4	0.4	0.4	0.4	0.42	63
Zn	1.6	3.2	6.5	12.0	21.0	172	630

[a]Number of analyses carried out.

eutrophication level of water bodies can be defined rapidly and reliably by using chlorophyll measurements.

The eutrophication status of lakes has for several years been estimated by using a chlorophyll-based classification system (Premazzi and Chiaudani, 1992). The OECD had in 1982 already drawn up an eutrophication classification of lakes, which is very centrally based on measured chlorophyll values. This scheme divides lakes into five categories on the basis of the data presented in Table 1.5.3.

Periphytic growth on artificial substrates measured as chlorophyll *a* $(mg\,m^{-2})$ has been found to be a good indicator for an early warning of eutrophication, especially in rivers, but also in lakes. There are cases where there have been no differences in the eutrophication between two areas estimated by phosphorus and pelagic chlorophyll *a*, while the difference in periphytic growth has been tenfold.

The devices which are used in monitoring periphytic growth in lakes are quite simple. Ordinary plastic plates are used for the incubation of periphyton, with an incubation time of, for example, three weeks at a depth of one metre. After the incubation, the periphyton growth is detached by scraping, and after homogenization the chlorophyll content is determined.

An interesting example of the use of periphytic growth is presented in Figure 1.5.3. In Lake Pyhäjärvi in central Finland, the northern part of the lake is eutrophied because of sewage, while the southern part seems to be quite oligotrophic according to the phosphorus concentrations. However, local fishermen have reported that the wastewaters also influence the research area shown as '2' in the figure, even though it has not been possible to identify the effect by using current chemical variables. The species of the phytoplankton present had previously given indications of the eutrophic influence of the wastewaters, with finally measurements of the periphytic growth confirming the views of the fishermen.

The mussel-incubation method (Herve, 1991) has been used in Finland since 1984 for monitoring specific organic compounds. Using this method, common lake mussels (*Anodonta piscinalis*) obtained from quite unpolluted watercourses were first pre-incubated in aquariums for two weeks to guarantee the cleanness

Table 1.5.3 The OECD eutrophication classification of lakes.

Trophic category	Chlorophyll content $(mg\,m^{-3})$	Maximum chlorophyll content $(mg\,m^{-3})$
Ultra-oligotrophic	< 1	< 2.5
Oligotrophic	< 2.5	< 8.0
Mesotrophic	2.5–8	8–25
Eutrophic	8–25	25–75
Hyper-eutrophic	> 25	> 75

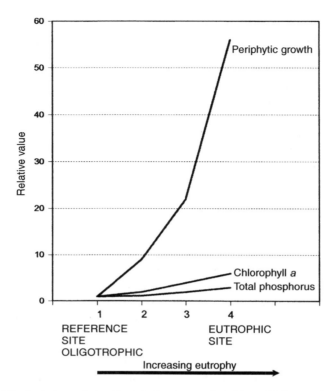

Figure 1.5.3   Relative values of total phophorus and chlorophyll *a* concentrations, and periphytic growth, in different parts of Lake Pyhäjärvi (central Finland).

of the mussels and to eliminate weak individuals. After the pre-incubation, the mussels were transferred to the monitoring sites and housed in plastic cages anchored in the epilimnion at a depth of one metre. The incubation period in the monitoring areas has been exactly the same every year, i.e. four weeks from · the beginning of August to the beginning of September.

After incubation, the mussels were then transferred to the laboratory, where composite samples of the soft tissues were analysed for different compounds (Herve, 1991). The results have been expressed in fat (lipid) weight (Figure 1.5.4). For more information on the chemical and environmental fate of these compounds, see also Chapter 3.2.

The chlorophenolic compounds were divided into two groups, as follows (Paasivirta *et al.*, 1980):

(a) chlorophenols originating from wood preservation, combustion, chlorination, and pesticide use;
(b) chlorophenols originating mainly from bleaching processes.

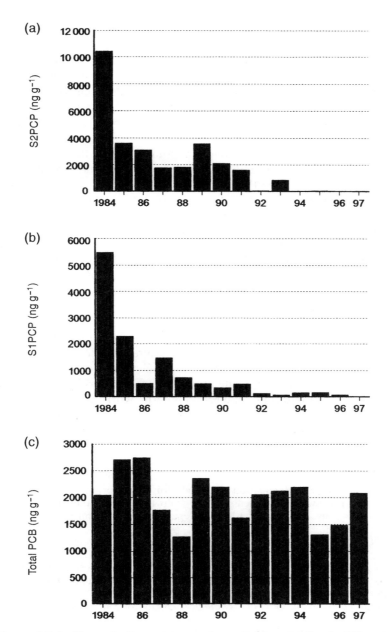

Figure 1.5.4 Concentrations (expressed as lipid weight) of different types of chlorophenolics (a,b) and polychlorinated biphenyls (c) in the recipient watercourse of a Finnish pulp mill, measured over the period 1984–1997.

The concentration of the chlorophenols resulting from the bleaching processes in incubated mussels, when monitoring first started in 1984, was very high. The improved treatment of wastewaters, and many process modifications, especially in pulp bleaching, have had a rather rapid effect on the receiving water bodies. The situation has radically improved, so that nowadays it is very difficult to identify these compounds, even with the very sensitive mussel-incubation method. Some small concentrations found in mussels could originate from old sediments, which can be resuspended in the water phase by various mechanisms.

The concentrations of chlorophenolic compounds originating from wood preservation, combustion, chlorination and pesticide use have significantly decreased since 1984–1985 (See Figure 1.5.4).

Polychlorinated biphenyls are persistent and ubiquitous contaminants which are frequently found in small amounts in lakes, especially in the inland waters of industrialized countries. These compounds are widely used for different purposes in electrical capacitors and transformers, and in some lubricating and cutting oils. Monitoring activity has shown that their disappearance is very slow (see Figure 1.5.4), and will probably take several tens of years.

1.5.7   CONCLUSIONS

The ecosystem is all the time searching for a balance between the biotope and the biocoenosis. The variations are often quite natural and depend wholly on the season and weather conditions. All of these factors must be considered carefully in advance when a monitoring programme for lakes is being designed. The success of the monitoring requires long-term work and gathering of material so that the variations or trends possibly caused by man can be discerned from the intrinsic variations and development trends of nature.

REFERENCES

Herve, S., 1991. Mussel Incubation Method for Monitoring Organochlorine Compounds in Freshwater Recipients of the Pulp and Paper Industry, Research Report No. 36. University of Jyväskylä, (Department of Chemistry) Jyväskylä, Finland.

Hutchinson, G. E., 1957. *A Treatise on Limnology, Volume I: Geography, Physics, and Chemistry*, John Wiley & Sons, New York.

Paasivirta, J., Särkkä, J., Leskijärvi, T. and Roos, A., 1980. Transportation and enrichment of chlorinated phenolic compounds in different aquatic food chains, *Chemosphere*, **9**, 441–456.

Premazzi, G. and Chiaudani, G., 1992. Ecological Quality of Surface Waters, Quality Assessment Schemes for European Community Lakes, European Communities Commision, EUR 14563 – Ecological Quality of Surface Waters, Environment Quality of Life Series, Environment Institute, University of Milan, Milan, Italy.

Wetzel, R. G., 1983. *Limnology*, 2nd Edn, Saunders College Publishing, Philadelphia.

Part Two
Biocoenosis in Evaluation of the Ecological State of the Lake

Chapter 2.1

Phytoplankton in Water Quality Assessment – An Indicator Concept

EVA WILLÉN

Hydrological and Limnological Aspects of Lake Monitoring
Edited by Pertti Heinonen, Giuliano Ziglio and André Van der Beken
©2000 John Wiley & Sons, Ltd, ISBN 0 471 89988 7

2.1.1 INTRODUCTION

Microalgae have been used as a biological classification tool for lakes and slow-flowing waters since the late 19th century. Well-known examples are given by Apstein (1896) and Kolkwitz and Marsson (1902, 1908) from Germany, by Huitfeldt-Kaas (1906) from Norway, and by Naumann (1919) and Teiling (1916, 1955) from Sweden. As fundamental components of the water food web (primary producers of organic matter, oxygen producers, food resources for grazers, compartments of the microbial loop, etc.) they are among the first to react to water quality changes and by this they initiate a chain reaction which is successively reflected within other groups of organisms (zooplankton, benthic fauna, fish, birds, etc.). The great value of a thorough knowledge of the structure (taxa, abundances and biomass) and function (response to environmental change) of the algal community in a water body is much more far-sighted than to be surprised by sudden and noxious waterblooms, by a distaste in fish or raw water, by sudden changes in the food web compartments and the succeeding reaction of other organisms, or by destruction of the water for practical use.

2.1.2 WHAT IS MEANT BY ALGAE?

For a long time, the definition of algae has been *plantlike organisms without roots, vascular tissues and leavy shoots, which constitute a heterogenous assembly of oxygen-producing, photosynthetic organisms* (South and Whittick, 1987). Some algae are not able to photosynthesize, but are still classified as algae due to a close resemblance to photosynthetic forms. Algae are more regarded as a functional group of organisms than a true systematic unity. Several classification systems have been proposed for algae but at present there is no general consensus about this. Supported by molecular systematics, the planktic algae are separated in various classes, including prokaryotes (cyanobacteria/blue-green algae), protists (chrysophytes, diatoms, prymnesiophytes, tribophytes, dinophytes, cryptophytes, euglenoids, etc.) and plants (i.e. chlorophytes/green algae). Based on the variety of phyla considered as algae, the term *phytoplankton* may be inappropriate as it focuses on the kingdom of plants, although this term is so well established that it will be hard to replace. Planktic algae could be an alternative expression to use.

2.1.3 EXPERIENCES OF INDICES AND INDICATORS

Experiences from the Nordic countries and from Germany have formed a school for the establishment of lake typology schemes, and for water quality assessments based on dominant species in the algal community. In Sweden the geographical position of a lake, i.e. in lowland or highland areas, was used for separation into Baltic and Caledonian types (Teiling, 1916). These two types were soon characterized as eutrophic and oligotrophic (Naumann, 1919). Soon after this, the humic lake was also classified as a special component of low productive waters with a high content of coloured matter (Thienemann, 1921), with typical dominant taxa for these lake types being distinguished (Table 2.1.1).

Somewhat later, the relationship between algal groups prevailing in nutrient-rich and nutrient-poor lakes was formalized in indices for use mainly in those particular countries where the indices were established, i.e. in Finland (Järnefelt, 1952, 1956), in Denmark (Nygaard, 1949), and in Sweden (Thunmark, 1945). Cyanobacteria, chlorophytes, centric diatoms, and euglenoids were related to the prevailing group of desmids in nutrient-poor lakes. In other indices, separate orders within one and the same algal class (i.e. among diatoms or green algae) were also related to each other (Table 2.1.2).

Even at an early stage, criticism was directed on the use of indices based on algal classes (Teiling, 1955). Many classes turned out to be very heterogenous, consisting of species occurring in both oligo- and eutrophic lakes. The criticism of the use of a comprehensive index for lake characterization could, to a great

Table 2.1.1 Compilation of the phytoplankton characteristics given to oligotrophic and eutrophic lake types in the late 19th and early 20th century (Apstein, 1896; Huitfeldt-Kaas, 1906; Teiling, 1916; Naumann, 1919).

Oligotrophic plankton formation	Eutrophic plankton formation
Desmidiacean type	Eutrophic peridinean type (*Ceratium* and *Peridinium*)
Chlorophycean type (*Botryococcus* and *Sphaerocystis*)	Eutrophic diatomean type (*Melosira* and *Stephanodiscus*)
Chrysophycean type (*Dinobryon* and *Mallomonas*)	Pediastrum type
Oligotrophic peridinean type (*Ceratium* and *Peridinium willei*)	Cyanophycean type (Chroococcales)
Oligotrophic diatomean type (*Tabellaria* and *Cyclotella*)	
Teiling's Caledonian lake type	Teiling's Baltic lake type
Small algal biomass with netplankton characterized by desmids	Large algal biomass with varying portions of water-bloom forming blue-green algae
Large species richness	Small species richness
Apstein's Dinobryon lake type	Apstein's chlorophycean lake type
Huitfeldt-Kaas' chlorophycean lake type (*Botryococcus* and *Sphaerocystis*)	Huitfeldt-Kaas' schizophycean lake type (*Anabaena* and *Coelosphaerium*)

Table 2.1.2 Examples of indices used for the trophic evaluation of lakes based on occurring planktic algae (species numbers, abundances or biomass).

Index	Calculation	Result	Reference
Thunmark's chlorophycean index	*Chlorococcales* spp./ *Desmidiales* spp.	$\leqslant 1$ = oligotrophy > 1 = eutrophy	Thunmark (1945)
Nygaard's myxophycean index	*Cyanophyceae* spp./ *Desmidiales* spp.	$\leqslant 1$ = oligotrophy > 1 = eutrophy	Nygaard (1949)
Nygaard's diatom index	*Centrales* spp./ *Pennales* spp.	$\leqslant 1$ = oligotrophy $\geqslant 1$ = eutrophy	Nygaard (1949)
Nygaard's euglenophycean index	*Euglenophyceae/ Cyanophyceae + Chlorophyceae* spp.	$\leqslant 1$ = oligotrophy > 1 = eutrophy	Nygaard (1949)
Nygaard's compound index	*Cyanophyceae + Chlorococcales + Centrales + Euglenophyceae* spp./ *Desmidiales* spp.	$\leqslant 1$ = oligotrophy 1–3 = mesotrophy > 3 = eutrophy	Nygaard (1949)
Palmer's genus index of organic pollution	$\sum I_{\text{genus}}{}^{a}$	> 20 = heavy pollution 15–19 = intermediate pollution < 15 = low pollution	Palmer (1969)
Palmer's species index of organic pollution	$\sum I_{\text{species}}{}^{b}$	> 20 = heavy pollution 15–19 = intermediate pollution < 15 = low pollution	Palmer (1969)
Hörnström's trophic index	$I_{\text{L}} = \sum (f I_{\text{s}})/\sum f^{c}$	10–100, high values indicate successively higher pollution	Hörnström (1981)
Brettum's trophic index based on phyto-plankton volumes	$I_{\text{T}} = \sum (v I_{\text{s}})/\sum v^{d}$	An index value for each trophic level from ultraoligotrophy to hypertrophy	Brettum (1989)
Finnish trophic index based on occurrences of species	spp. indicating eutrophy/spp. indicating oligotrophy	$\leqslant 8$ indicates oligo-trophic conditions	Järnefelt (1952, 1956), Heinonen (1980)
Finnish trophic index based on volumes of species	Total volume of spp. indicating eutrophy/total volume of spp. indicating oligotrophy	$\leqslant 35$ indicates oligo-trophic conditions	Järnefelt (1952, 1956) Heinonen 1980)

[a]Genus index value = 1–5
[b]Species index value = 1–6
[c]I_{L}, trophic index of lake; f, species frequency in a 5-degree scale; I_{s}, trophic index of species
[d] I_{T}, index of trophic level T; v, volume of species per litre; I_{S}, trophic index of species

extent, be explained by the lack of uniform and standardized methods for sampling, counting procedures and taxonomic resolution. The species used in most of the mentioned indices were mainly based on analyses of net samples, usually with nets of a rather coarse mesh size (50–150 μm). Such a technique concentrated large species, while the smaller ones, which often are dominating elements in the plankton community, were disregarded. When the Utermöhl method for plankton countings was introduced, the possibility of obtaining quantitative information on all planktic algae, whether large or small, was facilitated (Utermöhl, 1958). Since this technique has been successively more used, a detailed separation of species abundances/biomasses is possible in relation to environmental variables. Examples of large-scale regional studies based on the Utermöhl counting technique, where indicators and indicator indices are used for evaluation of a trophic and/or acidic range, have been published by Heinonen (1980), Hörnström (1981) and Brettum (1989).

Heinonen (1980) has listed c. 100 species indicating eutrophy and 25 indicating oligotrophy, on the bases of experiences gained from Finnish lakes (Table 2.1.3). The indicator species are chosen based on their frequency in eutrophic or oligotrophic lakes. Species are classified as eutrophic indicators when their occurrence ratio, i.e. eutrophic lakes/oligotrophic lakes, is >2. The corresponding ratio for the selection of oligotrophic indicators is 0.7. The trophic state of the lakes for the Finnish material is based on the species number or the biomass of the eutrophic indicators in relation to the oligotrophic ones (Table 2.1.2). Brettum has presented a tentative system in which 150 species are scored based on their proportion of the total biomass for a number of classes of environmental variables (total P content, total N content, N/P, pH, etc.). A final trophic index can then be constructed for trophic levels covering the range from ultraoligotrophy to hypertrophy (Table 2.1.2). An overview of some bordering levels of total volumes of planktic algae used for classification of lakes in a trophic gradient is presented in Table 2.1.4.

An index developed for the rating of organic pollution has been designed by Palmer (1969). This is based on a scoring of 1–6 for 10 genera and 20 species in accordance with statements from a large number of references (Table 2.1.5). High numbers are given to species with a special affinity to organically polluted waters. This index is founded on the presence/absence data of the involved species or genera.

Finally, I will say a few words about diversity indices; these have been widely used, especially in monitoring studies in the USA. Some of these indices are sensitive to the number of species, while others more reflect the evenness with which the individuals are distributed (Washington, 1984). When dealing with phytoplankton, the taxonomic competence of the analyser decides the outcome of the index value, which may be one reason for their restricted use in Europe. Another reason could be that its use as an early warning technique is limited as the index value may be the same in both species-rich and species-poor lakes,

Table 2.1.3 Indicator species of nutrient-rich and nutrient-poor Finnish lakes, according to Järnefelt (1952, 1956) and Heinonen (1980). Data taken from Tikkanen and Willén (1992).

Indicators of nutrient-rich lakes	
Actinastrum hantzschii	*Kirchneriella obesa*
Anabaena circinalis	*Lagerheimia* spp.
Anabaena planctonica	*Lepocinclis* spp.
Anabaena spiroides	*Melosira varians*
Aphanizomenon gracile	*Micractinium pusillum*
Aphanocapsa delicatissima	*Microcystis aeruginosa*
Aphanothece elabens	*Microcystis viridis*
Arthrodesmus octocornis	*Monoraphidium contortum*
Asterionella formosa	*Nitzschia acicularis*
Aulacoseira granulata	*Oocystis solitaria*
Aulacoseira islandica	*Ophiocytium* spp.
Centritractus spp.	*Pandorina charkowiensis*
Characiopsis longipes	*Pandorina morum*
Chroococcus dispersus	*Pediastrum angulosum*
Chroococcus minutus	*Pediastrum biradiatum*
Closteriopsis longissima	*Pediastrum boryanum*
Closterium aciculare	*Pediastrum duplex*
Closterium gracile	*Pediastrum duplex* var. *gracillimum*
Closterium macilentum	*Pediastrum tetras*
Closterium pronum	*Peridiniopsis penardiforme*
Closterium venus	*Peridiniopsis polonicum*
Coelastrum cambricum	*Peridinium bipes*
Coelastrum microporum	*Peridinium umbonatum*
Coelastrum reticulatum	*Peridinium willei*
Cosmarium humile	*Phacus* spp.
Cosmarium meneghinii	*Planktolyngbya contorta*
Cosmarium regnellii	*Planktolyngbya subtilis*
Cosmarium regnesii	*Polyedriopsis spinulosa*
Cyclostephanus dubius	*Pseudanabaena limnetica*
Cyclotella meneghiniana	*Pseudostaurastrum limneticum*
Diatoma tenuis	*Scenedesmus acuminatus* var. *minor*
Dictyosphaerium ehrenbergianum	*Scenedesmus armatus*
Dictyosphaerium elegans	*Scenedesmus bicaudatus*
Dimorphococcus lunatus	*Scenedesmus denticulatus*
Enallax acutiformis	*Scenedesmus hystrix*
Entomoneis spp.	*Scenedesmus magnus*
Euglena spp.	*Scenedesmus obtusus*
Fragilaria berolinensis	*Scenedesmus opoliensis*
Fragilaria capucina	*Scenedesmus sempervirens*
Fragilaria crotonensis	*Scenedesmus subspicatus*
Fragilaria ulna	*Selenastrum bibraianum*
Franceia ovalis	*Selenastrum gracile*
Goniochloris fallax	*Staurastum avicula*
Gonium pectorale	*Staurastrum paradoxun* var. *parvum*
Kirchneriella contorta	*Staurastrum tetracerum*
Kirchneriella lunaris	*Staurodesmus convergens*

Continued

Table 2.1.3 *Continued.*

Indicators of nutrient-rich lakes	
Staurodesmus dejectus	*Trachelomonas hispida*
Strombomonas verrucosa	*Trachelomonas intermedia*
Synura uvella	*Trachelomonas kelloggii*
Teilingia granulata	*Trachelomonas oblonga*
Tetraedriella regularis	*Trachelomonas planctonica*
Tetraedron spp.	*Trachelomonas varians*
Tetrastrum spp.	*Trachelomonas volvocina*
Trachelomonas abrupta	*Trachelomonas volvocinopsis*
Trachelomonas acanthostoma	*Westella botryoides*
Trachelomonas armata	*Volvox aureus*

Indicators of nutrient-poor lakes	
Bitrichia spp.	*Euastrum elegans*
Chroococcus turgidus	*Kephyrion* spp.
Cosmarium contractum	*Mallomonas allorgei*
Crucigeniella rectangularis	*Mallomonas akrokomos*
Cyclotella kuetzingiana	*Merismopedia glauca*
Diatoma vulgaris	*Ouadrigula lacustris*
Dinobryon acuminatum	*Rhabdogloea ellipsoidea*
Dinobryon bavaricum	*Staurodesmus incus*
Dinobryon cylindricum	*Stichogloea olivacea*
Dinobryon divergens	*Stichogloea doederleinii*
Dinobryon sertularia	*Willea irregularis*
Euastrum bidentatum	

depending on the equitability of the species involved. The ecological interpretation of these types of indices has been considered as obscure (Kalff and Knoechel, 1978).

2.1.4 PHYTOPLANKTON AND THE SEASONAL DEVELOPMENT

As planktic algae have very short generation times, they also react rapidly to shifts in the environment. Changes in the physical and/or chemical status of the water are traced after some weeks through alterations of the species and their abundances. Long-lasting water quality changes may be reflected among the planktic algae from one growth season to another (Reynolds, 1987). Knowledge about the compositional changes in the phytoplankton community during the course of a growth season is very important, e.g. when interpreting deviations in species composition and biomass, when choosing the period of

Table 2.1.4 Examples of total volumes of planktic algae, given at various trophic levels.

Oligotrophy		Mesotrophy		Eutrophy				
ultra-	oligo-	meso-1	meso-2	eu-1	eu-2	hyper-	Reference	Comments
0.12	0.4	0.6	1.5	2.5	5	>5	Brettum (1989)	Norwegian lakes; growth-season mean values
0.2	0.5	1	2.5	—	10	>10	Heinonen (1980)	Finnish lakes; July–August values
—	1	—	2	—	5	>10	Rosén (1981)	Swedish lakes; August values
1	5	—	15	25	5	>5	Willén (1999)	Swedish lakes; growth season mean values
—	0.5	—	2	>2	—	—	Rott (1984)	Austrian lakes; growth-season mean values
—	1	—	1–5	—	5–10	>10	Trifonova (1989)	South-western Russian lakes; growth-season mean values
—	1	—	3	5	10	>10	Wetzel (1975)	No season reported

investigation in relation to a certain goal, and when choosing the relevant number of samples and seasons for water quality assessment (Figure 2.1.1).

The phytoplankton pass through the same successional stages throughout a year as a forest in the northern latitudes will do in a century. This succession follows a course decided by alterations between the mixing and stratification periods. Usually, 30–100 generations are developed during a growth season, depending on the species occurring. The seasonal progression runs from pioneer stages, then over a number of intermediate phases, and finally to an equilibrium phase. This latter stage is, however, rarely reached in any lake, since the planktic community is more or less constantly reverted to an earlier stage as a result of the usually frequent events of wind stress, which causes at least a certain mixing of the surface layers (Sommer, 1991; Padisák, 1992; Reynolds, 1993).

In the late winter period in temperate lakes, and even beneath an ice cover, certain small flagellated species are already developing, and sometimes in large amounts if sufficient light is available. In spring, particularly in small wind-protected forest lakes with a very short mixing period, many different flagellated species of dinoflagellates, chrysophytes and cryptophytes develop. In larger lakes with longer mixing periods, diatoms grow to substantial biomasses, especially if the water is nutrient-rich. After this 'spring-bloom', a clearwater phase occurs, with an increase of different zooplankton grazing on

Table 2.1.5 Palmer's indicators of organic pollution elaborated for the USA (Palmer, 1969).

Genus	Pollution index[a,b]	Species	Pollution index[a,c]
Anacystis	1	*Ankistrodesmus falcatus*	3
Ankistrodesmus	2	*Arthrospira jenneri*	2
Chlamydomonas	4	*Chlorella vulgaris*	2
Chlorella	3	*Cyclotella meneghiniana*	2
Closterium	1	*Euglena gracilis*	1
Cyclotella	1	*Euglena viridis*	6
Euglena	5	*Gomphonema parvulum*	1
Gomphonema	1	*Melosira varians*	2
Lepocinclis	1	*Navicula cryptocephala*	1
Melosira	1	*Nitzschia acicularis*	1
Micractinium	1	*Nitzschia palea*	5
Navicula	3	*Oscillatoria chlorina*	2
Nitzschia	3	*Oscillatoria limosa*	4
Oscillatoria	5	*Oscillatoria princeps*	1
Pandorina	1	*Oscillatoria putrida*	1
Phacus	2	*Oscillatoria tenuis*	4
Phormidium	1	*Pandorina morum*	3
Scenedesmus	4	*Scenedesmus quadricauda*	4
Stigeoclonium	2	*Stigeoclonium tenue*	3
Synedra	2	*Synedra ulna*	3

[a]1 = moderate pollution, 6 = high pollution.
[b]A sum value of > 20 on the genus level indicates a heavily polluted environment
[c]A sum value of > 15 on species level indicates a clean-water environment

the planktic algae. Following a number of intermediate stages, with the development of many green algae, chrysophytes and some dinoflagellates, a mid- or late-summer phase appears, with the dominance of large, slow-growing species. This latter successional stage is usually recognized as the equilibrium phase, displaying the appearance of cyanobacterial water-blooms in nutrient-rich lakes, while other algal classes prevail in nutrient-poorer lakes. The late-summer period gradually proceeds to the next large-scale mixing period in the autumn, as the water temperature decreases.

2.1.5 THE REACTION OF PLANKTIC ALGAE TO ENVIRONMENTAL CHANGES

The following changes in the planktic algal community are expected in a trophic gradient of increased nutrient availability:

- increasing biomasses;
- a prolonged waterblooming season;

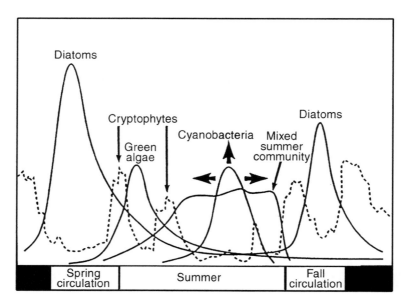

Figure 2.1.1 Model showing the development of varous planktic algal groups throughout the growth season. (Modified after Stewart and Wetzel 1986.)

- increasing biomass variations within the growth season of a year;
- a change in size structure of the community, in which the proportion of large species increases – many of these are considered as stress tolerants;
- a decreasing evenness, in which a few species become dominants;
- a decrease in species richness, especially in hypertrophic environments;
- a changed structure among the algal classes;
- a changed structure in the species composition.

The following changes are expected in a gradient of increasing acidity:

- decreasing biomasses;
- decreasing species richness;
- decreasing evenness;
- increasing trivialization of the plankton flora, in which only a few algal classes occur, e.g. (a) most planktic diatoms disappear at a pH ⩽ 5.5, and (b) the proportion of planktic cyanophytes is substantially reduced at a pH ⩽ 5.5;
- the number of life forms decreases, and in very acid waters flagellated species prevail.

The well-known fact that the planktic algal biomass rises with increasing nutrient availability, and especially with an increase in the phosphorus concentration, is illustrated worldwide. This particular condition is behind the introduction of various phosphorus-control measures in order to revert eutrophication signs and reduce waterblooms. According to Swedish experiences, disturbing amounts of planktic algae are reached at total phosphorus concentrations of $50 \mu g \, l^{-1}$ (Willén, 1992a). However, even at a concentration level of $25 \mu g \, l^{-1}$, complaints cannot be ignored due to the occurrence of algae reaching a biomass of $5 \, mm^3 \, l^{-1}$ which is a level where many people get an unaesthetic impression of the water. At a peak biomass level of $10 \, mm^3 \, l^{-1}$, complaints of an algal 'peasoup' are so unanimous that the situation has to be regarded as highly disturbing (Figure 2.1.2). It is, however, probable that the opinion of disturbance levels of plankton biomasses varies with latitude. In those parts of the world where the majority of lakes are in an oligotrophic or mesotrophic state, as in most of the Nordic countries, the threshold of tolerance may be much lower than in regions which are used to richer vegetations due to other temperature and nutrient conditions.

The relationship between the growth-season mean values of the total phosphorus concentration ($\mu g \, l^{-1}$) and the phytoplankton biomass ($mm^3 \, l^{-1}$) in Swedish lakes is calculated by using the following:

total planktic algal volume$_{May-Oct.}$ =
$$0.05 \text{ total phosphorus concentration}_{May-Oct.} - 0.2 \qquad (1)$$

where R^2 is 0.63 (R is the correlation coefficient) and n (number of seasons from 60 lakes) is 344.

Shallow wind-mixed lakes (mean depth $\leqslant 3$ m) have twice as large a biomass per unit phosphorus, according to the following expression:

total planktic algal volume$_{May-Oct.}$ =
$$0.055 \text{ total phosphorus concentration}_{May-Oct.} + 0.132 \qquad (2)$$

where R^2 is 0.73 and n (number of seasons from 10 lakes) is 56.

The relationship between a growth-season mean biomass and a late-summer biomass is calculated by using the following:

$$\text{total volume}_{Aug.} = 1.65 \text{ total volume}_{May-Oct.} - 0.26 \qquad (3)$$

where R^2 is 0.75 and n (number of seasons from 60 lakes) is 391.

In Swedish assessment criteria for lakes, various water quality states, based on the biomass of planktic algae, shown in Table 2.1.6, have been suggested (Willén, 1999). The total phosphorus ranges (May–October, mean values) of the different classes are shown in Table 2.1.7 (Persson, 1999). The changing proportion of algal classes, as a function of increased phosphorus concentrations and increasing biomasses of planktic algae, is illustrated in Figure 2.1.3. The successive increase of cyanobacteria and the evident decrease of chrysophytes are the most conspicuous features. Another sign is a growing

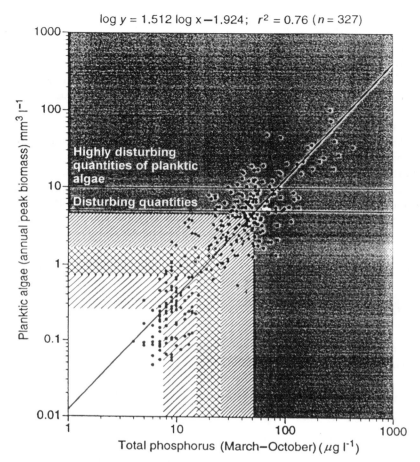

Figure 2.1.2   Annual peak volumes of planktic algae in relation to mean concentrations of total phosphorus for the period March–October. The levels of disturbing and highly disturbing biomasses are indicated. The different contours delimit the trophic evaluations set by the Swedish Environment Protection Agency in 1991.

proportion of diatoms in lakes with rising nutrient levels. The first sign of an incipient eutrophication process in oligotrophic lakes is often an increase in the spring biomass of diatoms. Very small amounts of planktic diatoms occur in depauperized lakes.

In Sweden, as well as in the Nordic countries Norway and Finland, the tribophycean species, *Gonyostomum semen* (Ehr.) Diesing (systematics according to Christensen, 1994), is successively invading more lakes during its immigration from the rest of Europe. As this species is comparatively large (up to 100 μm in length), it also occupies a large portion of the total biomass in moderate amounts. The main habitat is a small ($<1\,\mathrm{km^2}$) brown-water lake of a mesotrophic or eutrophic character.

Changes in biomass and algal class structures are robust measures for interpretation along the gradients of either eutrophication or oligotrophication processes. Interpretation of changes on the species level, however, owe more to the experience of the investigator. Knowledge about the characteristic algal associations along a trophic spectrum in relation to the lake type is very important. Due to the stochasticity of species occurrence, which to a large extent depends on the weather conditions, changes in the dominance structure is the easiest to verify statistically.

2.1.6 PHYTOPLANKTON CHARACTERISTICS AND LAKE TYPES

The characteristic assemblages of planktic algae in a trophic gradient of temperate lakes have been outlined, among others, by Reynolds (1980, 1984), by Rott (1984) (especially for Tyrolean lakes), by Willén (1992b) for oligotrophic Swedish lakes, and by Olrik (1993) for Danish lakes. Some

Table 2.1.6 The Swedish eutrophication classification system according to the biomass of planktic algae.

Class	Designation	Mean biomass $(\mathrm{mm^3\,l^{-1}})^a$	Biomass $(\mathrm{mm^3\,l^{-1}})^b$	Trophic state
1a	Particularly small biomass	$\leqslant 0.1$	$\leqslant 0.1$	Ultraoligotrophy
1b	Very small biomass	0.1–0.5	0.1–0.5	Oligotrophy
2	Small biomass	0.5–1.5	0.5–2	Mesotrophy
3	Moderately large biomass	1.5–2.5	2–4	Eutrophy I
4	Large biomass	2.5–5	4–8	Eutrophy II
5	Very large biomass	>5	>8	Hypertrophy

[a]During May–October.
[b]During August.

Table 2.1.7 The ranges of total phosphorus concentrations shown for the different classes of lake given in Table 2.1.6.

Class	P range $(\mu\mathrm{g\,l^{-1}})$
1a	$\leqslant 6$
1b	6–12.5
2	12.5–25
3	25–50
4	50–100
5	>100

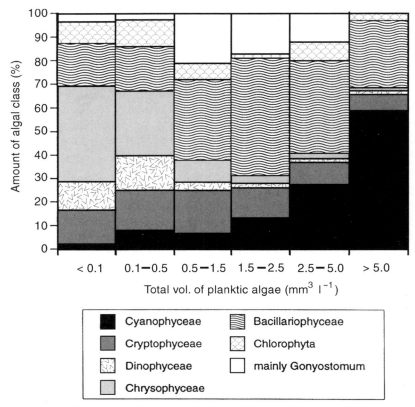

Figure 2.1.3 The proportion of algal classes in a gradient of rising nutrients as indicated by rising total volumes of planktic algae; the total phosphorus concentrations of the volume intervals are 6, 12.5, 25, 50, 100 and $> 100 \, \mu g \, l^{-1}$. The data represent mean values for May–October from 100 Swedish lakes, with each sampled for a period of 4 years.

phycologists had already tried in the 1940s to set up plankton associations in accordance with the then prevailing school of floristic sociology (Du Rietz, 1936; Teiling, 1942; Thunmark, 1945; Nygaard, 1949). When comparing, for example, the late summer period in these outlines, dinoflagellates (*Peridinium, Ceratium*) are common components in oligotrophic lakes, sometimes along with desmids (*Closterium, Cosmarium, Staurastrum, Staurodesmus*) and such colonial chrysophytes as *Dinobryon* and *Uroglena*. At slightly higher nutrient levels, diatoms and certain cyanophyte genera types of *Gomphosphaerioid* forms can develop.

In eutrophic stages, the water-bloom-forming cyanophytes occur in assemblages with large dinoflagellates and diatoms. The last-mentioned group is prevalent if the water is exposed to mixing. Turbulence is thus another important selection factor for the occurrence of a certain set of species. Further discussions on this aspect have been presented by Reynolds (1993), who places various lakes and their seasonal planktic dominants along a scale ranging from continuous mixing to an amictic state. By choosing some common cyanophyte genera, the following rough outline can be given:

Anabaena

Aphanizomenon   Microcystis   Planktothrix

————————————————————————→

Increasing water turbulence

The selection of separate species for a certain kind of lake along a trophic or acidic gradient has been elucidated empirically, as mentioned above, by Heinonen (1980) and by Brettum (1989) after analyses of large data sets (cf. Tables 2.1.2 and 2.1.3). In Sweden, an August sampling of 1250 lakes was performed in the early 1970s and the planktic algae were analysed quantitatively by the use of Utermöhl's technique (Rosén, 1981). As a result, frequently occurring taxa were reflected in relation to seven environmental variables, including total salt content (conductivity), nutrients (nitrogen and phosphorus), light penetration (Secchi depth), and coloured matter (colour). The main lake types and their occurring species, evaluated after this survey, are given in Tables 2.1.8 and 2.1.9. With the presently available multivariate methods, it is possible to carry out further work in this field and combine a classification of algal assemblages and a number of statistically significant environmental variables. An attempt in this direction, involving more than 300 Swedish lakes distributed all over the country, is currently under preparation.

2.1.7 WATER-BLOOMING AND TOXIN-PRODUCING CYANOBACTERIA

In suggested classifications, based on the biota, of ecological water quality in Europe, mass-developing cyanobacteria are almost always considered. When these water-blooms occur during longer periods, they are used as an indicator of a poor or a bad water quality (Nixon *et al.*, 1996). A massive water-bloom is not only an ecologically inferior phenomenon, but it also has a negative impact on the use of water by restraining open-air activities and by decreasing the production of fish used for consumption. In addition, it affects the use of water in drinking-water supplies.

About 60% of occurring water-blooms contain toxic cyanobacterial strains, which is a figure that seems relevant for many countries. This fact places a

Table 2.1.8 Characteristic species found in different kinds of Swedish oligotrophic lakes in a late summer succession stage (Rosén, 1981). Nomenclature used is mainly according to *Süsswasserflora von Mitteleuropa* (Komárek and Anagnostidis, 1986, 1989; Anagnostidis and Komárek, 1988).

Species	Acid/acidified lakes[a]	Forest lakes[b]	Clear-water forest lakes[c]	Forest lakes with elevated nutrient levels[d]
Asterionella formosa			×	
Bitrichia chodatii		×	×	
Botryococcus spp.				×
Chlamydocapsa spp.				×
Chroococcus limneticus			×	
Chrysidiastrum catenatum			×	
Chrysochromulina spp.			×	
Cryptomonas spp.				×
Cyclotella bodanica			×	
Cyclotella comta			×	
Dinobryon borgei		×		
Dinobryon crenulatum	×	×		
Dinobryon cylindricum var. *alpinum*			×	
Dinobryon divergens				×
Dinobryon pediforme	×			
Dinobryon sociale var. *americanum*	×			
Dinobryon spp.				×
Dinobryon suecicum		×		
Elakatothrix spp.			×	
Gymnodinium fuscum			×	
Gymnodinium uberrimum	×		×	
Istmochloron trispinatum	×			
Kephyrion, Pseudo-kephyrion, Kephyriopsis		×		
Merismopedia tenuissima		×		
Monoraphidium dybowskii		×	×	
Monoraphidium griffithii		×		
Monoraphidium spp.			×	
Mougeotia spp.	×			
Oocystis spp.			×	
Oocystis submarina	×			
Peridinium inconspicuum	×		×	
Pseudopedinella spp.			×	
Quadrigula spp.			×	×
Rhabdogloea ellipsoidea			×	
Rhodomonas lacustris				×
Small naked *chryso-* and *dinoflagellates*	×	×	×	
Snowella, Woronichinia			×	
Sphaerocystis spp.			×	

Continued

Table 2.1.8 *Continued.*

Species	Acid/ acidified lakes[a]	Forest lakes[b]	Clear- water forest lakes[c]	Forest lakes with elevated nutrient levels[d]
Spiniferomonas spp.			×	
Staurastrum anatinum			×	
Staurastrum longipes			×	
Staurastrum spp.				×
Staurodesmus crassus			×	
Staurodesmus extensus var. *joshuae*			×	
Staurodesmus sellatus		×	×	
Staurodesmus triangularis var. *limneticus*			×	
Stichogloea doederleinii			×	
Staurodesmus spp.				×
Tabellaria fenestrata			×	
Tabellaria flocculosa var. *asterionelloides*			×	
Tabellaria flocculosa var. *teilingii*			×	
Willea irregularis			×	
Woloszynskia ordinata			×	

[a]Total phosphorus $\leqslant 5\,\mu g\,l^{-1}$; pH < 5; extremely species-poor
[b]Total phosphorus $\leqslant 7\,\mu g\,l^{-1}$; pH 6–6.5; colour $\leqslant 25\,mg\,Pt\,l^{-1}$; low conductivity
[c]Total phosphorus 7–10 $\mu g\,l^{-1}$; total nitrogen 300–500 $\mu g\,l^{-1}$; conductivity $> 100\,\mu S\,cm^{-1}$
[d]Total phosphorus 10–12 $\mu g\,l^{-1}$; conductivity $> 100\,\mu S\,cm^{-1}$

special pressure on authorities to regulate for toxin content in water supplies. Even in countries with advanced cleaning processes, hepatotoxins can pass through various filters, and are traced in the drinking water – a highly unsatisfactory condition. As hepatotoxins are found to be liver-cancer promotors, they have been objects for special consideration, and routine test methods have been developed (Carmichael, 1997). Many countries have now started programmes on cyanotoxin-based issues. Some examples of these are inventories of toxin concentrations in raw water supplies, advice on treatment processes in waterworks, the initiation of various research programmes of methodological or ecological relevance, and the support of advisory services to the public.

The common cyanobacterial toxins are characterised in three groups, namely neurotoxins, hepatotoxins and protracted toxic substances. Further documentation on toxic cyanobacteria, occurence, health effects related to the toxins, chemical structure and quantitative determination of toxins is presented in Chapter 2.2. Neurotoxins cause rather rapid reactions which affect the nervous system by over-exciting the muscle cells, so that they stop operating. Signs of toxicoses among domestic and wild animals include staggering, muscle fasciculations, gasping, convulsions, and also, in the case of birds,

Table 2.1.9 Characteristic species found in different kinds of Swedish meso- and eutrophic lakes in a late summer succession stage (Rosén, 1981). Nomenclature used is mainly according to *Süsswasserflora von Mitteleuropa* (Komárek and Anagnostidis 1986, 1989; Anagnostidis and Komárek, 1988).

Species	Humic lakes[a]	Meso-trophic lakes[b]	eutrophic lakes[c]
Anabaena spp.		×	×
Aphanizomenon flos-aquae		×	×
Aphanocapsa spp.			×
Asterionella formosa	×	×	
Aulacoseira alpigena	×		
Aulacoseira granulata		×	×
Botryococcus spp.	×	×	
Ceratium hirundinella		×	×
Chlorococcal genera		×	×
Chrysococcus spp.	×		
Closterium acutum var. *variabile*		×	
Closterium spp.			×
Crucigenia tetrapedia	×		
Cryptomonas spp.	×		×
Cyclotella spp.		×	
Cyclotella stelligera	×		
Dinobryon bavaricum	×	×	
Dinobryon borgei	×		
Dinobryon crenulatum	×	×	
Dinobryon divergens		×	
Dinobryon suecicum	×		
Eudorina spp.			×
Euglena spp.			×
Fragilaria crotonensis		×	×
Gonyostomum semen	×		
Gymnodinium uberrimum	×		
Kephyrion, Pseudo-kephyrion, Kephyriopsis	×		
Mallomonas caudata		×	
Merismopedia tenuissima	×		
Microcystis spp.		×	×
Monoraphidium dybowskii	×		
Monoraphidium griffithii	×		
Oocystis borgei	×		
Oocystis marssonii	×		
Oocystis submarina	×		
Peridiniopsis polonicum			×
Peridinium bipes		×	×
Peridinium cinctum		×	×
Peridinium inconspicuum	×		

Continued

Table 2.1.9   *Continued.*

Species	Humic lakes[a]	Meso-trophic lakes[b]	Eutrophic lakes[c]
Phacus spp.	×		×
Planktothrix agardhii		×	
Quadrigula pfitzeri	×		
Rhizosolenia longiseta	×	×	
Rhodomonas lacustris	×		
Scenedesmus spp.		×	
Small naked *chryso-* and *dinoflagellates*	×		
Staurastrum anatinum		×	
Staurastrum chaetoceras			×
Staurastrum cingulum		×	
Staurastrum luetkemuelleri		×	
Staurastrum lunatum var. *planctonicum*		×	
Staurastrum paradoxum var. *parvum*			×
Staurastrum pingue		×	
Staurastrum planctonicum		×	×
Staurastrum smithii			×
Staurastrum tetracerum			×
Staurodesmus cuspidatus var. *curvatus*		×	
Staurodesmus extensus		×	
Staurodesmus patens			×
Stephanodiscus rotula		×	×
Synedra acus var. *angustissima*			×
Synedra spp.		×	
Tabellaria flocculosa var. *asterionelloides*		×	
Trachelomonas spp.	×		×
Uroglena spp.		×	
Willea irregularis		×	
Woronichinia compacta			×

[a]Total phosphorous $> 10\,\mu g\,l^{-1}$; pH > 6; colour $> 25\,mg\,Pt\,l^{-1}$
[b]Total phosphorus $12–17\,\mu g\,l^{-1}$
[c]Total phosphorus $> 17\,\mu g\,l^{-1}$

Table 2.1.10   Water-blooming cyanobacteria in August.

Class	Denomination	Biomass $(mm^3\,l^{-1})$
1	Very small biomass	$\leqslant 0.5$
2	Small biomass	$0.5–1$
3	Intermediate biomass	$1–2.5$
4	Large biomass	$2.5–5$
5	Very large biomass	> 5

opistothonos. Death is often caused by respiratory arrest. No mortality occurring among humans has yet been notified resulting from exposure to these neurotoxins, but several types of diffuse symptoms such as headache, stomach disorder, dizziness, feeling of sickness, etc., have been reported. Among domestic animals, deaths occur annually in all of the Nordic countries. Hepatotoxins act more slowly, and affect the liver by hemorrhage due to distortion of the cytoskeleton of the liver cells. Death occurs within hours to a few days after consumption and may be preceded by coma, muscle tremor and forced expiration of air. The liver-cancer promotion caused by hepatotoxins is (among others) due to their mode of triggering early cell changes in human beings. They may certainly be dangerous in cases of long-term consumption of badly treated drinking water. Hepatotoxins are the most commonly encountered toxins produced by cyanobacteria (Lampert *et al.*, 1994; Carmichael, 1997). Protracted toxins cause the deaths of experimental animals (mice) after a longer period of time than for the other toxins, but without any significant organ damage. The active components of these toxins are still unknown.

The following planktic taxa have been verified by mouse bioassays, by high performance liquid chromatography (HPLC) analysis or by immunological tests to be toxic in freshwaters in the Nordic countries (Sivonen *et al.*, 1990; Kaas *et al.*, 1996; Skulberg, 1996; Willén and Mattsson, 1997).

The genus *Anabaena*:
  Anabaena circinalis Rab.
  Anabaena crassa (Lemm.) Kom.-Legn et Cronb.
  Anabaena farciminiformis Cronb. et Kom.-Legn.
  Anabaena flos-aquae Bréb.
  Anabaena lemmermannii P. Richter
  Anabaena mendotae Trel.
  Anabaena solitaria Kleb.
  Anabaena spiroides Kleb.

The genus *Aphanizomenon*:
  Aphanizomenon flos-aquae (L.) Ralfs
  Aphanizomenon gracile (Lemm.) Lemm.
  Aphanizomenon klebahnii (Elenk.) Pechar et Kalina
  Aphanizomenon yezoense Watanabe

The genus *Gloeotrichia*:
  Gloeotrichia echinulata (J.E. Smith) Richter

The genus *Microcystis*:
  Microcystis aeruginosa Kütz.

Microcystis botrys Teil.
Microcystis flos-aquae (Wittr.) Kirchn.
Microcystis viridis (A.Br.) Lemm.
Microcystis wesenbergii (Kom.) Kom. in Kondr.

The genus *Planktothrix*:
Planktothrix agardhii (Gom.) Anagn. et Kom. (Syn.: *Oscillatoria agardhii* Gom.)
Planktothrix mougeotii (Bory ex. Gom.) Anagn. et Kom. (Syn.: *Oscillatoria agardhii* var. *isothrix* Skuja)
Planktothrix prolifica (Gom.) Anagn. et Kom. (Syn.: *Oscillatoria prolifica* Gom.)
Planktothrix rubescens (DC ex. Gom) Anagn. et Kom. (Syn.: *Oscillatoria rubescens* Gom.)

The genus *Snowella*:
Snowella lacustris (Chod.) Kom. et Hind.
The genus *Woronichinia*:
Woronichinia naegeliana (Unger) Elenk. (Syn.: *Gomphosphaeria*) *naegeliana* (Unger) Lemm.

In the above-mentioned Swedish assessment criteria for lakes and water-courses, both the biomass of water-blooming cyanobacteria and the number of toxin-producing taxa have been considered as water quality indicators (Willén 1999). The ranking of biomass values may, however, be highly specific for different physical geographical regions. For Sweden, the biomass classes shown in Tables 2.1.10 and 2.1.11 have been suggested.

Table 2.1.11 Number of potential toxin-producing genera of cyanobacteria in August.

Class	Denomination	No. of genera
1	None or few	$\leqslant 2$
3	Intermediate number	3–4
5	Large number	> 4

2.1.8   CONCLUSIONS

The above discussion of the various tools which make use of planktic algae in water quality assessments, provides just an outline. However, a more detailed knowledge for practical use can be obtained by using the given references. An urgent requirement for the near future is to combine the experiences obtained from various physical geographical regions, and compare the scales used for assessment of water quality based on planktic algae. Another line of approach is to agree on the relevant 'background values' in various water quality states for different kinds of lakes and ecoregions. Such 'background values', combined with realistic and acceptable cultural aspects could then be used to set achievable goals for nutrient discharge in order to receive the suggested biological quality.

REFERENCES

Anagnostidis, K. and Komárek, J., 1988. Modern approach to the classification system of cyanophytes, 3: Oscillatoriales, *Archiv für Hydrobiologie*, Suppl. 80(1–4), Algological Studies 50–53, 327–472.

Apstein, C., 1896. *Das Süsswasserplankton, Methode und Resultate der Quantitativen Untersuchung*, Kiel and Leipzig.

Brettum, P., 1989. Alger som Indikator på Vannkvalitet i Norske Innsjøer: Planteplankton, Norsk Institutt for Vannforskning (NIVA), Report O-86116: 2344, Oslo, Norway (in Norwegian).

Carmichael, W., 1997. The cyanotoxins, *Advances in Botanical Research*, 27, 211–256.

Christensen, T., 1994. *Algae, a Taxonomic Survey*, AIO Print Ltd, Odense, Denmark.

Du Rietz, E., 1936. Classification and nomenclature of vegetation units 1930–1935, *Svensk Botanisk Tidskrift*, 30.

Heinonen, P., 1980. Quantity and composition of phytoplankton in Finnish inland waters, *Publications of the Water Research Institute (Helsinki)*, 37, 1–91.

Huitfeldt-Kaas, H. 1906. *Planktonundersøgelser i Norske Vand*, Christiania, National-trykkeriet, Kiel and Leipzig.

Hörnström, E., 1981. Trophic characterization of lakes by means of qualitative phytoplankton analysis, *Limnologica (Berlin)*, 13, 246–261.

Järnefelt, H., 1952. Plankton als Indikator der Trophiegruppen der Seen, *Annales Academiæ Scientiarium Fennicæ, Ser. A, IV, Biologica*, 18.

Järnefelt, H., 1956. Zur Limnologie einige Gewässer Finnlands, 16, *Annales Societatis Vanamo*, 17, 1–201.

Kaas, H., Henriksen, P. and Jensen, J. P, 1996. Blågrønalgetoksiner i Bade-og Drikkevand, Report from Miljø-og Energiministeriet, Danmarks Miljøundersøgelser, Copenhagen, Denmark (in Danish).

Kalff, J. and Knoechel, R., 1978. Phytoplankton and their dynamics in oligotrophic and eutrophic lakes, *Annual Review of Ecology and Systematics*, 9, 475–495.

Kolkwitz, R. and Marsson, M. 1902. Grundsätze für die biologische Beurteilung des Wassers nach seiner Flora und Fauna, *Mitteilung der Königliche Prüfungsanstalt für Wasserversorgung und Abwasserbeseitigung*, 1, 33–72.

Kolkwitz, R. and Marsson, M., 1908. *Ökologie der Pflanzlichen Saprobien*, Berichte der Deutschen Botanische Gesellschaft, Berlin.

Komárek, J. and Anagnostidis, K., 1986. Modern approach to the classification system of cyanophytes, 2: Chroococcales, *Archiv für Hydrobiologie*, Suppl. 73(2), Algological Studies 43, 157–226.

Komárek, J. and Anagnostidis, K., 1989. Modern approach to the classification system of cyanophytes, 4: Nostocales, *Archiv für Hydrobiologie*, Suppl. 82(3), Algological Studies 56, 247–345.

Lampert, T. W., Holmes, C. F. B. and Hrudey, S. E., 1994. Microcystin class of toxins: health effects and safety of drinking water supplies, *Environmental Review*, 2, 167–186.

Naumann, E., 1919. Några synpunkter angående limnoplanktons ökologi, Med särskild hänsyn till fytoplankton, *Svensk Botanisk Tidskrift*, **13**, 51–58 (in Swedish).

Nixon, S. C., Mainstone, C. P., Milne, I., Moth Iversen, T., Kristensen, P., Jeppesen, E., Friberg, N., Papathaanassiou, E., Jensen, A. and Pedersen, F., 1996. The Harmonised Monitoring and Classification of Ecological Quality of Surface Waters in the European Union, Draft Final Report No. CO 4096, Water Research Centre, Marlow, UK.

Nygaard, G., 1949. Hydrobiological studies on some Danish ponds and lakes, II: The quotient hypothesis and some little known or new phytoplankton organisms, *Kunglige Danske Vidensk, Selskab*, **7**, 1–242.

Olrik, K., 1993. Planteplankton–Oekologi. Miljøprojekt 243. Miljøministeriet, Miljøstyrelsen, Copenhagen, Denmark (in Danish).

Padisák, J., 1992. Spatial and temporal scales in phytoplankton ecology, *Abstracta Botanica*, **16**, 15–23.

Palmer, C. M., 1969. A composite rating of algae tolerating organic pollution, *Journal of Phycology*, **5**, 78–82.

Persson, G. 1999. Växtnäringsämnen/Eutrofiering, in the Background Report on Chemical and Physical Parameters used for the Assessment of the Water Quality of Lakes and Watercourses (Report 4920), Swedish Environmental Protection Agency Stockholm, Sweden.

Reynolds, C., 1980. Phytoplankton assemblages and their periodicity in stratifying lake systems, *Holarctic Ecology*, **3**, 141–159.

Reynolds, C., 1984. Phytoplankton periodicity: the interactions of form, function and environmental variability, *Freshwater Biology*, **14**, 111–142.

Reynolds, C., 1987. Community organization in the freshwater plankton, in Gee, J. and Giller, P. S. (Eds), *Organization of Communities, Past and Present*, Blackwell Scientific, Oxford, 297–325.

Reynolds, C., 1993. Scales of disturbance and their role in plankton ecology, *Hydrobiologia*, **249**, 157–171.

Rosén, G., 1981. Phytoplankton indicators and their relations to certain chemical and physical factors, *Limnologica (Berlin)*, **13**, 263–290.

Rott, E., 1984. Phytoplankton as biological parameter for the trophic characterization of lakes, *Internationale Vereinigung für Theoretische und Angewandte Limnologie, Verhandlungen*, **22**, 1078–1085.

Sivonen, K., Niemelä, S. I., Niemi, R. M., Lepistö, L., Luoma, T. H. & Räsänen, L.A, 1990. Toxic cyanobacteria (blue-green algae) in Finnish fresh and coastal waters, *Hydrobiologia*, **190**, 267–275.

Skulberg, O., 1996. Toxins produced by cyanophytes in Norwegian inland waters – health and environment, in *Chemical Data as a Basis of Geomedical Investigations*, Norwegian Academy of Science and Letters, Oslo, Norway, 197–216.

Sommer, U., 1991. Phytoplankton: directional succession and forced cycles, in Remmert, H. (Ed.), *The Mosaic-Cycle Concept of Ecosystems*, Ecological Studies 85, Springer-Verlag, Heidelberg.

South, G. R. and Whittick, A., 1987. *Introduction to Phycology*, Blackwell Scientific, Oxford.

Stewart, A. and Wetzel, R., 1986. Cryptophytes and other microflagellates as couplers in planktonic community dynamics, *Archiv für Hydrobiologie*, **106**, 1–19.

Teiling, E., 1916. En kaledonisk fytoplanktonformation, *Svensk Botanisk Tidskrift*, **10**, 506–519.

Teiling, E., 1942. *Schwedische Planktonalgen 4: Phytoplankton aus Roslagen*, Botaniska Notiser, Lund, Sweden.

Teiling, E., 1955. Some mesotrophic phytoplankton indicators, *Proceedings of the International Association of Theoretical and Applied Limnology*, **XII**, 212–215.

Thienemann, A., 1921. Seetypen, *Naturwissenschaften*, **9**.

Thunmark, S., 1945. Zur Soziologie des Süsswasserplanktons: Eine methodologisch-ökologische Studie, *Folia Limnologica Scandinavica*, **3**, 1–66.

Tikkanen, T. and Willén, T., 1992. *Växtplanktonflora*, Naturvårdsverket, Stockholm, Sweden (in Swedish).

Trifonova, I. S., 1989. Changes in community structure and productivity of phytoplankton as indicators of lake and reservoir eutrophication, *Archiv für Hydrobiologie, Beihefte: Ergebnisse der Limnologie*, **33**, 363–371.

Utermöhl, H., 1958. Zur Vervollkommnung der quantitativen Phytoplanktonmethodik, *Internationale Vereinigung für Theoretische und Angewandte Limnologie*, Mitteilungen, **9**, 1–38.

Washington, H. G., 1984. Diversity, biotic and similarity indices: A review with special relevance to aquatic ecosystems, *Water Research*, **18**, 653–694.

Wetzel, R., 1975. *Linmology*, W.B. Saunders Company, Philadelphia, PA, USA.

Willén, E., 1992a. Long-term changes in the phytoplankton of large lakes in response to changes in nutrient loading, *Nordic Journal of Botany*, **12**, 575–587.

Willén, E., 1992b. Planktonic green algae in an acidification gradient of nutrient-poor lakes, *Archiv für Protistenkunde*, **141**, 47–64.

Willén, E. and Mattsson, R., 1997. Water-blooming and toxin-producing cyanobacteria in Swedish fresh and brackish waters, 1981–1995, *Hydrobiologia*, **353**, 181–192.

Willén, E., 1999. Planktiska Alger, in the Background Report on Biological Parameters used for the Assessment of the Water Quality of Lakes and Watercourses (Report 4921), Swedish Environmental Protection Agency, Stockholm, Sweden.

Chapter 2.2
Toxic Cyanobacteria

KAARINA SIVONEN

Hydrological and Limnological Aspects of Lake Monitoring
Edited by Pertti Heinonen, Giuliano Ziglio and André Van der Beken
©2000 John Wiley & Sons, Ltd, ISBN 0 471 89988 7

2.2.1 INTRODUCTION

Cyanobacteria which form so-called water blooms may produce a wide variety
of toxins. Hepatotoxins are the most frequently found cyanobacterial toxins in
fresh and brackish waters worldwide. These are cyclic hepta- or pentapeptides,
known as microcystins and nodularins, respectively. More than 60 different
microcystins and few nodularin structures are known. Toxic blooms are one of
the most unwanted consequences of eutrophication, and have caused several
animal poisonings and are a health risk to humans when water containing toxic
cyanobacteria is used for drinking, food production, recreation and medical
treatments such as hemodialysis. There are several bioassays and chemical
detection methods available for cyanobacterial toxins.

2.2.2 MASS OCCURRENCES OF TOXIC CYANOBACTERIA

Water blooms of cyanobacteria frequently occur, in particular in eutrophic
waters. The toxicity of such cyanobacterial blooms first became known from
reports on animal poisoning cases and later from systematic surveys. These
surveys have revealed an unexpectedly high frequency of toxic blooms (Table
2.2.1).

Mass occurrences of hepatotoxic cyanobacteria are more common than
neurotoxic ones. Hepatotoxic blooms have been reported from all of the
continents and from almost every part of the world where they have been
studied. Mass occurrences of neurotoxic cyanobacteria have been reported
from certain parts of America, Europe, Australia and Asia. Cyanobacterial
blooms containing neurotoxins or hepatotoxins have caused numerous animal
poisonings worldwide (for examples in Europe, see Table 2.2.2).

The bloom-forming season of cyanobacteria depends on the climatic
conditions of the region. In the northern hemisphere, mass occurrences of
cyanobacteria are most prominent during the late summer and early autumn,
lasting for 2–4 months. Mass occurrences of cyanobacteria in Scandinavian
lakes have been occasionally found under the ice. The more south we go, then
the longer is the bloom-forming season of cyanobacteria, e.g. in France four
months, while in Portugal and in Spain six months is common (see also
Chapter 2.1 in this volume for a discussion of toxin-producing cyanobacteria
and other related aspects).

2.2.3 CYANOBACTERIAL HEPATOTOXINS: MICROCYSTINS
AND NODULARINS

A knowledge of the cyanobacterial genera and species which produce
microcystins has been accumulating over the years, based on the occurrence

Table 2.2.1 Frequencies of toxic mass occurrences of cyanobacteria reported in European freshwaters[a].

Country	Number of samples	Toxic samples (%)	Type of toxicity[b]	Reference
Denmark	296	82	Hepatotoxic SDF Neurotoxic	Henriksen, 1996
Germany (GDR)	10	70	Hepatotoxic SDF	Henning and Kohl, 1981
Germany	532	72[c]	Hepatotoxic	Fastner, 1997
Germany	329	21[c]	Neurotoxic/ Anatoxin-a	Bumke-Vogt, 1997
Greece	18	?	Hepatotoxic	Lanaras et al., 1989
Finland	215	44	Hepatotoxic Neurotoxic	Sivonen, 1990
France, Brittany	25	70[c]	Hepatotoxic	Vezie et al., 1997
Hungary	50	66	Hepatotoxic	Törökné, 1991
Norway	64	92	Hepatotoxic Neurotoxic SDF	Skulberg et al., 1994
Portugal	30	60	Hepatotoxic	Vasconcelos, 1994
Scandinavia (Norway, Sweden (Finland))	81	60	Hepatotoxic	Berg et al., 1986
Sweden	53	43	Hepatotoxic Neurotoxic	Mattsson and Willén, 1986
Sweden	331	47	Hepatotoxic Neurotoxic	Willén and Mattsson, 1997
The Netherlands	13	92	Hepatotoxic	Leeuwangh et al., 1983
UK	21	95	Hepatotoxic SDF	Richard et al., 1983
UK	78	68	Hepatotoxic Neurotoxic	Pearson, 1990
UK	50	48	Hepatotoxic	Codd and Bell, 1996

[a]Toxicity was determined by mouse bioassay if not indicated otherwise.
[b]SDF, slow death factor; mice die after 4 h, but within 24 h, possibly due to low amounts of toxins or unknown slow-acting toxic components.
[c]High performance liquid chromatography (for hepatotoxins) or gas chromatography with an electrocapture detector (for anatoxin-a) were the procedures used to detect toxins in these samples.

of such species in toxic blooms and by isolating the cyanobacterial strains and showing their toxin production. Species of *Microcystis, Anabaena, Oscillatoria/ Planktothrix, Nostoc,* and *Anabaenopsis* genera produce microcystins, while nodularins have been found only from *Nodularia spumigena.*

Table 2.2.2 Some recent animal poisoning cases that have occurred in Europe due to cyanobacterial toxins.

Country (location, year)	Animals affected	Toxin	Referemce
Sweden (Baltic Sea coast, 1982)	9 dogs	Nodularin	Edler *et al.*, 1985
Germany (Baltic Sea coast, 1983)	16 young cattle	Nodularin	Gussmann *et al.*, 1985
Finland (Baltic Sea coast, 1984)	1 dog, 3 puppies	Nodularin	Persson *et al.*, 1984
Denmark (Lake Knud sø, 1993 and 1994)	Birds	Anatoxin-a(S)	Henriksen *et al.*, 1997
Finland (Åland Island, 1986)	Fish, birds, muskrats	Hepatotoxins	Eriksson *et al.*, 1986
Finland (Lake Sääskjärvi, 1985)	2 cows	Anatoxin-a	Sivonen *et al.*, 1990
Finland (Lake Vanajavesi, 1985)	2 cows	Hepatotoxins	Sivonen *et al.*, 1990
Finland (Lake Säyhteenjärvi, 1986)	3 cows	Anatoxin-a	Sivonen *et al.*, 1990
Ireland (Caragh Lake, 1992–1994)	Dogs, cats	Anatoxin-a	James *et al.*, 1997
UK (Loch Insh, 1990 and 1991)	Several dogs	Anatoxin-a	Edwards *et al.*, 1992
UK (Rutland Water, 1989)	20 sheep, 14 dogs	Hepatotoxins	Pearson, 1990

To date, two isolated strains of *Nostoc*, one from a Finnish lake (Sivonen, 1990; Sivonen *et al.*, 1992) and the other from the UK (Beattie *et al.*, 1998), have been found to produce different microcystin analogs. *Anabaenopsis milleri*, from Greece (Lanaras and Cook, 1994) and a terrestrial cyanobacterium, *Haphalosiphon hibernicus* (Prinsep *et al.*, 1992), were also found to contain (most likely) microcystin analogs. The isolation and characterization of toxins from certain species has verified them as toxin producers. It is likely, however, that the list of verified toxic species will increase in future due to new isolates and improved isolation and culturing methods.

The first chemical structures of cyanobacterial hepatotoxins were solved during the last decade and the number of characterized toxin variants has greatly increased during this period. Cyanobacterial hepatotoxins occurring in freshwaters were found to be cyclic heptapeptides with the general structure of cyclo(D-Ala^1-X^2-D-MeAsp^3-Z^4-Adda^5-D-Glu^6-Mdha^7), in which X and Z are variable L-amino acids (Carmichael *et al.*, 1988). The amino acid, Adda, is the most unique structure in cyanobacterial hepatotoxins. Variations in the chemical structures of microcystins is very common, with about 60 different varieties of microcystins having been characterized from bloom samples and isolated strains of cyanobacteria thus far (for a list of the different microcystin variants, see Sivonen and Jones, 1999). In brackish waters, an identically acting and structurally very similar cyclic pentapeptide occurs. This is called nodularin, according to the producer, *Nodularia spumigena*. A few naturally occurring variations of nodularins have been found, including a non-bioactive nodularin. The most prominent areas of occurrence of nodularin-containing blooms are the Baltic Sea and the saline lakes and estuaries in Australia and New Zealand.

2.2.3.1 Toxicity

The LD_{50} (lethal dose for 50% of test animals) values of different toxins vary from highly toxic ($50 \mu g \, kg^{-1}$) to non-toxic (see Sivonen and Jones (1999) for a list of the LD_{50} values of different microcystin variants). The linear microcystins (non-bioactive microcystines) are thought to be either microcystin precursors or breakdown products. The mechanisms of the toxic reactions of cyanobacterial hepatotoxins are currently rather well understood. Microcystins and nodularin are hepatotoxins since they can not enter other types of cells due to their non-lipophilic nature. With hepatocytes, they enter by using bile-acid carriers. Early this decade, microcystins and nodularins were found to be inhibitors of the protein phosphatases types 1 and 2A. This inhibition in hepatocytes leads to hyperphosphorylation of the cytoskeletal microfilaments. The damaged hepatocytes integrate with each other, thus filling the liver with blood and causing death by lack of blood to other organs, which leads to haemorrhagic shock (Carmichael, 1994).

Cyanobacterial hepatotoxins in lower doses have been found to be tumor promoters (Fujiki *et al.*, 1996), while one study carried out by Ohta *et al.* (1994) has suggested nodularin to be carcinogen.

2.2.4 CYLINDROSPERMOPSINS

In the tropical and subtropical waters of Australia, the alkaloid cytotoxin, cylindrospermopsin, has caused health problems via drinking water (Hawkins *et al.*, 1985). This is a cyclic guanidine alkaloid cytotoxin with a molecular weight of 415 (Ohtani *et al.*, 1992). Cylindrospermopsin has effects on the liver, the kidneys, the thymus and the heart (Hawkins *et al.*, 1985; Terao *et al.*, 1994). The LD_{50} in mice is $2.1 \, mg \, kg^{-1}$ at 24 h and $0.2 \, mg \, kg^{-1}$ at 5–6 days. This toxin is produced by *Cylindrospermopsis raciborskii* in the tropical and subtropical waters of Australia. In Europe, it has been found to be present (and to be toxic) in Lake Balaton, Hungary. In Japan, the similar toxin was isolated and characterized from *Umezakia natans* (Harada *et al.*, 1994), and in Israel from *Aphanizomenon ovalisporum* (Banker *et al.*, 1997). Recently, these latter species have also been found to be common producers of cylindrospermopsin in Australian waters (Shaw, 1998). New structural variants of cylindrospermopsin have also been recognized. Studies with cultured rat hepatocytes showed that cylindrospermopsin inhibits glutathione synthesis (Runnegar *et al.*, 1994, 1995). Electron microscopy studies on experimentally poisoned mice showed cylindrospermopsin to be a potent inhibitor of protein synthesis (Terao *et al.*, 1994).

2.2.5   CYANOBACTERIAL NEUROTOXINS

Mass occurrences of neurotoxic cyanobacteria have been reported from America, Europe and Australia, where they have caused several animal poisonings. Using chemical analysis, cyanobacterial neurotoxins have been recently detected in Japan and North Korea (Harada *et al.*, 1993; Park *et al.*, 1998). In mouse bioassays, the death by respiratory arrest occurs quickly (within 2–30 min). Signs of poisonings include respiratory distress, tremors, convulsions, fasciculations, and with some toxins, salivation. Cyanobacterial neurotoxins block the neurotransmission system, which does not allow any apparent organ damage to be discovered in autopsy (Beasley *et al.*, 1989).

2.2.5.1   Anatoxin-a

This is a secondary amine (with low molecular weight) alkaloid (Devlin *et al.*, 1977). It blocks the neurotransmission system by mimicking the effect of acetylcholine (Swanson *et al.*, 1986). Anatoxin-a is produced by *Anabaena flos-aquae*, *Anabaena* spp. (*flos-aquae/lemmermannii/circinalis*-group), *Anabaena planktonica*, *Aphanizomenon Cylindrospermum* and by benthic *Oscillatoria*. *Homoanatoxin-a* is an anatoxin-a homologue determined from an *Oscillatoria formosa* strain. The LD_{50} values of anatoxin-a and homoanatoxin-a are 200–250 $\mu g\,kg^{-1}$ (Skulberg *et al.*, 1992).

2.2.5.2   Anatoxin-a(S)

Anatoxin-a(S) is a unique phosphate ester of a cyclic *N*-hydroxyguanine. This ester is an anticholinesterase (Carmichael *et al.*, 1990) and its LD_{50} is 20 $\mu g\,kg^{-1}$ (mouse). Anatoxin-a(S) was first characterized from an Canadian *Anabaena flos-aquae* strain, NRC 525-17 (Matsunaga *et al.*, 1989) and has caused animal poisonings in North America (Cook *et al.*, 1989). More recently, it has been found from blooms and *Anabaena lemmermannii* strains in Denmark connected to bird kills (Henriksen *et al.*, 1997). Structural variations of anatoxin-a(S) have not yet been detected.

2.2.5.3   Saxitoxins

Saxitoxins, which are also called Paralytic Shellfish Poisons (PSPs), are a group of alkaloid neurotoxins; they can be either non-sulfated, singly sulfated (gonyautoxins) or doubly sulfated. These PSPs block the sodium channels (Sasner *et al.*, 1984). The toxicity of the PSPs varies, with saxitoxin being the most potent (LD_{50} $10\,\mu g\,kg^{-1}$, i.p. mouse). Saxitoxins are produced by cyanobacteria in freshwaters and by marine dinoflagellates (red tide organisms) (Anderson, 1994). They have also been found in *Aphanizomenon flos-aquae*, *Anabaena circinalis*, *Lyngbya wollei* and *Cylindrospermopsis raciborskii*. In North America, the benthic freshwater cyanobacterium, *Lyngbya wollei*, has produced three known and six new saxitoxins (Onodera *et al.*, 1998). *Cylindrospermopsis raciborskii* in Brazil was found to contain mostly neosaxitoxins and smaller amount of saxitoxins (Lagos *et al.*, 1997). Very recently, *Oscillatoria/Phormidium* strains from an Italian lake were shown to produce saxitoxins (Pomati *et al.*, 1998). For the different saxitoxin variants found from cyanobacteria, see the recent review by Sivonen and Jones (1999).

2.2.6   DETECTION METHODS OF CYANOBACTERIAL TOXINS

In the earliest studies, mouse bioassays were used to determine the toxicity of samples by evaluating the LD_{50} or MLD_{100} (the dose killing all the test animals) levels. Recently, alternative bioassays have been developed and are currently used to screen the toxicity of samples such as the brine shrimp (*Artemia salina*) (Kiviranta *et al.*, 1991), *Daphnia* (Baird *et al.*, 1989), and Thamnotox (Törökné, 1997), along with other insect larvae tests (Kiviranta *et al.*, 1993). Of these, the one most often used is the *Artemia salina* test which is comparatively easy to perform. For the screening of cyanobacterial hepatotoxins, primary hepatocytes have been successfully used in the bioassay (Heinze, 1996). The very sensitive screening methods available for cyano-bacterial hepatotoxins include the commercially available ELISA test (Chu *et al.*, 1990) and the colorimetric or radioactive protein phosphatase inhibition assays (Lambert *et al.*, 1994). Similarly, detection of anatoxin-a(S) is based on the acetylcholinesterase inhibition assay (Carmichael *et al.*, 1990).

Several chemical separation methods based on liquid chromatography have been developed to quantitate different microcystins and saxitoxins. Capillary electrophoresis or thin layer chromatography can also achieve separation. The liquid chromatography detection of microcystins and nodularin is usually based on UV detection, with fluorescence being used for saxitoxins. Both liquid- and gas-chromatography-based methods have been used to quantitate anatoxin-a. A combination of chromatographic separating techniques and mass spectrometry would improve the identification of cyanobacterial toxins but will not solve the problems completely unless tandem mass spectrometry (MS/MS) is employed. The lack of analytical standards for anatoxin-a(S), and of a full range of saxitoxin and microcystin variants has made the chemical analysis of these toxins difficult. For a current summary of the detection methods that are used see the review by Harada *et al.* (1999).

The development of accurate detection methods for cyanobacterial toxins has made it possible to study the environmental concentrations of these toxins, as well as the toxin production of cyanobacteria and the fate of the cyanobacterial toxins in environmental and water treatment processes.

The quantitative determination of toxin concentrations in environmental samples is mostly carried out on the concentrated, lyophilized bloom samples (particulate material, which contains not only cyanobacterial cells but also usually some zooplankton), with results being expressed as mg or μg per g of dry weight. The highest amounts of cyanobacterial toxins that have been found in the environment are several mg of toxins per g of dry weight (Sivonen and Jones, 1999). Cyanobacterial mass occurrences can be monocultural, where one single species is dominating, or can be composed of a variety of toxic and non-toxic species. In the former case, it is easy to attribute the toxicity to one particular species. Toxic and non-toxic cyanobacteria, even from the same species, can co-occur in blooms, and then can not be separated by microscopic identification (Vezie *et al.*, 1997). It has been observed that some strains are more toxic than others, so that one heavily toxic organism, even when occurring in minor amounts, may cause the bloom sample to be toxic. The co-occurrence of toxic and non-toxic species and strains is possibly the main cause for the high variation of toxin concentrations in environmental samples.

Further development of the analytical techniques and the screening methods has enabled the measurement of very low concentrations of toxins and made it possible to also determine toxins from water (μg/ng per L). Concentrations reported in the literature for dissolved toxins in water, vary from trace concentrations up to 1800 μg L^{-1} or higher (Sivonen and Jones, 1999).

2.2.7 CYANOBACTERIAL TOXIN PRODUCTION

The development of quantitative methods for determining cyanobacterial toxins has also enabled studies on cyanobacteria toxin production. It has been found that cyanobacterial strains may produce several microcystin or saxitoxin variants simultaneously.

Toxic strains of cyanobacteria seem to be constantly toxic; in different conditions the toxin content may vary but the toxin production seems not to cease totally. It is well established that the toxin production of cyanobacteria is strain- and not species-specific. The toxic and non-toxic strains of the same species can not be separated by microscopy. Currently, several laboratories are using molecular biological methods to characterize toxic and non-toxic planktic cyanobacteria (Neilan, 1996; Rudi *et al.*, 1997), work which will help to clarify the taxonomic status of these organisms. Dittmann *et al.* (1997) showed that microcystins in *Microcystis aeruginosa* are produced non-ribosomally by a large enzyme complex, namely microcystin synthetase. In the near future, this may lead to the development of specific methods to differentiate hepatotoxic strains from non-toxic ones. However, this task is complicated by the fact that cyanobacteria produce several other cyclic/linear peptides, involving similar mechanisms.

The effects of several environmental factors on the growth and toxin production of cyanobacteria have been studied mostly in batch, but also in continuous cultures. Sivonen and Jones (1999) have recently reviewed these studies. There have been studies on the effect of culture age in batch cultures and the effects of temperature, light, nutrients, salinity, CO_2, pH and micronutrients, including iron and zinc, on the growth and toxin production of cyanobacteria. Studies have been performed with hepatotoxic *Microcystis*, *Oscillatoria*, *Anabaena*, *Nodularia*, anatoxin-a-producing *Anabaena*, *Aphanizomenon* and *Oscillatoria*, and saxitoxin-producing *Aphanizomenon* and *Anabaena circinalis* (Sivonen and Jones, 1999). Hepatotoxins are mostly kept within cells when the conditions for the growth of the organism are favourable. The amount of hepatotoxins and saxitoxins increases during the (logarithmic) growth-phase, being highest in the late (logarithmic) phase. Maximum anatoxin-a production is found in the same conditions during the growth-phase.

In general, most studies carried out to date have shown that the organisms produce most toxins in the same conditions, i.e. those which are also favourable for their growth. Two- to threefold differences in toxin production have been reported in relation to the light conditions. *Oscillatoria* preferred low light intensities for growth, while *Anabaena* and *Aphanizomenon* preferred moderate and high levels, respectively. All strains produced the most toxins when growing under their optimum light conditions. The strains and species also differed slightly in their optimum growth temperatures. The toxin production in most studies was highest at temperatures between 18 and 25 °C, whereas low (10 °C) or high temperature (30 °C) decreased the toxin production. Temperature conditions caused two- to threefold differences in toxin production. Different microcystins might also be produced under different temperatures and light conditions.

From the management point of view, the response of toxic cyanobacteria to nutrients is an important issue. It was found that hepatotoxic strains produced more toxins under high concentrations of phosphorus, while for anatoxin-a production, phosphorus had no effect. The differences induced by low and high phosphorus concentrations varied between two- to fourfold in laboratory and field studies. Interestingly, the positive correlation of total phosphorus with microcystin in cells of *M. aeruginosa*, or in bloom material of *Microcystis* spp. (Lahti *et al.*, 1997) has also been found in field studies. Species such as *Microcystis* and *Oscillatoria* produce more toxins under high nitrogen concentrations. The nitrogen-fixing species are not dependent on the nitrogen in the media for their toxin production. Orr and Jones (1998) found a linear correlation between the cell division and microcystin production rates. They concluded that the factors increasing the growth rate also promote toxin production and that cyanobacterial toxins may not be true secondary metabolites.

There have only been a few studies reported on the effects of micronutrients on cyanobacterial growth and toxin production and the results obtained have sometimes been conflicting.

Physiological differences between toxic and non-toxic strains are not yet fully understood. More studies are also needed to reveal the cascades leading to the development of toxic blooms. Molecular ecological methods will for the first time allow us to study natural waters at population/genotype levels and are likely to increase our understanding on the factors regulating the occurrence of different cyanobacterial populations.

2.2.8   WATER TREATMENT AND FATE OF CYANOBACTERIAL TOXINS

Cyanobacterial toxins are largely kept within well-growing cells and released to the water by cell lysis. Thus, the removal of cyanobacterial cells in water-treatment processes is as important as removal of the toxins liberated in the water. It has been shown that chemical flocculation effectively removes the cyanobacterial cells (Lepistö *et al.*, 1994). For the removal of toxins from water, activated-carbon and ozonation treatments are needed (Drikas, 1994). There may be a risk, however, that activated-carbon treatment is not always sufficient to remove the toxins completely (Lambert *et al.*, 1996). At the chlorine concentrations normally used in water-treatment processes in Finland, toxins were not removed (Keijola *et al.*, 1988). In an Australian study, the high amounts of chlorine used were shown to oxidize the toxins (Drikas, 1994). Trace amounts of microcystins have been detected in the observation tubes and pumping wells of waterworks which produce drinking water by the artificial recharge of groundwater and bank filtration (Lahti *et al.*, 1998).

In Brazil, it was reported that over 50 patients had died in a hemodialysis clinic due to the presence of microcystins in the water used for their treatment (Jochimsen *et al.*, 1998). This emphasizes the importance of water quality control in medical treatments.

Microcystins and nodularin are cyclic peptides and are very resistant to degradation and the boiling of water does not destroy them. The same is true for food-containing microcystins/nodularin. It has been recommended that viscera, especially liver of fish or waterfowl, should not be used for human consumption during periods of heavy cyanobacterial blooms. Biodegradation of toxins can occur, and bacteria able to degrade both microcystins and nodularin have been isolated (Bourne *et al.*, 1996). The biodegradation rate seems to be highly dependent on the water temperature (Jones and Orr, 1994; Lahti *et al.*, 1997).

Cyanobacterial neurotoxins degrade more easily than hepatotoxins; in fact, neurotoxins such as anatoxin-a and anatoxin-a(S) decompose under alkaline conditions (Stevens and Krieger, 1991). The breakdown of anatoxin-a is further accelerated by sunlight (Stevens and Krieger, 1991).

Saxitoxins are known to accumulate in feeders, where the latter filter large amounts of water containing mussels. It is also known that in biological systems some of these toxins may transform into more poisonous variants (Jones and Negri, 1997). Fresh water mussels are usually not used food for humans, which lessens the danger of poisonings compared to seafood. More studies are needed on the fate and biodegradation of saxitoxins.

2.2.9 CONCLUSIONS

The protection of waters of eutrophication is the most important way to prevent pristine waters becoming nutrient-rich and thus starting to produce harmful cyanobacterial blooms. It seems that phosphorus not only increases the growth of cyanobacteria but also the amount of toxins produced. On the other hand, high levels of nitrogen compounds in water may change the cyanobacterial population in a reservoir from nitrogen-fixing species to non-nitrogen-fixing ones. Such species include the frequently toxic *Microcystis* or *Oscillatoria*.

If a water reservoir contains massive blooms, the use of water for any purpose should be avoided. In many countries, intensive monitoring programmes are carried out during the summer in important recreational areas and beaches where cyanobacterial blooms are present are closed. When such lakes or reservoirs are used for the production of drinking water, efficient methods to remove the cyanobacterial cells and toxins should be used. As the incident in Brazil showed, such treatment should be used not only for drinking water but also in particular for water which is used for medical treatment. The World Health Organization has suggested a guide-line value for the presence of microcystins in drinking water. This may promote more intensive monitoring of drinking waters in the future and also help to minimize the risks.

REFERENCES

Anderson, D.M., 1994. Red tides, *Scientific American*, August, 52–58.

Baird, D. J., Soares, A. M. V. M., Girling, A. E., Barber, I., Bradley, M. C. and Calow, P., 1989. The long-term maintenance of *Daphnia magna* for use in ecotoxicity tests: problems and prospects, in *Proceedings of the First Conference on Ecotoxicology*, Lokke, H., Tyle H. and Bro-Rasmussen, F. (Eds), Lungby, Denmark, 144–148.

Banker, P. D., Carmeli, S., Hadas, O., Teltsch, B., Porat, R. and Sukenik, A., 1997. Identification of cylindrospermopsin in *Aphanizomenon ovalisporum* (Cyanophyceae) isolated from Lake Kinneret, Israel, *Journal of Phycology*, **33**, 613–616.

Beasley V. R., Dahlem A. M., Cook W. O., Valentine W. M., Lowell R. A., Hooser S. B., Harada K.-I., Suzuki M. and Carmichael W. W., 1989. Diagnostic and clinically important aspects of cyanobacterial (blue-green algae) toxicoses, *Journal of Veterinary Diagnostic and Investigation*, **1**, 359–365.

Beattie, K. A, Kaya, K., Sano, T. and Codd, G. A., 1998. Three dehydrobutyrine (Dhb)-containing microcystins from the cyanobacterium *Nostoc* spp., *Phytochemistry*, **47**, 1289–1292.

Berg, K., Skulberg, O. M., Skulberg, R., Underdal, B. and Willén, E., 1986. Observations of toxic blue-green algae (cyanobacteria) in some Scandinavian lakes, *Acta Vet. Scand.*, **27**, 440–452.

Bourne, D., Jones, G. J. Blakeley, R. L., Jones, A., Negri, A. P. and Riddles, P., 1996. Enzymatic pathway for the bacterial degradation of the cyanobacterial cyclic peptide toxin microcystin LR, *Applied and Environmental Microbiology*, **62**, 4086–4094.

Bumke-Vogt, C., 1997. Cyanobakterielle Neurotoxine: Untersuchungen zum Vorkommen von Anatoxin-a in einigen deutschen Gewässern im Verlauf der Jahre 1995/1996, in *Toxische Cyanobakterien in deutschen Gewässern. Verbreitung, Kontrollfaktoren und ökologische Bedeutung*, Institut für Wasser-, Boden- und Lufthygiene des Umweltbundesamtes, Berlin, 35–39 (in German).

Carmichael, W. W., 1994. The toxins of cyanobacteria, *Scientific American*, January, 64–72.

Carmichael, W.W., Beasley, V. R., Bunner, D. L., Eloff, J. N., Falconer, I., Gorham, P., Harada, K.-I., Krishnamurthy, T., Yu, M.-J., Moore, R. E., Rinehart, K., Runnegar, M., Skulberg, O. M. and Watanabe, M., 1988. Naming of cyclic heptapeptide toxins of cyanobacteria (blue-green algae), *Toxicon*, **26**, 971–973.

Carmichael, W. W., Mahmood, N. A. and Hyde, E. G., 1990. Natural toxins from cyanobacteria (blue-green algae), in *Marine Toxins, Origin, Structure and Molecular Pharmacology*, ACS Symposium Series No. 418, Hall, S. and Strichartz, G. (Eds), American Chemical Society, Washington DC, USA, 87–106.

Chu, F. S., Huang, X. and Wei, R. D., 1990. Enzyme-linked immunosorbent assay for microcystins in blue-green algal blooms, *Journal of AOAC*, **73**, 451–456.

Codd, G. A. and Bell, S. G., 1996. The occurrence and fate of blue-green algae in freshwaters, National Rivers Authority Research and Development Report 29, Her Majesty's Stationery Office, London.

Cook, W. O., Beasley, V. R., Lovell, R. A., Dahlem, A. M., Hooser, S. B., Mahmood, N. A. and Carmichael, W. W., 1989. Consistent inhibition of peripheral cholinesterases by neurotoxins from the freshwater cyanobacterium *Anabaena flos-aquae*: studies on ducks, swine, mice and steer, *Environmental Toxicology and Chemistry*, **8**, 915–922.

Devlin, J. P., Edwards, O. E., Gorham, P. R., Hunter, M. R., Pike, R. K. and Stavric, B., 1977. Anatoxin-a, a toxic alkaloid from *Anabaena flos-aquae* NCR-44h, *Canadian Journal of Chemistry*, **55**, 1367–1371.

Dittman, E., Neilan, B. A., Erhard, M., v. Döhren, H. and Börner, T., 1997. Insertional mutagenesis of a peptide synthetase gene which is responsible for hepatotoxin production in the cyanobacterium *Microcystis aeruginosa* PCC 7806, *Molecular Microbiology*, **26**, 779–787.

Drikas, M., 1994. Removal of cyanobacterial toxins by water treatment processes, in *Cyanobacteria – A Global Perspective*, Australian Centre for Water Quality Research, Adelaide, Australia, 30–44.

Edler L., Fernö, S., Lind, M. G., Lundberg, R. and Nilsson, P. O., 1985. Mortality of dogs associated with bloom of the cyanobacterium *Nodularia spumigena* in the Baltic Sea, *Ophelia*, **24**, 103–109.

Edwards, C., Beattie, K. A., Scrimgeour, C. M. and Codd, G. A., 1992. Identification of anatoxin-a in benthic cyanobacteria (blue-green algae) and in associated dog poisonings at Loch Insh, Scotland, *Toxicon*, **30**, 1165–1175.

Eriksson, J., Meriluoto, J. and Lindholm, T., 1986. Can cyanobacterial peptide toxins accumulate in aquatic food chains?, in *Perspectives in Microbial Ecology*, Megusar, F., and Gantar, M. (Eds), Proceedings of the Fourth International Symposium on Microbial Ecology, Slovene Society for Microbiology, Ljubljana, Slovenia, 655–658.

Fastner, J., 1997. Microcystinvorkommen in 55 deutschen Gewässern, in *Toxische Cyanobakterien in deutschen Gewässern. Verbreitung, Kontrollfaktoren und ökologische Bedeutung*, Institut für Wasser-, Boden- und Lufthygiene des Umweltbundesamtes, Berlin, 27–34 (in German).

Fujiki, H., Sueoka, E. and Suganuma, M., 1996. Carcinogenesis of microcystins, in *Toxic Microcystis*, Watanabe, M. F., Harada, K.-I., Carmichael, W. W. and Fujiki, H. (Eds), CRC Press, Boca Raton, Fla., USA, 203–232.

Gussmann, H., Molzahn, M. J. and Bicks, B., 1985. Vergiftungen bei Jungrindern durch die Blaualge Nodularia spumigena, *Mh. Vet. Med.*, **40**, 76–79.

Harada, K.-I., Nagai, H., Kimura, Y., Suzuki, M., Park, H.-D., Watanabe, M. F., Luukkainen, R., Sivonen, K. and Carmichael, W. W., 1993. Liquid chromatography/ mass spectrometric detection of anatoxin-a, a neurotoxin from cyanobacteria, *Tetrahedron*, **49**, 9251–9260.

Harada, K.-I., Ohtani, I., Iwamoto, K., Suzuki, M., Watanabe, M. F., Watanabe, M. and Terao, K., 1994. Isolation of cylindrospermopsin from a cyanobacterium *Umezakia natans* and its screening method, *Toxicon*, **32**, 73–84.

Harada, K.-I., Kondo, F. and Lawton, L.A., 1999. Laboratory analysis of cyanotoxins, in *Toxic Cyanobacteria in Water: a Guide to Public Health Significance, Monitoring and Management*, Chorus, I. and Bertram, J. (Eds), WHO, 369–405.

Hawkins, P. R., Runnegar, M. T. C., Jackson, A. R. B. and Falconer, I. R., 1985. Severe hepatotoxicity caused by the tropical cyanobacterium (blue-green alga) *Cylindrospermopsis raciborskii* (Woloszynska) Seenaya and Subba Raju isolated from a domestic water supply reservoir, *Applied and Environmental Microbiology*, **50**, 1292–1295.

Heinze, R., 1996. A biotest for hepatotoxins using primary rat hepatocytes, *Phycologia*, **35**, 89–93.

Henning, M. and Kohl, J.-G., 1981. Toxic blue-green algae water blooms found in some lakes in the German Democratic Republic, *Int. Rev. Hydrobiol.*, **66**, 553–561.

Henriksen, P., 1996. Toxic cyanobacteria/blue-green algae in Danish fresh waters. Ph.D thesis. Dept. of Phycology, Botanical Institute, University of Copenhagen. 43p.

Henriksen, P., Carmichael, W. W., An, J. and Moestrup, Ø., 1997. Detection of an anatoxin-a(s)-like anticholinesterase in natural blooms and cultures of cyanobacteria/ blue-green algae from Danish lakes and in the stomach contents of poisoned birds, *Toxicon*, **35**, 901–913.

James, K. J., Sherlock, I. R. and Stack, M. A., 1997. Anatoxin-a in Irish freshwater and cyanobacteria, determined using a new fluorimetric liquid chromatographic method, *Toxicon*, **35**, 963–971.

Jochimsen, E. M., Carmichael, W. W., An. J., Cardo, D. M., Cookson, S. T., Holmes, C. E. M., Antunes, M. B., Melo Filho, D. A., Lyra, T. M., Barreto, V. S. T., Azevedo, S. M. F. O. and Jarvis W. R., 1998. Liver failure and death after exposure to microcystins at a hemodialysis center in Brazil, *The New England Journal of Medicine*, **338**, 873–878.

Jones, G. J. and Negri, A. P., 1997. Persistence and degradation of cyanobacterial paralytic shellfish poisons (PSPs) in freshwaters, *Water Research*, **31**, 525–533.

Jones, G. J. and Orr, P. T., 1994. Release and degradation of microcystin following algicide treatment of a *Microcystis aeruginosa* bloom in a recreational lake, as determined by HPLC and protein phosphatase inhibition assay, *Water Research*, **28**, 871–876.

Keijola, A.-M., Himberg, K., Esala, A.-L., Sivonen, K. and Hiisvirta, L., 1988. Removal of cyanobacterial toxins in water treatment processes – laboratory and pilot-scale experiments, *Tox. Assess.*, **3**, 643–656.

Kiviranta, J., Sivonen, K., Niemelä, S. I. and Huovinen, K., 1991. Detection of toxicity of cyanobacteria by *Artemia salina* bioassay, *Environ. Toxicol. Water Qual.*, **6**, 423–436.

Kiviranta J., Abdel-Hameed, A., Sivonen, K., Niemelä, S. I. and Carlberg, G., 1993. Toxicity of cyanobacteria to mosquito larvae – screening of active compounds, *Environ. Toxicol. Water Qual.*, **8**, 63–71.

Lagos, N., Liberona, J. L., Andrinolo, D., Zagatto, P. A., Soares, R. M. and Azevedo, S. M. F. Q., 1997. First evidence of paralytic shellfish toxins in the freshwater cyanobacterium *Cylindrospermopsis raciborskii* isolated from Brazil, Abstracts of the VIII International Conference on Harmful Algae, June 25–29, 1997, Vigo, Spain, 115.

Lahti, K., Rapala, J., Färdig, M., Niemelä, M. and Sivonen, K., 1997. Persistence of cyanobacterial hepatotoxin, microcystin-LR, in particulate material and dissolved in lake water, *Water Research*, **31**, 1005–1012.

Lahti, K., Vaitomaa, J., Kivimäki, A.-L. and Sivonen, K., 1998. Fate of cyanobacterial hepatotoxins in artificial recharge of groundwater and in bank filtration, in *Artificial Recharge of Groundwater*, Peters *et al.* (Eds), Balkema, Rotterdam, 211–216.

Lambert, T. W., Boland, M. P., Holmes, C. F. B. and Hrudey, S. E., 1994. Quatitation of the microcystin hepatotoxins in water at environmentally relevant concentrations with the protein phosphatase bioassay, *Environ. Sci. Technol.*, **28**, 753–755.

Lambert, T. W., Holmes, C. F. B. and Hrudey, S. E., 1996. Adsorption of microcystin-LR by activated carbon and removal in full scale water treatment, *Water Research*, **30**, 1411–1422.

Lanaras, T. and Cook, C. M., 1994. Toxin extraction from an *Anabaenopsis milleri* – dominated bloom, *Science of Total Environment*, **142**, 163–169.

Lanaras, T., Tsitsamis, S., Chlichlia, C. and Cook, C. M., 1989. Toxic cyanobacteria in Greek freshwaters, *J. Appl. Phycol.*, **1**, 67–73.

Leeuwangh, P., Kappers, F. I., Dekker, M. and Koerselman, W., 1983. Toxicity of cyanobacteria in Dutch lakes and reservoirs, *Aquat. Toxicol.*, **4**, 67–73.

Lepistö, L., Lahti, K., Niemi, J. and Färdig, M., 1994. Removal of cyanobacteria and other phytoplankton in four Finnish waterworks, *Archiv für Hydrobiologie, Algological Studies*, **75**, 167–181.

Matsunaga, S., Moore, R. E., Niemczura, W. P. and Carmichael, W. W., 1989. Anatoxin-a(s), a potent anticholinesterase from *Anabaena flos-aquae*, *J. Am. Chem. Soc.*, **111**, 8021–8023.

Mattson, R. and Willén, T., 1986. Toxinbildande Blågröna Alger i Svenska Insjöar 1985 (Toxin-producing Blue-green Algae in Swedish Lakes in 1985), Naturvårdsverket Rapport 3096 (2)/1986, Laboratoriet för Miljökontroll, Uppsala, Sweden (in Swedish, with English summary).

Neilan, B. A., 1996. Detection and identification of cyanobacteria associated with toxic blooms: DNA amplification protocols, *Phycologia*, **35**, 147–155.

Ohta, T., Sueoka, E., Iida, N., Komori, A., Suganuma, M., Nishiwaki, R., Tatematsu, M., Kim, S.-J., Carmichael, W. W. and Fujiki, H., 1994. Nodularin, a potent inhibitor of protein phosphatases 1 and 2A, is a new environmental carcinogen in male F344 rat liver, *Cancer Research*, **54**, 6402–6406.

Ohtani, I., Moore, R. E. and Runnegar, M. T. C., 1992. Cylindrospermopsin, a potent hepatotoxin from the blue-green alga *Cylindrospermopsis raciborskii*, *Journal of the American Chemical Society*, **114**, 7941–7942.

Onodera, H., Satake, M., Oshima, Y., Yasumoto, T. and Carmichael, W. W., 1998. New saxitoxin analogues from the freshwater filamentous cyanobacterium *Lyngbya wollei*, *Natural Toxins*, **5**, 146–151.

Orr, P. T. and Jones, G. J., 1998. Relationship between microcystin production and cell division rates in nitrogen-limited *Microcystis aeruginosa* cultures, *Limnology and Oceanography*, **43**, 1604–1614.

Park, H.-D., Kim, B., Kim, E. and Okino, T., 1998. Hepatotoxic microcystins and neurotoxic anatoxin-a in cyanobacterial blooms from Korean lakes, *Environ. Toxicol. Water Qual.*, **13**, 225–234.

Pearson, M. J., 1990. Toxic Blue-green Algae, Report of the National Rivers Authority, Water Quality Series No. 2, Stanley L. Hunt, Rushden, Northants, UK.

Persson, P.-E., Sivonen, K., Keto, J., Kononen, K., Niemi, M. and Viljamaa, H., 1984. Potentially toxic blue-green algae (cyanobacteria) in Finnish natural waters, *Aqua Fenn.*, **14**, 147–154.

Pomati, F., Sacchi, S., Rossetti, C. and Neilan, B. A., 1998. And now saxitoxin-producing cyanobacetria in Europe, Abstracts of the 4th International Conference on Toxic Cyanobacteria, Beaufort, NC, USA, 44.

Prinsep, M. R., Caplan, F. R., Moore, R. E., Patterson, G. M. L., Honkanen, R. E. and Boynton, A. L., 1992. Microcystin-LA from a blue-green alga belonging to the Stignonematales, *Phytochemistry*, **31**, 1247–1248.

Richard, D. S., Beattie, K. A. and Codd, G. A., 1983. Toxicity of cyanobacterial blooms from Scottish freshwaters, *Environ. Technol. Lett.*, **4**, 377–382.

Rudi, K., Skulberg, O. M., Larsen, F. and Jakobsen, K. S., 1997. Strain characterization and classification of oxyphotobacteria in clone cultures on the basis of 16S rRNA sequences from the variable regions V6, V7, and V8, *Applied and Environmental Microbiology*, **63**, 2593–2599.

Runnegar M. T., Kong S., Zhong Y.-Z., Ge, J.-L. and Lu, S. C., 1994. The role of glutathione in the toxicity of a novel cyanobacterial alkaloid cylindrospermopsin in cultured rat hepatocytes, *Biochemical and Biophysical Research Communications*, **201**, 235–241.

Runnegar, M. T., Kong, S.-M., Zhong, Y.-Z. and Lu, S. C., 1995. Inhibition of reduced glutathione synthesis by cyanobacterial alkaloid cylindrospermopsin in cultured rat hepatocytes, *Biochem. Pharmacol.*, **49**, 219–225.

Sasner J. J. Jr., Ikawa M. and Foxall T. L., 1984. Studies on *Aphanizomenon* and *Microcystis* toxins, in *Seafood Toxins*, ACS Symposium Series, No. 262, Regelis, E. P. (Ed.), American Chemical Society, Washington, DC, USA, 391–406.

Shaw, G. R., 1998. Cylindrospermopsin production by two cyanobacterial species in Australia, Abstracts of the 4th International Conference on Toxic Cyanobacteria, Beaufort, NC, USA, 4.

Sivonen, K., 1990. Toxic Cyanobacteria in Finnish Fresh waters and the Baltic Sea, Reports from the University of Helsinki (Department of Microbiology), 39/1990.

Sivonen, K. and Jones, G., 1999. Cyanobacterial toxins, in *Toxic Cyanobacteria in Water: a Guide to Public Health Significance, Monitoring and Management*, Chorus, I. and Bartram, J. (Eds), WHO, 41–111.

Sivonen, K., Niemelä, S. I., Niemi, R. M., Lepistö, L., Luoma, T. H. and Räsänen, L. A., 1990. Toxic cyanobacteria (blue-green algae) in Finnish fresh and coastal waters, *Hydrobiologia*, **190**, 267–275.

Sivonen, K., Namikoshi, M., Evans, W. R., Färdig, M., Carmichael, W. W. and Rinehart, K. L., 1992. Three new microcystins, cyclic heptapeptide hepatotoxins, from *Nostoc* spp. strain 152, *Chemical Research in Toxicology*, **5**, 464–469.

Skulberg, O. M., Carmichael, W. W., Anderson, R. A., Matsunaga, S., Moore, R. E. and Skulberg, R., 1992. Investigations of a neurotoxic Oscillatorialean strain (cyanophyceae) and its toxin. Isolation and characterization of homoanatoxin-a, *Environmental Toxicology and Chemistry*, **11**, 321–329.

Skulberg, O. M., Underdal, B. and Utkilen, H., 1994. Toxic water blooms with cyanophytes in Norway – current knowledge, *Algological Studies*, **75**, 279–289.

Stevens, D. K. and Krieger, R. I., 1991. Stability studies on the cyanobacterial nicotinic alkaloid anatoxin-a, *Toxicon*, **29**, 167–179.

Swanson, K. L., Allen, C. N., Aronstam, R. S., Rapoport, H. and Albuquerque, E. X., 1986. Molecular mechanisms of the potent and stereospecific nicotinic receptor agonist (+)-anatoxin-a, *Molecular Pharmacology*, **29**, 250–257.

Terao, K., Ohmori, S., Igarashi, K., Ohtani, I., Watanabe, M. F., Harada, K.-I., Ito, E. and Watanabe, M., 1994. Electron microscopic studies on experimental poisoning in mice induced by cylindrospermopsin isolated from blue-green alga *Umezakia natans*, *Toxicon*, **32**, 833–843.

Törökné, A. K., 1991. Toxin-producing Cyanobacteria in Hungarian Freshwaters, Doctoral Dissertation, University of Budapest, Budapest, Hungary.

Törökné, A. K., 1997. Interlaboratory trial using Thamnotox kit for detecting cyanobacterial toxins, Abstracts of the VIII International Conference on Harmful Algae, June 25–29, 1997, Vigo, Spain, 114.

Vasconcelos, V. M., 1994. Toxic cyanobacteria (blue-green algae) in Portuguese fresh waters, *Archiv. Hydrobiol.*, **130**, 439–451.

Vezie, C., Brient, L., Sivonen, K., Betru, G., Lefeuvre, J.-C. and Salkinoja-Salonen, M., 1997. Occurrence of microcystins containing cyanobacterial blooms in freshwaters of Brittany (France), *Archiv für Hydrobiologie*, **139**, 401–413.

Willén, E. and Mattsson, R., 1997. Water-blooming and toxin-producing cyanobacteria in Swedish fresh and brackish waters, 1981–1995, *Hydrobiologia*, **353**, 181–192.

Chapter 2.3

Use of Littoral Algae in Lake Monitoring

PERTTI ELORANTA

Hydrological and Limnological Aspects of Lake Monitoring
Edited by Pertti Heinonen, Giuliano Ziglio and André Van der Beken
©2000 John Wiley & Sons, Ltd, ISBN 0 471 89988 7

2.3.1   INTRODUCTION

Lake monitoring is one sector of the applied research fields in limnology which has many aims. It can be divided into different groups, according to the main purposes, as follows:

(a) monitoring of seasonal dynamics in lakes, including hydrological measurements (water level, flows, etc.), thermal and chemical stratification, and dynamics of other physical, chemical and biological variables;
(b) monitoring and evaluation of effects by different loads, i.e. point sources or/and diffuse sources;
(c) monitoring of long-term lake development, such as eutrophication, saprobity and acidification;
(d) monitoring of lake water quality for different water use, namely compliance to characteristics for human consumption or for other water uses.

The intensity of temporal and spatial monitoring varies very much, depending on the methods used, purpose and resources.

Lake monitoring has traditionally been carried out by using measurements and sampling from the pelagial zone at the lake deeps. In addition to several sets of physical and chemical variables, biological data have also been collected for monitoring purposes. The biological variables used most often in lake monitoring have been phytoplankton biomass (fresh mass, biovolume or chlorophyll *a*), primary production, and analyses of zoobenthos from lake deeps. In addition, fish and zooplankton have also been included in the programmes.

All of these analyses can give proper information on lake conditions in general, and by using several sampling stations the horizontal variation in lake water quality may also be followed. Practically all of the lake typological classifications are based on the data obtained from the pelagial zone. It has been assumed that the conditions at the mid-lake station illustrates the lake as a whole. This is, however, not necessarily true. The lake morphology, bathymetric conditions, currents and location of different loads may cause situations when littoral observations could also be very useful, and sometimes even necessary.

2.3.2   ALGAL COMMUNITIES IN LAKES

The only algal community in the pelagial zone is phytoplankton. In open-shore lakes with steep littoral zones, phytoplankton can also be used for narrow-shore monitoring purposes. Typical littoral algal communities are formed and named according to the substrata available in each place. On the sediment surface, epipelic algae live, with epilithic and epiphytic algae living on rocks and plants, respectively. In addition, distinct algal communities (epipsammic) live on sand grains. Although each substratum has its own characteristic communities, there are so many indicator taxa in each case that any of these communities could be used for monitoring purposes.

2.3.3   SAMPLING STRATEGIES FOR LITTORAL ALGAE

Sampling should be performed during the monitoring from similar substrata, and from time to time at one place, in order to keep results comparable at one sampling location. The monitoring sampling can be carried out from the sediment surface or from the surface of some more permanent substratum such as rocks or pier poles. Aquatic macrophytes are not very suitable as a sampling substratum due to the large changes in the epiphytic community during the seasons and also to the rather short annual period when they occur and during which the algae may colonize. Due to the heterogeneity of the local communities, the sampling should be carried out from any substratum and from several sublocations at each station. The collected samples should be kept separate from each other whenever some statistical comparisons between different stations are planned.

Natural substrata rather rarely afford possibilities to use quantitative methods for algological analyses. Spatial variation is very wide and it is very difficult to get comparable and representative quantitative samples from time to time at one location. Therefore, many different methods have been developed and used for quantitative sampling. The most common method is the exposure of different man-made materials in the water for use in algal colonizing. After certain exposure periods, sampling is then carried out from these substrata.

All kind of materials have been tested, e.g. glass slides, acrylic plastic plates, plastic or aluminum foils, wooden plates, ceramic plates, and even glass fibre filters used normally in the laboratory. The idea of the use of artificial substrata is to obtain comparable quantitative samples after the same colonizing time and from any location independent from the local available natural substrata. However, the locations should have the same depth, current and light conditions in order to obtain comparable results on the local loading that is being monitored. If the artificial substrata have been used for monitoring, the exposure time should be long enough to also give enough time for filamentous algae to colonize the substrata. If the short-term effects need to be studied, then periods of two to four weeks could be used, but in these cases the majority of algae would be diatoms.

2.3.4   VARIABLES MEASURED FROM LITTORAL ALGAE

If quantitative samples have been collected, the useful biomass estimates would be chlorophyll *a* content, fresh mass, dry mass, organic content and ash-free dry mass (AFDM). From these, chlorophyll alone describes algal mass only, whereas the other variables also include other organic material (detritus, animals and other heterotrophic organisms) or even inorganic particles (clay and sand particles) mixed with algae. The degree of autotrophy has been described by using the autotrophic index (AFDM/chlorophyll *a*) (Weber, 1973). According to Weitzel (1979), index values > 100 indicate heterotrophic conditions and pollution, while values < 100 indicate unpolluted waters.

Physiological variables could also be used to evaluate the growth potential of the littoral algae to (a) test the possible toxic effects of some temporary effluent load or (b) test the growth-limiting factors in the water. The most often measured variables used in these studies are the rate of the primary production or the nitrogen fixation *in situ* or *in vitro* in closed chambers in the field or in the laboratory (Loeb, 1981; Loeb and Reuter, 1981). If the littoral habitates are comparable between the stations (currents, exposure, vegetation, etc.), then the colonizing and growth rate on the artificial substrata also give some comparable values of the growth potential (Eloranta and Kunnas, 1982).

In the lake littorals, such as in the running waters, qualitative samples and community analyses are useful for the littoral monitoring purposes. Typical variables calculated from these samples are the species richness, diversity and especially the contributions of different taxa belonging to different indicator groups. In addition, several indices have been developed to describe the levels of organic load (saprobity) (see e.g. Sládeček, 1973, 1986; Descy, 1979; Prygiel and Coste, 1993; Lecointe *et al.*, 1993), nutrient load (Kelly and Whitton, 1995) and trophy (Schiefele, 1987; Steinberg and Schiefele, 1988; Schiefele and Schreiner, 1991; Hofmann, 1994). Although these indices are mainly used in rivers, they can very well be adopted to the lake littoral studies as well. In the acidification studies, equations have been developed to calculate the water pH by using diatoms as indicators (e.g. Eloranta, 1990; Huttunen and Turkia, 1990). Diatoms may also indicate currents, nitrogen-uptake metabolism (Cholnoky, 1968; Van Dam *et al.*, 1994), conductivity and salinity (Hustedt, 1953, 1957; Simonsen, 1962; Ziemann, 1991; Van Dam *et al.*, 1994) and oxygen requirements (Van Dam *et al.*, 1994), and they can be described and grouped according to their original habitats (e.g. Denys, 1991a,b; Van Dam *et al.*, 1994).

2.3.5   PURPOSES OF MONITORING LITTORAL ALGAE

Littoral monitoring by using algal communities can be highly recommended as a method of evaluating the quantity and quality of the diffuse load from different parts of a lake (Figure 2.3.1).

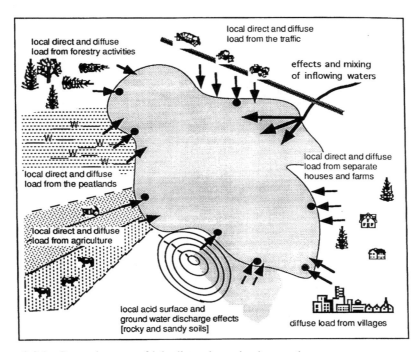

Figure 2.3.1   Example cases of lake littoral monitoring needs.

The method can also be used to compare the trophy or saprobity of different parts of the lake (Figure 2.3.2). In this case, the sample substratum should be taken into account, i.e. it should be similar between the different sampling locations in order to make comparisons more relevant.

Littoral conditions may differ very drastically from those in the pelagial due to the vicinity and direct connection with the terrestrial ecosystem, and often on account of the littoral vegetation making an important buffer zone between the shore and open water. In many lakes, the diffuse load from nearby drainage basins may be an important source of nutrients from fields, and humic substances from peatlands or from the other loads. In addition, the effects of point-load sources are best seen at the shores close to sewers (Figure 2.3.3). Therefore, by using littoral communities in monitoring it is possible to monitor and show the local sources of different type of loads. Algae are good tools for this monitoring due to their sensitivity and wide ecological range in general, but also because among the algae there are numerous good indicators for the many ecological factors mentioned before. Algae exist everywhere, and they can be collected from natural substrata at any time in the open water season.

A shortcoming in the use of littoral methods is the lack of any quantitative classifications similar to the trophy classifications, for example, for chlorophyll

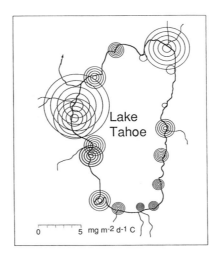

Figure 2.3.2   Periphyton growth rate in the lake littoral as a measure of local loads at Lake Tahoe (California–Nevada) (redrawn from Goldman 1974).

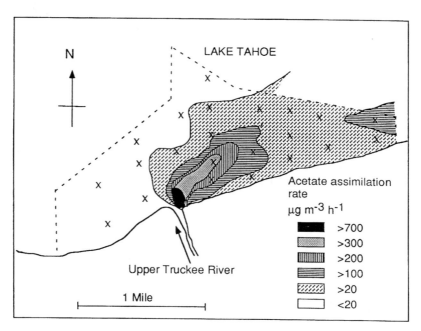

Figure 2.3.3   Heterotrophic activity as a measure of local organic load by inflowing waters and the spatial pattern of their mixing and dilution with lake waters (at Lake Tahoe) (redrawn from Goldman 1974).

a concentrations in pelagial water. The littoral communities are very different from those in the pelagial phytoplankton. This may cause one limit for the use of littoral algae due to the lack of taxonomic expertise of personnel in the monitoring laboratories. Although all algal groups may give some information on the environmental conditions, the diatoms are the most useful group for monitoring. Diatoms occur everywhere, and always exist, plus the group is rich in species and their ecology is very well known. Diatoms have been used in river monitoring and also in lake saprobity monitoring since the beginning of this century.

The lack of any absolute quantitative classifications is compensated very well by the fact that for evaluation of the trophic state, the many indices mentioned above, e.g. saprobity, pH, salinity, currents, nitrogen loads, etc., have been developed to make temporal monitoring and comparison between stations more precise. Of great help in these applications is the software and database program OMNIDIA (Lecointe *et al.*, 1993), which includes all freshwater diatoms and their ecological information obtained from the literature. This software can rapidly calculate many trophy and saprobity indices, and pH values, and can produce many ecological groupings of community structure for describing the state and possible changes in the littoral environment.

REFERENCES

Cholnoky, B. J., 1968. *Die Ökologie des Diatomeen in Binnengewässern*, Cramer, Vaduz Liechtenstein.

Denys, L., 1991a. A Check-List of the Diatoms in the Holocene Deposits of the Western Belgian Coastal Plain with a Survey of their Apparent Ecological Requirements. I: Introduction, Ecological Code and Complete List, Ministère des Affaires Economiques, Service Géologique de Belgique, Brussels, Belgium.

Denys, L., 1991b. A Check-List of the Diatoms in the Holocene Deposits of the Western Belgian Coastal Plain with a Survey of their Apparent Ecological Requirements. II: Centrales, Ministère des Affaires Economiques, Service Géologique de Belgique, Brussels, Belgium.

Descy, J. P., 1979. A new approach to water quality estimation using diatoms, *Nova Hedwigia*, **64**, 305–323.

Eloranta, P., 1982. Periphyton growth and diatom community structure in a cooling water pond, *Hydrobiologia*, **96**, 253–265.

Eloranta, P., 1990. Periphytic diatoms in the acidification project lakes, in Kauppi, P., Anttila, P. and Kenttämies, K. (Eds), *Acidification in Finland*, Springer-Verlag, Berlin, Heidelberg, 985–994.

Eloranta, P. and Kunnas, S., 1982. Periphyton accumulation and diatom communities on artificial substrates in recipients of pulp mill effluents, *Biol. Res. Rep. Univ. Jyväskylä*, **9**, 19–33.

Goldman, C. R., 1974. Eutrophication of Lake Tahoe Emphasizing Water Quality, EPA-660/3-74-034, Environmental Protection Agency, Corvallis, OR, USA.

Hofmann, G., 1994. Aufwuchs-Diatomeen in Seen und ihre Eignung als Indikatoren der Trophie, *Bibliotheca Diatomologica*, **30**.

Hustedt, F., 1953. Die Systematik der Diatomeen in ihren Beziehungen zur Geologie und Ökologie nebst einer Revision des Halobiensystems, *Sv. Bot. Tidskr.*, **47**, 509–519.

Hustedt, F., 1957. Die Diatomeenflora der Flusssystems der Weser im Gebiet der Hansestadt Bremen, *Abh. Naturw. Verein. Bremen*, **34**, 181–440.

Huttunen, P. and Turkia, J., 1990. Surface sediment diatom assemblages and lake acidity, in Kauppi, P., Anttila, P. and Kenttämies, K. (Eds), *Acidification in Finland*, Springer-Verlag, Berlin, Heidelburg, 995–1008.

Kelly, M. G. and Whitton, B. A., 1995. The Trophic Diatom Index: a new index for monitoring eutrophication in rivers, *J. Appl. Phycol.*, **7**, 433–444.

Lecointe, C., Coste, M. and Prygiel, J., 1993. 'OMNIDIA': software for taxonomy, calculation of diatom indices and inventories management, *Hydrobiologia*, **269/270**, 509–513.

Loeb, S. L., 1981. An *in situ* method for measuring the primary productivity and standing crop of the epilithic periphyton community in lentic systems, *Limnol. Oceanogr.*, **26**, 394–399.

Loeb, S. L. and Reuter, J. E., 1981. The epilithic periphyton community: a five-lake comparative study of community productivity, nitrogen metabolism and depth distribution of standing crop, *Verh. Int. Verein. Limnol.*, **21**, 346–352.

Prygiel, J. and Coste, M., 1993. Utilisation des indices diatomiques pour la mesure de la qualité des eaux du bassin Artois-Picardie: bilan et perspectives, *Ann. Limnol.*, **29**, 255–267.

Schiefele, S., 1987. Indikationswert benthischer Diatomeen in der Isar zwischen Mittenwald und Landshut, Diploma Thesis, Universität München.

Schiefele, S. and Schreiner, C., 1991. Use of diatoms for monitoring nutrient enrichment, acidification and impact of salt rivers in Germany and Austria, in Whitton, B. A., Rott, E. and Friedrich, G. (Eds), *Use of Algae for Monitoring Rivers*, Institut für Botanik, Universität Innsbruck, 103–110.

Simonsen, R., 1962. Untersuchungen zur Systematik und Ökologie der Bodendiatomeen der westlichen Ostsee, *Int. Rev. Hydrobiol. Syst. Beih.*, **1**, 1–144.

Sládecek, V., 1973. System of water quality from the biological point of view, *Arch. Hydrobiol., Beih. Ergebn. Limnol.*, **7**.

Sládecek, V., 1986. Diatoms as indicators of organic pollution, *Acta Hydrochim. Hydrobiol.*, **14**, 555–566.

Steinberg, C. and Schiefele, S., 1988. Biological indication of trophy and pollution of running waters, *Z. Wasser-Abwasser-Forschung*, **21**, 227–234.

Van Dam, H., Mertens, A. and Sinkeldam, J., 1994. A coded checklist and ecological indicator values of freshwater diatoms from the Netherlands, *Netherlands J. Aquatic Ecol.*, **28**, 117–133.

Weber, C. I., 1973. Biological Field and Laboratory Methods for Measuring the Quality of Surface Waters and Effluents, EPA-670/4-73-001, Environmental Protection Agency, Corvallis, OR, USA.

Weitzel, R. L., 1979. Periphyton measurements and applications, in Weitzel, R. L. (Ed.), *Methods and Measurements of Periphyton Communities: A Review*, STP 690, American Society for Testing and Materials, Philadelphia, PA, USA, 3–33.

Ziemann, H., 1991. Veränderungen der Diatomeenflora der Werra unter dem Einfluss des Salzgehaltes, *Acta Hydrochim. Hydrobiol.*, **19**, 159–174.

Chapter 2.4

Use and Applicability of Zoobenthic Communities in Lake Monitoring

ESA KOSKENNIEMI

Hydrological and Limnological Aspects of Lake Monitoring
Edited by Pertti Heinonen, Giuliano Ziglio and André Van der Beken
©2000 John Wiley & Sons, Ltd, ISBN 0 471 89988 7

2.4.1 INTRODUCTION

Benthic invertebrates (zoobenthos) have been increasingly used in freshwater monitoring over the last few decades (Rosenberg and Resh, 1993). Due to its large species richness which covers all kinds of freshwater habitats and to an increased ecological knowledge on species responses to environmental conditions, zoobenthos can now be used for different monitoring purposes, such as eutrophication, acidification, changes in habitat structure and species diversity, and toxicity. Zoobenthos is rather easy to sample, and as sedentary and rather long-lived organisms they can reflect site-specific, long-term changes in nature. Hence, it is not surprising that zoobenthos studies also play an important role in current national and international freshwater monitoring. According to the European Community water policy directive proposals, zoobenthos seems to be in a key position for assessing and monitoring the ecological status of freshwaters in the future Europe.

In general, most attention has been paid to rivers for freshwater monitoring in Europe. This is understandable, since rivers have traditionally been important for man throughout the whole of Europe. Lakes are not so evenly distributed, but they are of increasing importance, even in areas where they are rare. Many large lakes in Europe can even be regarded as unique ecosystems. They are important in landscape formation and as recreation objects, but are also largely threatened in their specific ecosystem characteristics, thus calling for good international co-operation in monitoring and restoration (Tilzer and Bossard, 1992).

Lakes occur in the depressions of the catchment landscape, where a decreased velocity of water leads to sedimentation of alloctonic and autoctonic matter on to the bottom. In general, deeper profundal bottoms are accumulation areas, whereas shallower, littoral parts of the lake are a mosaic of various bottom types. All of the lake-bottom habitats can be used for zoobenthos monitoring, but usually for different purposes. In general, the importance of lake littoral has been underestimated in monitoring, partly due to its complex habitat structure and laborious sampling and sorting procedures.

2.4.2 ZOOBENTHOS MONITORING DESIGN

Zoobenthos communities are understood here to be a species set occurring at a site in a lake. Only macrozoobenthos, i.e. benthic macroinvertebrates is treated in this chapter. No simple definition is found for this aquatic animal group. Rosenberg and Resh (1993) defined them as organisms that inhabit the bottom substrates (sediments, debris, logs, macrophytes, filamentous algae, etc.) of freshwater habitats for at least a part of their life cycle. The sieve mesh size used for sample rinsing is usually between 200 and 500 μm.

In planning a monitoring study programme it is important that every part is carefully designed, thus trying to avoid errors and unnecessary risks in research, as well as pursuing cost-effective studies. A typical monitoring programme can be seen as a process consisting of temporally subsequent steps, as follows:

(1) Environmental problem and scientific questions arising.
(2) Planning the monitoring involving scientific, economical and organizational aspects.
(3) Sampling and treatment:
    (a) sample collection;
    (b) sample preservation and transport;
    (c) sample sorting and taxa identification.
(4) Field measurement and background data:
    (a) data recording;
    (b) data transcription and entry;
    (c) data analysis.
(5) Interpretation of results and conclusions.
(6) Publication and dissemination.
(7) Remedial measures (in nature, society, etc.) and improvements to the monitoring programme (returning back to stages (1) and (2)).

It goes without saying that careful planning and implementation of the above-mentioned steps can improve and guarantee the successfulness of the monitoring programme. In general, the equipment and procedures used in lake zoobenthos sampling and sorting are rather well standardized, but other 'steps' could include more individualism and specific geographical-cultural features in monitoring. The 'sample' stages (3a–3c) are, however, usually the most critical steps in zoobenthos studies.

In zoobenthos monitoring, as in monitoring in general, two main types can be found, namely surveillance monitoring and compliance (statutory) monitoring (Rosenberg and Resh, 1993; Powell, 1995). Regarding compliance, the monitoring objectives are usually more clearly defined by various norms and limits which are not allowed to be exceeded. Surveillance monitoring has more characteristics of a general environmental status assessment, repeated at certain time intervals. Typically, it should be a true long-term monitoring with a rather wide spectra of type parameters being measured.

In order to detect spatio-temporal differences in nature, we also need knowledge about the events of certain reference areas, to be able to compare the condition of the control areas before and after changes of impacted areas, i.e. as Resh and McElravy (1993) simply stated: 'The establishment of suitable spatial and/or temporal controls is an essential component of sampling design for any biomonitoring study . . .'. Very often we will start our studies without having any valid information before the impacted period, or with measurements in the control areas being largely neglected. Moreover, sampling design should avoid pseudoreplication – unfortunately, a very common problem in zoobenthos monitoring (Hurlbert, 1984). For example, it is still very common that profundal samples are taken 'around the boat', and not over a larger area of the bottom from randomly selected sampling points. Furthermore, the number of replicates taken seems to be more based on the tradition ('three replicates is enough') and formal compliance of existing standards, than on statistical assumptions and the monitoring aim (Veijola *et al.*, 1996; Johnson, 1998).

The monitoring programmes should be revised when necessary. This task is important, but is certainly very often largely neglected. Is our purpose in monitoring only to store accumulating data? Did we really answer questions relevant to our monitoring programme? Johnson (1998) recently stressed that in order to assess the impact quality and quantity in field monitoring, more focus should be placed on evaluating the robustness of the indicator metrics. Should the whole zoobenthic community be monitored or only some 'good' indicator species; is it better to measure biomass, density, taxa richness or biotic indices, and does our material allow the use of robust statistical tests in detection of human and natural impacts? These are all questions that should be asked.

Monitoring which covers several decades may have some problems concerning the continuity and result reliability. A large sampling programme repeated annually is rather expensive and is therefore sensitive to the economical situation and preferences in the society. Later on, it may become scientifically uninteresting in the long run and useful revisions of the programme will be neglected. Even changes in the political situation may affect the monitoring implementation. Maybe we are asking why do we need to monitor when no clear changes were detected in zoobenthos? It is very important to know what are the qualitative (metrics) and quantitative (expected effect size) environmental changes we are monitoring. The aim of the monitoring programme is always, however, to test the hypotheses clearly stated beforehand, and not to confirm or to defend uncertain assumptions.

2.4.3 LAKES AS OBJECTS IN ZOOBENTHOS MONITORING

In lake ecosystems, zoobenthos is an important energy link, constituting a number of species with different ecological traits and limits. Zoobenthos communities vary largely in their assemblage, both within a lake and among different lakes. The 'within-lake' aspect illustrates the bathymetric, vertical distribution of the species in the littoral–sublittoral–profundal areas, and the horizontal distribution of the species, with a large variation in community structure, especially in the different littoral habitats. True 'littoral' or 'profundal' species are not so easy to determine. Criteria for zone division, e.g. by penetration of light on to the bottom areas or by the thermal stagnancy pattern of the water column, are somewhat arbitrary for division of the zoobenthos into littoral or profundal species. Bottom-substrate characteristics are, for example, very important to the occurrence of zoobenthic species.

In general, oligochaetes, chironomids and small molluscs dominate in lake profundals (Brinkhurst, 1974) if oxygen is available, excluding shorter possible stagnation periods. Especially in the large northern lakes, glacial relict crustaceans may also be important (Särkkä *et al.*, 1990), and generally indicates high water quality. Towards the littoral, insect groups other than chironomids (e.g. caddis flies, mayflies, water beetles, dragonflies, etc.) and large molluscs become common. In strongly regulated lakes, this fauna is relatively poor in species, in abundance and in biomass (Grimås, 1961) and the amplitude of the regulation seems to be an important factor affecting zoobenthos biomass (Palomäki, 1994). The distribution of species may cover both littoral and profundal regions, for instance if soft bottoms also occur in the littoral, or if the species have seasonal migrations between the bottom areas, such as the large crustacean relicts.

Many cold-stenothermic species occur in the littoral of the northern subarctic/arctic lakes, but occur only in restricted profundal 'refugia' below the thermocline in the more southern lakes (Brundin, 1949). Some lake habitats much resemble those of running waters, i.e. the profundal soft bottoms or exposed stony littorals are rather similar when compared to pools or riffles in river conditions. This similarity is also partly included in some zoobenthos classification systems. Little is known about the ecotone in the land/water interface zones of lakes, and this is certainly an important future study object in lake management and restoration (Schiemer *et al.*, 1995).

In 'among-lake' variations, the profundal chironomid-fauna-based typology describes mostly those lakes along the oligotrophy–eutrophy axis (Brinkhurst, 1974). This trophic aspect of typology fits well with the zoobenthos monitoring theme of organic pollution in lakes. Saether (1979) widened the classification from Europe to North America and tried to find additional chironomid indicator species in the sublittoral and littoral lake-bottom areas. Furthermore, he divided lakes into 15 trophic classes based on the occurrence of profundal chironomid indicators and showed, with a data set on Palearctic and Nearctic non-humic lakes, a clear correlation between the class order and the ratios of both the average total phosphorus and the average chlorophyll *a* to the mean lake depth.

Especially in northern Europe, the water in lakes is often rich in humic substances, constituting the typological group of dystrophic lakes (Brundin, 1956). Large amounts of humic compounds in water hamper high phytoplankton production and sedimentation of fine organic matter is low. Several dipteran species are typical or at least occur commonly, in humus-rich water, but this specific lake type still calls for further research to be carried out (Johnson and Wiederholm, 1989; Saether, 1979).

In the case of Spanish reservoirs, Prat (1978) created a trophic classification and typology based on the profundal chironomid assemblages in 100 study objects over the period 1973–1975. Chlorophyll *a* and water depth were important factors in this typological classification, which partly followed Brundins' species lists. Later, Real and Prat (1993) used data from the same reservoirs, resampled between 1987 and 1988, to monitor possible changes in the water quality. Canonical correlation analysis was performed for the data set (71 reservoirs) and special attention was paid to functional feeding groups of the zoobenthos and environmental variables related to the trophic state in the reservoirs. A dichotomy classification 'key', based on oligochaete and chironomid occurences, was created for the reservoirs and the role of the different disturbance factors was discussed. Oxygen availability and water depth, and also chlorophyll content, were important factors for the zoobenthos, but the role of environmental variables remained largely unclear. Water-level fluctuations and inorganic/organic inputs were suggested to be important stress factors determining the zoobenthos community structure.

2.4.4   CHANGES IN THE TROPHIC STATE AND ZOOBENTHOS MONITORING

Eutrophication due to anthropogenic activity is certainly the most common study theme in lake monitoring. In cases where industrial or municipal point-source loading causes eutrophication in a lake, the sampling network is rather easy to design along the recipient. Here, the division of the network into sampling units can be outlined, e.g. along water physico-chemistry or water-movement distribution patterns.

Various total quantitative features of zoobenthos, such as abundance, biomass and production variables, can be used in monitoring. In general, high variable values indicate eutrophy. In long-term monitoring, selected data may show trends that are somewhat difficult to detect, due to high interannual fluctuations related to changes in weather conditions or in biotic interactions (Timm *et al.*, 1996). In particular, the profundal zoobenthos is used for eutrophication monitoring and many zoobenthic species reflect well the changes in the trophic state of the lake and have a good indicator value for this monitoring purpose (Johnson, 1995).

The Benthic Quality Index (*BQI*) scores reported by Wiederholm (1980) indicate well the trophic continuum in lakes and can be calculated on the chironomid or oligochaete indicator assemblages. The index value can be calculated by using the following equation:

$$BQI = \sum k_i n_i N^{-1} \tag{1}$$

where k_i is the weight of the indicator taxon, n_i is the number of individuals for each indicator species and N is the total number of individuals of all of the indicator species.

For chironomids containing altogether twelve indicator taxa in five classes, the highest weight values ($k = 5$) indicating extreme oligotrophy will be attained by the exclusive occurence of *Heterotrissocladius subpilosus* (Kieff.) larvae, whereas at the other end of the *BQI* scale, the occurrence of *Chironomus plumosus* (L.) will give the lowest weight values ($k = 1$).

For oligochaete *BQI* scores, only four species are included: namely *Stylodrilus heringianus* ($k = 1$), *Psammoryctes barbatus* ($k = 2$), *Potamothrix hammoniensis* ($k = 3$) and *Limnodrilus hoffmeisteri* ($k = 4$). Wiederholm's *BQI* score gradient seem to fit well with the phosphorus content in water (Johnson *et al.*, 1993). These species-level classifications can, however, be modified when introduced to other geographical areas. For Finnish lakes, Paasivirta (1987) and Kansanen *et al.* (1990) used modified *BQI* versions in their monitoring studies. Several new chironomid species were introduced into the modifications and the orders of *Potamothrix* and *Limnodrilus* were reversed. Especially for oligochaetes, Milbrink (1978) showed the trophic distribution of several species, while Lang (1989) developed a Swiss version of trophic classification based on oligochaete worms.

For Swiss lakes, Lang and Reymond (1996) have used oligochaete communities in the monitoring of eutrophication. Oligochaete worms seemed to interpret well the changes in the trophic lake status. In the studies carried out for Lake Geneva, oligochaete indicator metrics fitted well with eutrophication processes. Changes in the community structure, divided into three indicator species classes (oligo-, meso- and eutrophy) were in good concordance with temporal changes or the 'among-basin' differences of phosphorous concentration in water, whereas the total zoobenthic biomass seemed to have a poor indicator value. In the long-term zoobenthos monitoring of Lake Geneva several well-known but also somewhat surprising features became evident.

Many of the Swiss lakes are in stages of recovery from man-made eutrophication. In particular, the profundal regions will become colonized by species of the pre-impact period, this at last representing 'the landmark of the successful recovery of the whole lake'. Hence, sublittoral communities have reacted more rapidly towards lowered water column nutrient concentrations than those of the profundal. The differences in the responses of the water column and the sediment to changed nutrient inputs can cause misunderstandings in the perceived successfulness of remedial measures. In the case of Lake Geneva, the phosphorus and chlorophyll concentrations and plantonic primary production were all observed to decrease, but the oligochaete worm communities, however, 'refused' to display this in the 1980s. Later, however, an improvement in the bottom conditions was detected and confirmed in 1991 by changes in the oligochaete species assemblages. The monitoring of only oligochaete communities was not enough in all lakes of the area, and in some cases the total zoobenthic metrics (biomass) and other zoobenthic groups (chironomids, *Chaoborus*) needed also to be considered.

For the Finnish Vanajavesi lake system, Kansanen (1985) was able to show clear changes in the zoobenthos metrics during the pollution history and an ecological deterioration of the lake recipient since the pre-industrial area. Sample data for living bottom fauna reached back to the 1920s and these results served as reference communities before the rapid industrialization. The occurence of indicator species, such as *Monoporeia*, *Chaoborus* and several chironomid and oligochaete species mentioned above, were found to be useful in studying the eutrophication history and its spatial gradient within and between different lake basins. Results similar to those presented in Lang's papers were obtained. The zoobenthic sensitivity to pollution was observed to be higher in deeper profundal regions, as was the case for Lake Geneva. Kansanen adopted a very wide spatio-temporal approach in his studies, using water physico-chemistry, sediment characteristics and zoobenthos data. Kansanen *et al.* (1984, 1990) also used clustering and ordination techniques for lake classification and pollution assessment.

Sediment cores are very useful in palaeolimnological eutrophication history studies (Walker, 1993). Very often, data are lacking from the pre-impacted period, reaching perhaps back to a century ago. Subfossil remains, especially the chitinous chironomid larval head capsules, in the stratified, radioisotope (e.g. ^{210}Pb) – dated sediment layers can be identified and used as the basis for interpreting past communities (Carter, 1977). The palaeolimnological chironomid assemblage can be similarly analysed as the samples of living communities (Kansanen, 1985; Meriläinen and Hamina, 1993). Meriläinen and Hamina observed a well-documented, four-stage pollution history for the large Lake Päijänne (in Finland) in subfossil chironomid communities during the past 150 years by using the Wiederholm *BQI* analytical method. The first stage (the years 1838–1936), with very little human interference, showed oligotrophy with typically high *BQI* values (4.00–4.28), followed by the stage of increasing pollution (*c.* 1944–1973) and lowered *BQI* levels (2.75–3.50). The eutrophication peak during the 'black decade' (1973–*c.* 1983), with heavy pollution, had a *BQI* value of 2.15, and *Chironomus anthracinus* gr. and *Chironomus plumosus* gr. being the dominants. Thereafter (*c.* 1983 onwards), more effective wastewater purification in both the wood-processing industry and the municipal treatment plants markedly reduced organic loading and the reversal stage then had increased *BQI* values of 2.90–3.00, thus resembling the second stage chironomid communities. Notably, samples of living benthos during the fourth stage had similar *BQI* values to those developed from the subfossil data, so confirming the validity of the palaeolimnological study in the lake trophic history documentation.

2.4.5 ACIDIFICATION AND ZOOBENTHOS MONITORING

Anthropogenic acidification of freshwater lakes and rivers is a serious problem in north-western Europe. Typically, acid-rain-affected deterioration of lake ecosystems is common in areas where the bedrock and soil have a low buffering capacity. Notably, lakes in large geographical areas are simultaneously exposed, thus affecting the sampling strategy to be adopted. The lake groups selected for acidification-effect monitoring can, for example, be selected randomly along the classes of acidification-load gradient. Upper catchment (high-altitude), small lakes are usually important early warning signs, showing dramatic changes in their ecosystem structure. In mapping areas which are possibly influenced and endangered by acidification, data sets on rain and/or lake water physico-chemistry or geology can be used (Kristensen and Hansen, 1994; Marchetto *et al.*, 1994). In palaeolimnological studies on acidification, the chironomid fossil studies are not as powerful a tool in monitoring as they are in the case of eutrophication.

In contrast to eutrophication, the littoral zone is the most important acidification zoobenthos monitoring habitat, and includes a number of good indicator taxa, e.g. *Gammarus*, molluscs and mayflies (Meriläinen and Hynynen, 1990; Koskenniemi, 1995). For example, Raddum *et al.* (1988) used a four-category (a–d) classification based on the presence/absence of indicator organisms. If a community sample has only species tolerating a pH level >5.5 in its list, the highest category (a) with a score value of 1 (less acidified) is reached. The other categories (b to d) (scoring 0.5, 0.25 and 0) include the species occuring at pH levels of 5.0–5.5, 4.7–5.0 and <4.7, respectively.

Meriläinen and Hynynen (1990) have listed a number of zoobenthic taxa in Finnish forest lakes in relation to both the mean and minimum pH values of the waters. In regression analysis, the total zoobenthic biomass was shown to be in good concordance with earlier literature data, displaying no correlation with the pH metrics, whereas abundance values (total zoobenthos and many taxonomical groups) had correlation with the mean pH values. Gastropods and mayflies also showed correlations with the minimum water pH levels. The zoobenthos acidification indices recommended in monitoring (e.g. Fjellheim and Raddum, 1990) are largely based on species-specific responses observed in the field assessment studies, but an approach to combine laboratory/field experiment data to field observations is highly recommended and is already in use.

2.4.6   TOXICITY

Very often, toxic conditions exist in a heavily polluted lake recipient area of industry or municipalities loaded by metals, acids or organic poisons. Toxic water and sediment conditions may totally deteriorate the zoobenthos, which later on show a recovery and may have decreased species diversity, but with high abundance and biomass due to organic enrichment effects from the same effluents. Typically, only a few target taxa are used in monitoring. Toxicity studies are also more case-specific and have usually a relatively small impact area.

Toxic effects and their distribution can be studied in a lake, e.g. by deformities in the chironomid head capsule or oligochaete setae, or by measuring substance concentrations and survival in sentinel (bioaccumulative) organisms (Milbrink, 1983; Warwick, 1991). Deformity-index systems are still rather complicated for routine monitoring use and need further developmental work.

These aspects were also clearly stated by Milbrink (1983), who showed high deformity levels in a recipient of mercury-contaminated sediments in Lake Vänern, in southern Sweden. In this study, impact areas were stratified according to water physico-chemistry and sediment heavy-metal concentrations. The incidence of chaetae deformities correlated significantly with the distribution of mercury deposits in the lake bottoms.

2.4.7   RESERVOIR ZOOBENTHOS

The newly created reservoir (man-made lake) constitutes a new resource for the invading zoobenthic species. Rapid changes in the community structure at the beginning and the exceptional conditions, such as the bottom structure of terrestrial origin and artificial water level fluctuations, provide remarkable challenges for studying the zoobenthic succession in reservoirs (Koskenniemi, 1994). Monitoring of continous changes, however, will take years or even decades or centuries (Krzyzanek and Kownacki, 1986). Typically, reservoir succession studies have been of a highly descriptive character, including a wide spectrum of ecosystem characteristics (water physico-chemistry, vegetation, plankton, zoobenthos, fish, etc.).

Zoobenthos has proved to be a powerful tool to interpret both the early succession (invasion) and the later, mostly habitat-controlled faunal development with partial 'lacustrinezation' of the reservoir ecosystem (Koskenniemi,

1994). After the first 'filling-up', the dominance of *Chironomus* and other detritivore chironomids with a high biomass is usually observed, followed by species with lower dispersal and space competition ability. Later in succession, the biomass values tend to be lower, and other groups, in addition to chironomids, become common. Usually, several reservoir bottom habitats are included in the sampling programme, which is very intensive at the beginning. Coarse-bottom quality and insect-dominated faunal structure greatly affect the sampling methodology. In particular, deeper soft-bottom areas of older reservoirs have a faunal composition resembling that of natural lakes.

2.4.8 TOWARDS INTEGRATED MONITORING

Increased international co-operation, including several present and drafted environmental directives of the European Community, and the tendency to build up multi-purpose national monitoring, calls for re-evaluation of the different approaches for measuring environmental changes and to combine these in integrated monitoring programmes. Traditionally, the different sub-programmes of lake ecosystem characteristics (for example, water chemistry, plankton, zoobenthos and fish monitoring) will function separately, often indicating the same ecological quality criteria, and are largely overlapping in their monitoring purpose. Rather than ask if one really needs to measure eutrophication development 'in a thousand different ways', it is better to appreciate the complementary role of the different measurements and to widen the monitoring approach. New challenges in monitoring, including the biodiversity aspect and the possible climatic changes are of a highly international and global character. Rather newer aspects of monitoring also include ecosystem health and ecological integrity. Therefore, monitoring methods need still further development, but also require intensive co-operation in order to harmonize their use in the future Europe.

REFERENCES

Brinkhurst, R. O., 1974. The Benthos of Lakes, Macmillan Press, London.
Brundin, L., 1949. Chironomiden und andere Bodentiere der sdschwedischen Urgebirgsseen, *Inst. Freshwater Res. Drottningholm*, **30**, p. 576.
Brundin, L., 1956. Die bodenfaunistischen Seetypen und ihre Anwendbarkeit auf die Sdhalbkugel. Zugleich eine Theorie der produktionbiologischen Bedeutung der glazialen Erosion, *Rep. Inst. Freshwater Res. Drottningholm*, **37**, 186–235.
Carter, C. E., 1977. The recent history of the chironomid fauna of Lough Neagh, from the analysis of remains in sediment cores, *Freshwater Biol.*, **7**, 415–423.
Fjellheim, A. and Raddum, G. G., 1990. Acid precipitation: biological monitoring of streams and lakes, *Sci. Total Environ.*, **96**, 57–66.

Grimås, U., 1961. The bottom fauna of natural and impounded lakes in northern Sweden (Ankarvattnet and Blåsjön), *Rep. Inst. Freshwater Res. Drottningholm*, **46**, 31–48.

Hurlbert, S. H., 1984. Pseudoreplication and the design of ecological field experiments, *Ecol. Monogr.*, **54**, 187–211.

Johnson, R. K., 1995. The indicator concept in freshwater biomonitoring, in Cranston, P. (Ed.), *Chironomids: from Genes to Ecosystems*, Proceedings of the 12th International Chironomid Symposium, CSIRO Publications, Canberra, Australia, 11–27.

Johnson, R. K., 1998. Spatiotemporal variability of temperate lake macroinvertebrate communities: detection of impact, *Ecol. Appl.*, **8**, 61–70.

Johnson, R. K. and Wiederholm, T., 1989. Classification and ordination of profundal macroinvertebrate communities in nutrient poor, oligo-mesohumic lakes in relation to environmental data, *Freshwater Biol.*, **21**, 375–386.

Johnson, R. K., Wiederholm, T. and Rosenberg, D. M., 1993. Freshwater biomonitoring using individual organisms, populations and species assemblages of benthic macroinvertebrates, in Rosenberg, D. M. and Resh, V. H. (Eds), *Freshwater Biomonitoring and Benthic Macroinvertebrates*, Chapman & Hall, New York, 40–158.

Kansanen, P., 1985. Interpretation of changes caused by wastewater loading on the basis of zoobenthos and sediment analyses in Lake Vanajavesi, *Academic Dissertations*, 13p + 5 articles. Helsinki University Press, Helsinki, Finland.

Kansanen, P., Aho, J. and Paasivirta, L., 1984. Testing the benthic lake type concept based on chironomid associations in some Finnish lakes using multivariate statistical methods, *Ann. Zool. Fenn.*, **21**, 55–76.

Kansanen, P., Paasivirta, L. and Väyrynen, T., 1990. Ordination analysis and bioindices based on zoobenthos communities used to assess pollution of a lake in southern Finland, *Hydrobiologia*, **202**, 153–170.

Koskenniemi, E., 1994. Colonization, succession and environmental conditions of the macrozoobenthos in a regulated, polyhumic reservoir, Western Finland, *Int. Revue Ges. Hydrobiol.*, **79**, 521–555.

Koskenniemi, E., 1995. Benthic fauna, in Bergström, I., Mäkelä, K. and Starr, M. (Eds.), Integrated Monitoring in Finland, p. 110–111. UN ECE Report, the Ministry of the Environment, Forssa, Finland.

Kristensen, P. and Hansen, H. O., 1994. European Rivers and Lakes. Assessment of their Environmental State. EEA Environmental Monographs 1, European Environment Agency, Copenhagen, 1–122.

Krzyzanek, E. and Kownacki, A. (Eds), 1986. Development and structure of the Goczalkowice reservoir ecosystem, *Ekol. Polska*, **34**, 307–577.

Lang, C., 1989. Effects of small-scale sedimentary patchiness on the distribution of tubificid and lumbriculid worms in Lake Geneva, *Freshwater Biol.*, **21**, 477–481.

Lang, C. and Reymond, O., 1996. Reversal of eutrophication in four Swiss lakes: evidence from oligochaete communities, *Hydrobiologia*, **334**, 157–161.

Marchetto, A., Mosello, R., Psenner, R., Barbieri, A., Bendetta, G., Tait, D. and Tartari, G., 1994. Evaluation of the level of acidification and the critical loads for alpine lakes, *Ambio*, **23**, 150–154.

Meriläinen, J. J. and Hamina, V., 1993. Recent environmental history of a large, originally oligotrophic lake in Finland: a palaeolimnological study of chironomid remains, *J. Palaeolimnol.*, **9**, 129–140.

Meriläinen, J. J. and Hynynen, J., 1990. Benthic invertebrates in relation to acidity in Finnish forest lakes, in Kauppi, L., Anttila, P. and Kenttämies, K. (Eds) *Acidification in Finland*, Springer-Verlag, Berlin, 1029–1049.

Milbrink, G., 1978. Indicator communities of oligochaetes in Scandinavian lakes, *Verh. Int. Ver. Limnol.*, **20**, 2406–2411.

Milbrink, G., 1983. Characteristic deformities in tubificid oligochaetes inhabiting polluted bays of Lake Vänern, southern Sweden, *Hydrobiologia*, **106**, 169–184.

Paasivirta, L., 1987. Macrozoobenthos of Lake Pyhäjärvi (Karelia), *Finnish Fish. Res.*, **8**, 27–37.

Palomäki, R., 1994. Response by macrozoobenthos biomass to water level regulation in some Finnish lake littoral zones, *Hydrobiologia*, **286**, 17–26.

Powell, M., 1995. Building a national water quality monitoring programme, *Environ. Sci. Technol.*, **29**, 458–463.

Prat, N., 1978. Benthic typology of Spanish reservoirs, *Verh. Int. Ver. Limnol.*, **20**, 1647–1651.

Real, M. and Prat, N., 1993. Factors influencing the distribution of chironomids and oligochaetes in profundal areas of Spanish reservoirs, *Netherlands J. Aquat. Ecol.*, **26**, 405–410.

Resh, V. H. and McElravy, E. P., 1993. Contemporary quantitative approaches to biomonitoring using benthic macroinvertebrates, in Rosenberg, D. M. and Resh, V. H. (Eds), *Freshwater Biomonitoring and Benthic Macroinvertebrates*, Chapman & Hall, New York, 159–194.

Rosenberg, D. M. and Resh, V. H., 1993. Introduction to freshwater biomonitoring and benthic macroinvertebrates, in Rosenberg, D. M. and Resh, V. H. (Eds), *Freshwater Biomonitoring and Benthic Macroinvertebrates*, Chapman & Hall, New York, 1–9.

Saether, O. A., 1979. Chironomid communities as water quality indicators, *Holarct. Ecol.*, **2**, 65–74.

Schiemer, F., Zalewski, M. and Thorpe, J. E., 1995. Land/inland water ecotones: intermediate habitats critical for conservation and management, *Hydrobiologia*, **303**, 259–264.

Särkkä, J., Meriläinen, J. J. and Hynynen, J., 1990. The distribution of relict crustaceans in Finland: new observations and some problems and ideas concerning relicts, *Ann. Zool. Fenn.*, **27**, 221–225.

Tilzer, M. M. and Bossard, P. (Eds), 1992. Large lakes and their sustainable development, *Aquat. Sci.* **54**, 91–103.

Timm, T., Kangur, K. Timm, H. and Timm, V., 1996. Macrozoobenthos of Lake Peipsi-Pihkva: long-term biomass changes, *Hydrobiologia*, **338**, 155–162.

Veijola, H., Meriläinen, J. J. and Marttila, V., 1996. Sample size in the monitoring of benthic macrofauna in the profundal of lakes: evaluation of the precision of estimates, *Hydrobiologia*, **322**, 301–315.

Walker, I. R., 1993. Paleolimnological biomonitoring using freshwater benthic macroinvertebrates, in Rosenberg, D. M. and Resh, V. H. (Eds), *Freshwater Biomonitoring and Benthic Macroinvertebrates*, Chapman & Hall, New York, 306–343.

Warwick, W. F., 1991. Indexing deformities in ligulae and antennae of Procladius larvae (Diptera: Chironomidae): application to contaminant-stressed environments, *Can. J. Fish. Aquat. Sci.*, **48**, 1151–1166.

Wiederholm, T., 1980. Use of benthos in lake monitoring, *J. Water Pollut. Control Fed.*, **52**, 537–547.

Chapter 2.5

Botanical Aspects in Lake Monitoring and Assessment

HEIKKI TOIVONEN

Hydrological and Limnological Aspects of Lake Monitoring
Edited by Pertti Heinonen, Giuliano Ziglio and André Van der Beken
©2000 John Wiley & Sons, Ltd, ISBN 0 471 89988 7

2.5.1 INTRODUCTION

Often, only hydrophytes, i.e. those plants living their whole life cycle (except flowering and seed dispersal) in water, are referred to by the term, aquatic macrophytes. Hydrophytes live submerged on the bottoms, in the water column, or floating on the water surface. Helophytes or emergent macrophytes with aerial shoots are also included in this class. Many mire and shore plants can occasionally live in shallow-water habitats.

The species pool of vascular hydrophytes in Fennoscandian inland waters is more than 80, and including the species occurring in brackish coastal waters of the Baltic Sea, this figure is closer to 90 (Linkola, 1933; Jensén, 1994). The number of helophytes is more than 30. About 50–60 bryophytes can be recognized as being aquatic macrophytes (Koponen *et al.*, 1995). From the larger algae, only stoneworts (charophytes) are commonly included in this class.

2.5.2 BIOLOGICAL CHARACTERISTICS OF AQUATIC MACROPHYTES

Aquatic macrophytes have been divided into different life- and growth-forms. These classifications have been treated in detail by Hutchinson (1975) and Luther (1983). A combined life- and growth-form classification for Nordic aquatic macrophytes is given in Table 2.5.1.

The growth depth for aquatic macrophytes is largely determined by light, with the lower limit being about the depth of the Secchi transparency. Vascular aquatics reach in most lakes depths of 2–4 m, while in very transparent waters they can even reach 8–10 m; algae and bryophytes can occur at even greater depths. Most aquatic

Table 2.5.1 Life- and growth-forms of Nordic aquatic macrophytes (Mostly according to Luther (1983), Hutchinson (1975) and Toivonen and Huttunen (1995)).

PLEUSTOPHYTES
    Lemnids
    Ceratophyllids
    Aquatic mosses

HAPTOPHYTES
    Macroalgae
    Many aquatic mosses

RHIZOPHYTES
    Charids
    Elodeids
    Isoetids
    Nymphaeids
    Helophytes (emergent)

plants, especially when submerged, are very flexible in their habit, due to various ecological factors, in particular light, nutrients, water currents and stream velocity. On average, aquatic plants have wide geographical ranges, partly due to their effective means of dispersal. Many emergent and submerged macrophytes can produce large clones due to effective vegetative reproduction.

The taxonomical identification of many submerged plants is difficult, because many of them hybridize, or occur as atypical forms, while some can only be identified when in flower or bearing ripe fruits. The taxonomic problems for many aquatic mosses and charophytes are even more difficult. In spite of a relatively low number of species, a detailed study which includes macrophytes, thus requires special expertise. Some high-quality identification guides have, however, been published (Casper and Krausch, 1981; Moore, 1985; Preston, 1995; Koponen *et al.*, 1995).

2.5.3 ECOLOGICAL DETERMINANTS

The occurrence of aquatic macrophytes is determined by many factors which should be taken into account when studying species composition and vegetation of a particular water body (Table 2.5.2). Correlation of most of these factors is, however, strong, and species occurrences are largely determined by the water chemistry and transparency, bottom quality and the exposure of habitats. However, local impacts of herbivory or competition between plants should be taken into account (Gaudet and Keddy, 1995). Aquatic macrophytes are also often sensitive to local catastrophes, e.g. channelization, drainage, algal blooms due to eutrophication, and oxygen depletion caused by long-lasting ice cover.

Table 2.5.2 Factors influencing aquatic flora and vegetation.

REGIONAL FACTORS
  Latitude
  Elevation
  Biogeographic history

LOCAL FACTORS
  Bedrock and soil of catchment area
  Land use of catchment area
  Size and shape of the water body
  Shore morphometry
  Physical and chemical quality of the sediment
  Hydrology (ice cover and water-level fluctuation)
  Water colour and transparency
  Water chemistry (pH, alkalinity and nutrients)
  Competition
  Herbivory, etc.
  Natural or anthropogenic catastrophes

In Finnish inland lakes, the trophic state of the habitat largely determines the species richness of helophytes. For hydrophytes, the principal determinants are the trophic state, the water transparency and the bottom type (Toivonen and Huttunen, 1995). When considering the diversity and species composition of Norwegian lakes, the calcium content, alkalinity and specific conductivity are of primary importance (Brandrud and Mjelde, 1997).

Iversen (1929) had earlier pointed out the importance of the lake pH as a principal determinant of macrophyte species richness. Acid lakes (pH 4.5–5.5) can support dense bryophyte and isoetid communities, which use CO_2 as their carbon source, but only relatively few elodeids (Heitto, 1990). Species diversity of elodeids and charophytes increases clearly in lakes with pH values >6, and especially in lakes with pH values >7 (Brandrud and Mjelde, 1997). Because pH can show great temporal and spatial variations during the growing season (diurnal fluctuations of 1 pH unit are not rare), these variations should be taken into account when measuring the pH.

In larger water bodies, there is always local variation in water quality or hydrology, while the erosive effects of waves and wind can also vary greatly. Local heterogeneity in the lake flora probably results to a great extent from these variations in exposure gradients. Transparency and water colour are important for submerged aquatics. If relatively high transparency (3–5 m) is combined with meso- or eutrophy and larger lake areas, high species richness is achieved. This combination of habitat factors is typical of those lakes in NE Europe which have a high number of species (Rørslett, 1991). The high species number is primarily due to the coexistence of plants representing various submerged growth-forms.

The species richness and occurrence of various growth forms vary greatly, being lowest in acid brown-water oligotrophic lakes and highest in the clear-water meso-eutrophic and eutrophic waters. In hypertrophic lakes, the species richness decreases again (Toivonen and Huttunen, 1995). In true oligotrophic lakes, the number of species in each growth-form category is low, and the occurrence of pleustophytes and elodeids, in particular, is restricted by nutrients. In eutrophic lakes, isoetids and submerged bryophytes occur only rarely. In the course of heavy eutrophication, all submerged vegetation disappears (Phillips *et al.*, 1978; Blindow, 1992).

The impact of acidification on aquatic macrophytes have been studied intensively in the Netherlands, Norway, Sweden and Finland. Acidified lakes often have clear water, and large beds of a few vascular plants and aquatic mosses (especially *Juncus bulbosus* and *Sphagnum auriculatum*) can occur, but at the same time the cover of Isoetid vegetation can drastically decrease (Roelofs, 1983; Brandrud and Mjelde, 1997). In acidified running waters, bryophyte vegetation displays considerable changes, such as a marked decrease of *Fontinalis antipyretica*, *F. dalecarlica* and *Hygrohypnum ochraceum* in brooks with a pH <5.5 (Brandrud and Mjelde, 1997).

2.5.4   INDICATOR VALUE OF MACROPHYTES IN LAKE MONITORING

Many characteristics of aquatic macrophytes can be used as indicators (Table 2.5.3). However, due to the local heterogenity of habitats, generalizations of indicator values are difficult, and variation in responses should be interpreted carefully. The aquatic macrophytes used in monitoring and assessment can be grouped as follows:

(a) in plant tests carried out under controlled laboratory/field conditions;
(b) by using the chemical content of certain species as indicators of heavy metal and other toxic loads;
(c) by using species or species groups as indicators of water or habitat quality;
(d) by studying temporal (long-term) changes in flora and vegetation;
(e) by studying or assessing biodiversity in water bodies.

2.5.5   PLANT TESTS AND CHEMICAL CONTENTS

Aquatic macrophytes has been used as test organisms far less than algal organisms (Lewis, 1995). The most important species for bioassays have been

Table 2.5.3   Macrophyte responses to ecological factors.

PLANT
   Morphology and anatomy
   Shoot dimensions (length and weight)
   Vitality
   Chemical constituents
     macro- and micro-nutrients
     heavy metals
     organic compounds
     chlorophyll
   Physiology
     enzyme activity
     production and assimilation variables

VEGETATION AND SPECIES COMPOSITION
   Species number and composition
   Life- and growth-forms
   Species categories according to trophic state, sabrobity, etc.
   Physiognomy
   Horizontal (zonation) and vertical stucture
   Stand characteristics
     size
     shoot density
     height (mean and maximum)
     biomass
   Phytosociological associations, habitat types, etc.

duckweeds, as *Lemna* spp. (Sinha *et al.*, 1994). The contents of various chemicals, especially macro- and micro-nutrients, heavy metals, etc., have been studied quite intensively since the work of Hutchinson (1975). Kovacs *et al.* (1984) studied the concentration of 40 micro-nutrients in eight aquatic vasculars and the green alga *Cladophora*, and observed that macrophytes are very efficient in concentrating many elements. Bryophytes have been used in many monitoring studies, including mercury and heavy metals (Kelly and Whitton, 1989; Siebert *et al.*, 1996). This approach, as well as other transplant studies of aquatic macrophytes, will probably have a wider use in the future.

2.5.6   WATER AND HABITAT QUALITY

As the aquatic flora and vegetation in a water body are influenced simultaneously by several factors (see Table 2.5.2), macrophytes only rarely have a good indicative value when evaluating some specific environmental variables. Instead, they give a good general estimation of the trophic state and other main characteristics of the site. Their use as indicators is further motivated by their relative persistence in the site. Fluctuations in the population size (although sometimes remarkable) of aquatic macrophytes are also usually minor in comparison to many other groups of organisms.

The aquatic vegetation usually reflects conditions of its habitat, i.e. the littoral zone, characterizing those of integral, shallow small water bodies better than those of large lake systems. The dependency of Pleustophytes, such as Lemnids and weak-rooted hydrophytes (e.g. *Elodea*), on the water character-istics is much more pronounced than that of Helophytes. Elster *et al.* (1995) demonstrated a good correlation of the root length of *Spirodela polyrrhiza* to the trophic state. In Nordic countries, macrophytes has usually been divided into four to six broad categories according to their adaptation to the nutrient level of their site, ranging from low (oligotraphents) to high (eutraphents) nutrient concentration indicators (Linkola, 1933; Jensén, 1994; Toivonen and Huttunen, 1995). Plant species suffer or benefit from the eutrophication of their natural site. In their response to eutrophication or to water pollution, aquatic macrophytes have been divided into three to seven categories (from clean to highly polluted water indicators) (Kurimo, 1970).

Concerning the species indicating oligotrophic habitats (oligotraphents), the Finnish classifications roughly correspond to the central European studies (Seddon, 1972; Pietsch, 1980; Wiegbleb, 1978, 1981; Mäkirinta, 1989). Many species recognized as indifferent or meso-eutraphents in Nordic countries (where oligotrophic waters prevail) are considered as being oligo- or mesotraphents in Central Europe.

Indicator values of various environmental variables (e.g. nitrogen, salinity, light) for aquatic macrophytes are widely used in Central European countries

(Ellenberg *et al.*, 1992). Saprobity values have also been given to aquatic macrophytes in central Europe (Sládecek, 1973). Although macrophytes may be indicators of pollution, they are not indicators of saprobity, except in those situations where the saprobic and trophic scales are parallel. Only organisms taking part in the destruction of organic matter could be properly used as indicators of saprobity. As primary producers, macrophytes indicate the level of trophy (Schmedtje and Kohmann, 1987).

Physiognomic characteristics are often used in studying and assessing water bodies. These include, e.g. vertical and horizontal ranges of species and various growth forms (Eloranta, 1970). Ilmavirta and Toivonen (1986) presented growth-form spectra for 10 lakes, with the species number and a semiquantitative vegetation index for each growth-form category. The vegetation index was derived by combining the frequency and abundance for each species. Advances in computerization and numerical methods will make possible an increased use of such indices in the future. In addition, analyses of growth-form spectra, as well as ecological species groups, created by using numerical approaches, will be used increasingly more.

Biometric variables can be used in characterizing macrophyte stands. These include area, shoot density, mean and maximum heights and growth depths at the inner and outer borders of stands, and underground or above-ground phytomass (cf. Table 2.5.3). The determination of underground biomass is laborious and therefore not suitable for routine monitoring, while in addition, aerial biomass determinations can only be used in larger-scale studies when measurements can be made indirectly, e.g. by means of height–weight regression analysis.

2.5.7 CHANGES IN AQUATIC VEGETATION AND SPECIES COMPOSITION

Macrophytes are suitable for longer-term studies of changes in the littoral and open-water areas (Wallsten, 1981; Rintanen, 1996).

Macrophyte communities can be effectively used in characterizing and mapping littoral habitats of various water bodies. Because macrophyte communities are habitats for many other organisms, and affect microbial communities and processes in sediment, this approach is important in assessing littoral characteristics of water bodies. Effects of the water level regulations can be successfully studied and monitored by the use of macrophytic vegetation (Nilsson and Keddy, 1990; Hellsten and Riihimäki, 1996).

Experiences on the use of aerial photos in studies on emergent vegetation successions in small vegetation-rich water bodies have been good. A combination of aerial photography and field studies provides the best results in the vegetation mapping; in particular, stands of helophytes and floating-leaved plants can be delimited fairly exactly (e.g. Meriläinen and Toivonen, 1979; Toivonen and Nybom, 1989), but the success of aerial photographs for studying submerged vegetation is much lower, and depends largely upon the bottom quality and water transparency. Detailed aerial photography should be carried out at a sufficiently low altitude, e.g. 500 m, by using a fine-grain, black and white, or preferably colour positive film. For reedbeds and shore meadows, infrared films can also be used with success. Satellite images have been used for mapping littoral communities, but the results have been largely coarse, largely due to the large pixel size and the heterogenity of the site (Raitala *et al.*, 1985).

In more recent years, new geographic information systems (GIS) and remote-sensing methods have been successfully used in mapping aquatic vegetation and estimating biomass and distribution of macrophytes over the scale of whole lakes (Lehmann and Lachavanne, 1997; Lehmann *et al.*, 1997). Peñuelas *et al.* (1993) investigated the use of spectral reflectance methods for different types of aquatic vegetation. Malthus and George (1997) compared the images acquired by airborne remote sensing with spectroradiometric measurements of the reflectance characteristics of water and stands of emergent, floating-leaved and submerged aquatic plants.

The classification units used in vegetation mapping, and also in most ecological and botanical studies in central Europe, have been often based on the *Braun–Blanquet* phytosociological approach. In northern Europe, however, the dominant species and physiognomic characteristics of vegetation (growth-form, synusia, zonation, etc.) are more widely used. Dierssen (1996) has presented a comprehensive survey on Nordic studies and classifications of aquatic vegetation.

2.5.8 BIODIVERSITY ASSESSMENT

The use of macrophyte flora and vegetation in assessing the biodiversity of freshwater habitats for nature conservation is a recent development. In many countries of central and southern Europe, aquatic submerged macrophytes are as a group highly vulnerable, because of the widespread use of their habitats as recipients waters for domestic and industrial waste, and also because of nutrient loading from agriculture (Best *et al.*, 1993). Many wetland and freshwater habitats are threatened in Europe. In most cases, good water quality and well-developed and diverse macrophyte communities are interdependent. The restoration of macrophyte beds is therefore recommended, instead of removing aquatic vegetation (Kristensen and Hansen, 1994).

Only a few species of aquatic macrophytes are currently included in the international 'red lists', with less than 10 aquatic macrophytes being included in the EU Habitat Directive and the European 'red list' (Kristensen and Hansen, 1994). However, many of national lists include an even greater number of aquatic macrophytes, which reflects the regional differences in the status of macrophytes.

In Finland, about 20 vascular plants, 20 bryophytes and 10 charophytes of aquatic habitats are threatened (Report on the monitoring of threatened animals and plants in Finland, 1991). In addition, many other species are rare or threatened from a regional point of view. A great deal of these species are plants of small brooks, springs and flooded areas, but species of larger water bodies are also threatened (*Najas flexilis, N. tenuissima, Oenanthe aquatica, Pilularia globulifera, Alisma wahlenbergii, Potamogeton friesii, P. rutilus, Dichelyma capillaceum,* some *Chara* spp. and *Nitella braunii*). In larger water bodies, the main threats are eutrophication and pollution, and hydrological changes due to water level manipulation, dredging and construction activities. In small water bodies, hydrological changes, disturbances to the shoreline, and eutrophication are prevailing factors.

The EU Habitat Directive (1992/43/EEC; Annex 1) also includes freshwater biotopes. In 1997, the list in Annex 1 was completed with nature types from boreal biogeographic regions. The list of freshwater nature types given in the Habitat Directive (and its amendments) occurring in Nordic countries is as follows:

- oligotrophic waters containing very few minerals of sandy plains;
- oligotrophic to mesotrophic standing waters with vegetation of *Littorelletea uniflorae* and/or *Isoeto-Nanojuncetea*;
- hard oligo-mesotrophic waters with benthic vegetation of *Chara* spp;
- natural eutrophic lakes with *Magnopotamion-* or *Hydrocharition*-type vegetation;
- natural dystrophic lakes and ponds;
- fennoscandian natural rivers;
- Alpine rivers and the herbaceous vegetation along their banks;
- Alpine rivers and their ligneous vegetation with *Myricaria germanica*;
- watercourses of plain to montane levels with *Ranunculion fluitans* and *Callitricho-Batrachion* vegetation;
- Fennoscandian mineral-rich springs and springfens;
- petrifying springs with tufa formation.

Each member state of the EU is requested to carry out appropriate conservation measures for these habitats. The favourable conservation status, as well as changes in the representatives of these nature types, will be regularly monitored in the future.

2.5.9 CONCLUSIONS

Macrophytes can be successfully used in many ways in assessing and monitoring the trophic state, and to study longer-term (5–20 years) changes in their habitats. The proportions of species indicating various trophic states, growth-form spectra, vertical and horizontal changes in stands, etc., can be successfully used in comparative studies of various water bodies or in monitoring temporal changes.

However, the research methods applied to aquatic plants vary greatly according to both the aim of the study and the area or species being studied. In studies of aquatic macrophytes there is therefore a great need for standardization of the survey methods. Much of this work addresses itself to the treatment of the great local heterogeneity of plant communties. The following issues need harmonization:

- time of survey;
- size and selection of survey sites;
- selection of survey transects;
- recording scales for macrophytes (abundance, cover and frequency);
- biomass sampling and measurements;
- measurement of stand characters;
- numerical evaluation of the data;
- vegetation mapping methods;
- phytosociological classification;
- physiognomic classification;
- habitat classification.

In vegetation analysis, the location of vegetation transects, and their intervals, numbers or widths, will all depend to a great extent on the purpose of the study, the size and integrity of the area, and the length of the shoreline. The selection should be made in an objective way, preferably by means of stratified sampling which takes the different shore types into account.

REFERENCES

Best, E. P. H., Verhoeven, J. T. A. and Wolff, W. J., 1993. The ecology of The Netherlands wetlands: characteristics, threats, prospects and perspectives for ecological research, *Hydrobiologia*, **265**, 305–320.

Blindow, I., 1992. Decline of charophytes during eutrophication: comparison with angiosperms, *Freshwater Biology*, **28**, 9–14.

Brandrud, T. E. and Mjelde, M., 1997. 5.1. Makrofytter, in Brandrud, T. E. and Aagaard, K. (Eds), *Virkninger av Forurensing på Biologisk Manfold: Vann og Vassdrag i By- Och Tettstedsnaere Områder. En Kunskapsstatus*, NIVA Temahefte 13, 28–38.

Casper, S. J. and Krausch, H.-D., 1980, 1981. Pteridophyta und Anthophyta, in: Ettl, H., Gerloff, J. and Heying, H. (Eds), *Süsswasserflora von Mitteleuropa*. Vols. **23** (pp. 1–403) and **24** (pp. 405–942), Gustav Fischer Verlag, Stuttgart, New York.

Dierssen, K., 1996. *Vegetation Nordeuropas*, Eugen Ulmer Verlag, Stuttgart.

Ellenberg, H., Weber, H.E., Düll, R., Virth, V., Werner, W. and Paulissen, D. 1992. Zeigerwerte von Pflanzen in Mitteleurope, *Scripta Botanica*, **18**, 1–248.

Eloranta, P., 1970. Pollution and aquatic flora of waters by sulphite cellulose factory at Mänttä, Finnish Lake District, *Annales Botanici Fennici*, **7**, 63–141.

Elster, J., Kvet, J. and Hauser, V., 1995. Root length of duckweeds (Lemnaceae) as an indicator of water trophic status, *Ekologia (Bratislava)*, **14**, 43–59.

Gaudet, C. L. and Keddy, P. A., 1995. Competitive performance and species distribution in shoreline plant communties: A comparative approach, *Ecology*, **76**, 280–291.

Heitto, L., 1990. Macrophytes in Finnish forest lakes and possible effects of airborne acidification, in Kauppi, P., Anttila, P. and Kenttämies, K. (Eds), *Acidification in Finland*, Springer-Verlag, Berlin, 963–972.

Hellsten, S. and Riihimäki, J., 1996. Effects of lake water level regulation on the dynamics of aquatic macrophytes in northern Finland, *Hydrobiologia*, **340**, 85–92.

Hutchinson, G. E., 1975. *A Treatise on Limnology. III. Limnological Botany*, John Wiley & Sons, New York.

Ilmavirta, V. and Toivonen, H., 1986. Comparative studies on macrophytes and phytopankton in ten small, brown-water lakes of different trophic status, *Aqua Fennica*, **16**, 125–142.

Iversen, J., 1929. Studien über die pH-verhältnisse dänischer Gewässer und ihren Einfluss auf die Hydrophyten Vegetation, *Botanisk Tidskrift*, **40**, 227–326.

Jensén, S., 1994. Sjövegetation, in Påhlsson, L. (Ed.), *Vegetationtyper i Norden*, *TemaNord*, **665**, 458–531.

Kelly, M. and Whitton, B. A., 1989. Interspecific differences in zinc, cadmium and lead accumulation by freshwater algae and bryophytes, *Hydrobiologia*, **175**, 1–12.

Koponen, T., Karttunen, K. and Piippo, S., 1995. Suomen vesisammalkasvio (Summary: Aquatic bryophytes of Finland), *Bryobrothera*, **3**, 1–86.

Kovacs, M., Nyary, I. and Toth, L., 1984. Microelement content of some submerged and floating aquatic plants, *Acta Botanica Hungarica*, **30**, 173–185.

Kristensen, P. and Hansen, H. O. (Eds), 1994. *European Rivers and Lakes. Assessment of their Environmental State*, EEA Environmental Monographs 1, European Environment Agency, Silkeborg, Denmark.

Kurimo, U., 1970. Effect of pollution on the aquatic macroflora of the Varkaus area, Finnish Lake District, *Annales Botanici Fennici*, **7**, 213–254.

Lehmann, A. and Lachavanne, J.-B., 1997. Geographic information systems and remote sensing in aquatic botany, *Aquatic Botany*, **58**, 195–207.

Lehmann, A., Jaquet, J.-M. and Lachavanne, J.-B., 1997. A GIS approach of aquatic plant spatial heterogeneity in relation to sediment and depth gradients, Lake Geneva, Switzerland, *Aquatic Botany* **58**, 347–361.

Lewis, M. A., 1995. Use of freshwater plants for phytotoxicity testing: A review, *Environ. Pollut.*, **87**, 319–336.

Linkola, K., 1933. Regionale Artenstatistik der Süsswasserflora Finnlands, *Ann. Soc. Bot. Vanamo*, **3**, 3–13.

Luther, H., 1983. On life forms, and above-ground and underground biomass of aquatic macrophytes. A review, *Acta Bot. Fennica*, **123**, 1–23.

Malthus, T. J. and George, D. G., 1997. Airborne remote sensing of macrophytes in Cefni Reservoir, Anglesey, UK, *Aquatic Botany*, **58**, 317–332.

Mäkirinta, U. 1989. Classification of south Swedish isoetid vegetation with the help of numerical methods, *Vegetatio*, **81**, 145–157.

Meriläinen, J. and Toivonen, H., 1979. Lake Keskimmäinen, dynamics of vegetation in a small shallow lake, *Annales Botanici Fennici*, **16**, 123–139.

Moore, J. A., 1986. *Charophytes of Great Britain and Ireland*, BSBI Handbook No. 5, Botanical Society of the British Isles, London.

Nilsson, C. and Keddy, P. A., 1990. Predictibility of change in shoreline vegetation in a hydroelectric reservoir, northern Sweden, *Can. J. Fish Aquatic Sci.*, **45**, 1896–1904.

Peñuelas, J., Gamon, J. A., Griffin, K. L. and Field, C. B., 1993. Assessing community type, plant biomass, pigment composition, and photosynthetic efficiency of aquatic vegetation from spectral reflectance, *Remote Sensing Environ.*, **46**, 110–118.

Phillips, G. E., Eminson, D. F. and Moss, B., 1978. A mechanism to account to macrophyte decline in progressiviely eutrophicated freshwaters, *Aquatic Botany*, **4**, 103–126.

Pietsch, W., 1980. Zeigerwerte der Wasserpflanzen Mitteleuropas, *Feddes Repertorium*, **91**, 106–213.

Preston, C. D., 1995. *Pondweeds of Great Britain and Ireland.* BSBI Handbook No. 8, Botanical Society of the British Isles, London.

Raitala, J., Jantunen, H. and Lampinen, J., 1985. Application of Landsat satellite data for mapping aquatic areas in north-eastern Finland, *Aquatic Botany*, **21**, 285–294.

Report on the monitoring of threatened animals and plants in Finland, 1991. Committee Report No. 30, Ministry of the Environment, Helskinki, Finland (In Finnish).

Rintanen, T., 1996. Changes in the flora and vegetation of 113 Finnish lakes during 40 years, *Annales Botanici Fennici*, **33**, 101–122.

Roelofs, J. G. M., 1983. Impact of acidification and eutrophication on macrophyte communities in soft waters in the Netherlands I. Field observations, *Aquatic Botany* **17**, 139–153.

Rørslett, B., 1991. Principal determinants of aquatic macrophyte richness in northern European lakes, *Aquatic Botany*, **39**, 173–193.

Schmedtje, U. and Kohmann, F., 1987. Bioindikation durch Makrophyten – Indizieren Makrophyten Saprobie, *Archiv für Hudrobiologie*, **109**, 455–469.

Seddon, B., 1972. Aquatic macrophytes as limnological indicators, *Freshwater Biology*, **2**, 107–130.

Siebert, A., Bruns, J., Krauss, G. J., Miersch, J. and Markert, B., 1996. Fundamental investigations into heavy metal accumulation in Fontinalis antipyretica L. ex Hedw, *Science of Total Environment*, **177**, 137–144.

Sinha, S., Rai, U. N. and Chandra, P., 1994. Accumulation and toxicity of iron and manganese in Spirodela polyrrhiza (L.) Schleiden, *Bull. Environ. Contam. Toxicol.*, **53**, 610–617.

Sládecek, V., 1973. System of water quality from the biological point of view, *Arch. Hydrobiol. Beih. Ergebn. Limnol*, **7**, 1–218.

Toivonen, H. and Huttunen. P., 1995. Aquatic macrophytes and ecological gradients in 57 small lakes in southern Finland, *Aquatic Botany*, **51**, 197–221.

Toivonen, H. and Nybom, C., 1989. Aquatic vegetation and its recent succession in the waterfowl wetland Koijärvi, S. Finland, *Annales Botanici Fennici*, **26**, 1–14.

Wallsten, M., 1981. Changes of lakes in Uppland, central Sweden, during 40 years, *Symbolae Bot. Upsaliensis*, **23**(3), 1–84.

Wiegleb, G., 1978. Untersuchungen ber den Zusammenhang zwischen hydrochemischen Umweltfaktoren und Makrophytenvegetation in stehenden Gewässern, *Archivum für Hydrobiologie*, **83**, 443–484.

Wiegleb, G., 1981. Application of multiple discriminant analysis on the analysis of the correlation between macrophyte vegetation and water quality, *Hydrobiologia* **79**, 91–100.

Chapter 2.6

Fish as Components of the Lake Ecosystems

HANNU LEHTONEN

Hydrological and Limnological Aspects of Lake Monitoring
Edited by Pertti Heinonen, Giuliano Ziglio and André Van der Beken
©2000 John Wiley & Sons, Ltd, ISBN 0 471 89988 7

2.6.1 INTRODUCTION

Many studies have revealed that piscivorous fish can control planktivorous and benthivorous fish abundances and thus indirectly the zooplankton and phytoplankton dynamics and nutrient levels in water (top-down control). However, fish abundance and species composition are also controlled by lower trophic levels and abiotic factors (bottom-up control). Nutrient levels and primary productivity correlate with fish production, but these correlations are often weak.

Fish populations have been largely ignored in studies regarding the nutrient budgets of lakes, in spite of the fact that in many cases the majority of the phosphorus content has been found in fish tissues. Part of the phosphorus bound in fish recycles through excretion and egestion. There is evidence that the decomposition of dead fish may also be an important source of nutrients. An important role of fish is also bound to their mobility. Diel migrations of fish are among the features in lakes which translocate nutrients between different areas.

Trophic cascade regulation is strongest between the piscivorous fish and their prey and weakens towards the lower trophic levels. Removal of plankti- and benthivorous fish above the threshold levels from eutrophic lakes has in many cases led to improved water quality. The achieved effect is much dependent on whether the natural loading of the lake is internal or external. Predatory fish affect water quality by two means, i.e. (1) predation on planktivorous fish decreases the pressure towards herbivorous zooplankton, which in turn increases the abundance of phytoplankton, and (2) a decrease of benthivorous fish diminishes the internal load of nutrients from sediments and affects water clarity.

Lake ecosystems have organized in complex food webs in which fish live at the top or close to the top. Fish have a strong influence on other parts of food webs. They feed on a variety of organisms belonging to other trophic levels and fish are themselves food of other fish, many invertebrates, birds and mammals. Many other types of relationships between fish and their biotic and abiotic environments also exist. Recently, much attention has been paid to the effects of fish on water quality. It has been demonstrated that fish may determine to a great extent the conditions under which they live. Any changes in one system inevitably produce changes in the others.

The role of fish is thus of crucial importance for the whole lake ecosystem and water quality. When assessing the influence of fish assemblages, the concept of 'balance' has been developed. The density and size distribution of piscivorous fish and their prey must be balanced in such a way that both are capable of growing and providing ample replacement of the individuals that have been lost from the population (Van Densen, 1994). For example, in eutrophic lakes there often exist cases of insufficient predation: a situation with too few predators, excess of small-sized planktivorous and benthivorous fishes and strong grazing on herbivorous zooplankton. This disturbed situation has given rise to a number of studies with respect to biomanipulation, i.e. to affect fish assemblage structures in unbalanced lakes. In 'healthy' ecosystems, it is thought that a large part of the total biomass consists of large, old individuals (Van Densen, 1994).

2.6.2 FISH AND OTHER AQUATIC ORGANISMS

Fish production is based on the production of lower trophic levels in the food web. The biomass and productivity of invertebrates in turn is, to a high degree, dependent on fish predation and also on water quality. The basic elements required for the primary production are solar energy and amounts of nutrients. Energy flows into lakes through organic matter from catchment areas. In lakes, there exist complicated food webs before the energy reach the fish (Figure 2.6.1). Food webs in Nordic lakes are composed of hundreds of species, ranging from microbes, through plankton, benthic fauna, and finally to fish. Aquatic organisms have different generation times, varying from hours to years.

The effect of a fish on other organisms is dependent both on the fish species and the ontogenetic phase, which each have different nutritional requirements. Common to all fish species is that they feed on zooplankton during the larval

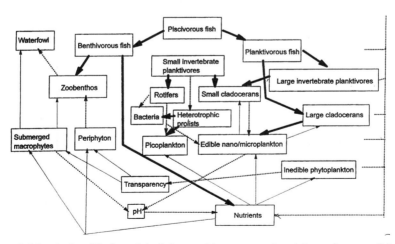

Figure 2.6.1 A simplified model of the trophic structure in a lake and some of the most important control directions; three important abiotic variables (transparency, pH and nutrients) have been included. The dashed arrow descending along the right-hand margin denotes the release of nutrients in excretion and faeces and through bacterial activities. For clarity, all competitive relationships have been ignored (redrawn from Sarvala, 1992).

phase. The size of prey plankton, however, varies depending on the species and ontogenetic phase. The smallest fish larvae are totally dependent on the smallest zooplankton, i.e. rotatorians and copepoda nauplii, for a short period after hatching. Their mouths are not yet large enough to swallow larger zooplankton. This period normally lasts for one to two weeks. Larger larvae can also feed on larger zooplankton. While all fish species are dependent on zooplankton at these early stages, most fish species then turn to other food items during their first months of life. There are, however, some species which continue mainly to feed on zooplankton during their whole life span (e.g. vendace (*Coregonus albula* (L.)), bleak (*Alburnus alburnus* (L.)) and blue bream (*Abramis ballerus* (L.)).

Food webs can be viewed from a variety of perspectives. When the emphasis is on fish effects, the control arrows point down. Different trophic levels influence the ecosystem in different ways. In general, the lower trophic levels enhance production, while fish predation has effects on the community structure, productivity and biomass (Crowder *et al.*, 1988). Depending on the direction of the control arrows, the terms 'top-down' and 'bottom-up' effects have been used.

It is often thought that in aquatic food webs, the highest trophic levels, with fish among them, are controlled by the production of the lower levels. This includes the idea that production of green plants regulates the whole function of the lake ecosystem. The direction in this case would therefore be bottom-up (McQueen *et al.*, 1989). Many studies have, however, revealed that the direction may also be reversed. Top-down control can cascade down the trophic levels from fish to phytoplankton (Carpenter *et al.*, 1985), and hence the community structure, productivity and biomass of the lower trophic levels are all controlled by fish (Figure 2.6.2). The existence of bottom-up control has been demonstrated in many studies (Helminen and Sarvala, 1993) and there is also strong evidence of top-down control (Rudstam *et al.*, 1994).

The top-down effects of piscivores on benthivores and planktivores reduce benthivory and planktivory, leading to increased sizes and numbers of zooplankton and zoobenthos, and thus to a reduced phytoplankton biomass. Top-down effects are highly dependent on the size and species structure of fish assemblages. Visually feeding fish suppress large, active prey, while olfactorily feeding fish suppress smaller, cryptic forms.

Piscivorous fish can thus have great impacts on the fish populations living at the lower trophic levels. However, the impact of piscivore predation on planktivores may in some cases be reduced due to the large sizes of prey fish. For example, bream (*Abramis brama* (L.)), blue bream and large roach (*Rutilus rutilus* (L.)) are outside the size categories of most piscivores.

Predatory interactions in lakes are likely to be among the most important factors that affect the biotic and physical environments. Both top-down and bottom-up controls should therefore be considered. So far, the relative effects

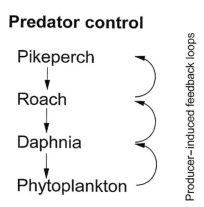

Figure 2.6.2 Top-down predator control and bottom-up consumer feedback loops (modified from Mills and Forney, 1988).

of top-down vs. bottom-up forces in the food-web function are not yet precisely known in spite of intensive studies.

2.6.3 FROM PRIMARY PRODUCTION TO FISH YIELD

There does exist a correlation between the production of the different trophic levels, i.e. the higher the trophic level in question, then the weaker it correlates with primary production or nutrient levels. McQueen *et al.* (1986) pointed out that top-down and bottom-up forces are different in eutrophic and oligotrophic lakes. The impact of fish on the lower trophic levels is strongest in eutrophic lakes, while primary productivity has the strongest control on food webs in oligotrophic lakes. An example of the relative strength of the different trophic levels in two lakes having different primary production is shown in Figure 2.6.3.

Fishery administrators have made numerous attempts to predict the potential levels of exploitable fish catches on the basis of measured phytoplankton biomass and primary production. Due to differences in fish species composition, the observed variation in fish production was high, varying between 1–6%, while the fish catch was only 0.02–0.2% from primary production (Morgan *et al.*, 1980). Globally, the estimated primary production (all waters) is estimated to be of the order of 100×10^9 (metric) tonnes of carbon per year, with the fish production being $250–280 \times 10^6$ tonnes and the potential catch in fisheries $120–150 \times 10^6$ tonnes (Moiseev, 1994). Some lakes at high latitudes may show exceptionally high fish production. An example has been found in Lake Pyhäjärvi (SW Finland) where the fish catch has been greater than 1% of the primary production. This is explained by the large

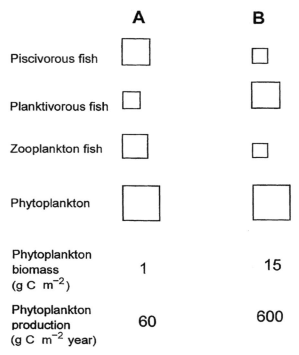

Figure 2.6.3 Relative biomass of different trophic levels (denoted by different-sized squares) in two model lakes (Riemann *et al.*, 1986) with a primary production of 60 (A) and 600 (B) g C m^{-2} per year (from Persson *et al.*, 1988).

euphotic zone and clear water which enables both planktivorous and benthivorous fish to utilize a large proportion of the food production (Sarvala *et al.*, 1984).

Some studies have revealed that a primary production increase of 5% leads to an increase of 6–9% in fish production (Nixon, 1988; Hardy and Gucinski, 1989). However, there exists only a few reliable estimates of fish production which support these estimates. Downing *et al.* (1990) compared the relationships between primary production and fish production on the basis of material collected from 19 lakes and found that planktonic primary production accounted for 79% of the fish production. These workers also found a significant correlation between fish biomass and fish production.

In some cases, climatological factors may influence fish production even more than nutrient levels. According to Regier *et al.* (1990), high water temperatures enhance the growth rates and decrease the age of maturation and development times of the egg and larval stages. It was also expected that at the species association/ecosystem level, temperature affects the productivity at all trophic levels to about 10–20% per 1 °C change in mean temperature. Meisner *et al.* (1987) predicted that an increase of 2% in the average annual air temperature would lead to about a 26% increase in aggregated maximum sustainable yields of commercially valuable species in the North American Great Lakes.

2.6.4 CAN FISH REGULATE PREY POPULATIONS?

The biological structure of the lake ecosystem is a complex and regulated hierarchically. The productivity of all trophic levels, including fish, is regulated by biological, chemical and physical factors, but the biological structure of the food web determines how the production is packaged. By assuming that predators and prey are regulated by different factors, it can be demonstrated that if there exist four trophic levels in a lake, then piscivorous fish control planktivorous fish by predation, zooplankton is controlled by competition and phytoplankton by zooplankton predation (Persson *et al.*, 1988, 1992). If the ecosystem consists of only three levels, with planktivorous fish being the highest, they are then controlled by competition, while zooplankton is controlled by planktivores and phytoplankton by competition. The general principle of this is shown in Figure 2.6.4.

An understanding of predator–prey interactions is a key factor in understanding the function of the whole ecosystem (Kitchell *et al.*, 1994). In most studies, the topic has been approached by using laboratory studies and small-scale experimental systems such as enclosures, ponds, etc. (Berryman, 1992). In these experiments, the impact of fish on their prey populations has almost always been drastic. In lakes, the situation is somehow different, and the predators seem in most cases to be fairly ineffective controllers of their prey (Table 2.6.1).

Trophic levels are usually correlated with size and show ontogenetic niche shifts. For example, piscivores such as pike, pikeperch and brown trout (the

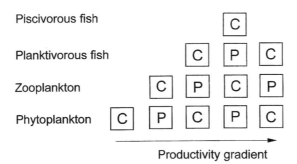

Figure 2.6.4 Number of trophic levels in a productivity gradient, showing the main structuring forces on different levels: C, competition; P, predation (redrawn from Persson *et al.*, 1988).

Table 2.6.1 Prey consumption by fish in different areas, expressed as a percentage of the total prey production.

Predator	Site	Prey	Consumption (%)	Reference
Benthivorous fish	SW Finland	Benthos	9–31	1
Yellowtail flounder	USA	Benthos	1–34	2
Benthivorous fish	USA	Benthos	30–50	3
Three fish species plus two crustaceans	Swedish west coast	Benthos	4–98	4
Shallow-water fish	Canada	Benthos	25–50	5
Eel	Lake Bodensee	Y-O-Y	2–17[a]	6
		Perch	10–100[b]	
Vendace (0+ and 1+)	Lake Pyhäjärvi	Zooplankton	9–14	7
Vendace	Lake Suomunjärvi	Zooplankton	9	8
Vendace	Lake Päijänne	Zooplankton	70	9
Planktivorous fish	Polish lakes	Zooplankton	8–20	10
Alewife	Lake Michigan	Zooplankton	2–20	11

[a]During normal-year periods.
[b]During weak-year periods.

dominant types of piscivorous fish in Nordic lakes), grow over six orders of magnitude in weight from larva to adult (a larva weighs a few mg, while an adult specimen may exceed 10 kg). Therefore, these fish species are small planktivores at the larval stage, benthivores in the fingerling stage and piscivores thereafter.

Figure 2.6.5 shows seven trophic levels which are typical of large Nordic lakes: (1) top piscivore (man), (2) large piscivores (pike and pikeperch), (3) benthivores and secondary piscivores (burbot (*Lota lota* (L.)), perch (*Perca fluviatilis* L.) and many cyprinids), (4) large planktivores (vendace, smelt and many cyprinids), (5) small planktivores (fish larvae, predatory zoobenthos and zooplankton), (6) zooplankton, and (7) phytoplankton. Many of these, however, feed on more than one trophic level and not all of the possible interactions are included in the figure. This type of multi-level omnivory is common in aquatic systems and results in a large number of interactions (Rudstam *et al.*, 1994).

2.6.5   FOOD-WEB MANIPULATION BY MAN

Many workers have applied trophic interactions between fish and other organisms as a tool in food-web manipulation for the purpose of water quality management. Such manipulation is based on two principles. First, piscivorous fish control the planktivorous fish which have a negative influence on herbivorous zooplankton. Secondly, benthivorous fish recycle bottom material by releasing sedimented nutrients and through excretion and egestion

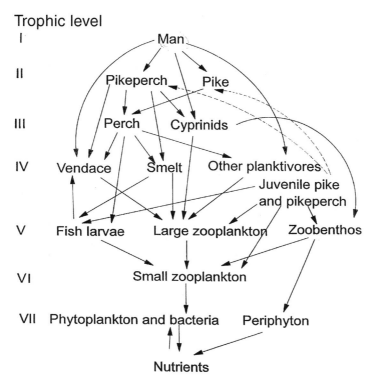

Figure 2.6.5 Schematic representation of the major potential top-down pathways and feedback loops (shown by dashed arrows) in a lake ecosystem (modified from Rudstam *et al.*, 1994).

(Horppila, 1994). Dense fish populations may also inhibit the growth of macrophytes through a reduction of the clarity of the water (Meijer *et al.*, 1990).

In eutrophic lakes, the balance of fish assemblages is disturbed and predators are not able to control the increase of cyprinid fish populations. The mass removal of cyprinid fishes has therefore been used in several lakes to balance the predator–prey relationship. Fishing can be considered as being a selective predator (Urho, 1994). The selective removal of planktivorous and benthivorous fish by both man and piscivorous fish affects the species composition and biomass of prey fish, and hence the species composition and biomass of phytoplankton and zooplankton (Tátrai and Istvánovics, 1986). The ultimate goal of food-web manipulation is usually to diminish the phytoplankton biomass which may be harmful for the use of water, and in some cases can also be toxic.

Simultaneously, it is possible to enhance the structure of fish populations. In eutrophic lakes, the function of food webs can be manipulated in two ways, first, by decreasing the external nutrient loading, or secondly, by decreasing the internal loading through a reduction in the density of the planktivorous fish, which in turn increases grazing on phytoplankton by zooplankton (Carpenter *et al.*, 1985). There are many examples of how removal of piscivores has affected the abundance of organisms belonging to the lower trophic levels and how this has affected water quality (Horppila, 1994).

2.6.6   CONCLUDING REMARKS

Fish can have both direct and indirect impacts on water quality, with direct impacts being usually much smaller than indirect ones. Examples of direct impacts are the diel migrations of fish and zooplankton which translocate nutrients, which may also affect the distribution of the latter. However, the effects of diel migrations on water quality are still poorly known.

Fish communities and their impacts on the structure and function of lake ecosystems have been subjects of research for many decades. Predator–prey interactions create feedback loops which regulate the trophic-level production. The strong top-down effects observed in manipulated ecosystems reflect the short time-scale of most studies. Over a longer period, bottom-up feedback loops may develop as species interactions lead to a reorganization of the ecosystem.

Manipulation of food webs has proven to be a suitable tool in the improvement of healthy ecosystems. The strong role of predatory fish indicates their strong top-down impact. Short-term introductions of piscivores have already demonstrated almost immediate effects on algal populations and water clarity. However, other observations made by Mills and Forney (1988) suggest that as predator and prey approach a balance, then interactions between the trophic levels become self-regulating.

So far, most of our understanding of trophic dynamics in lakes is based on short-term manipulations. High interannual variability in predator and prey abundances may obscure the biotic mechanisms which regulate community stability. In addition, the stability of ecosystems which have been altered by predator introductions remains unknown. Long-term studies will therefore be needed in order to understand the role of fish in ecological interactions.

REFERENCES

Berryman, A. A., 1992. The origins and evolution of predator-prey theory, *Ecology*, **73**, 1530–1535.
Boisclair, D. and Leggett, W. C. 1985. Rates of food exploitation by littoral fishes in a mesotrophic north-temperate lake. *Can. J. Fish. Aquat. Sci.*, **42**, 556–566.

Carpenter, S. R., Kitchell, J. F. and Hodgson, J. R., 1985. Cascading trophic interactions and lake productivity, *Bioscience*, **35**, 634–638.

Collie, J. S. 1987. Food consumption by yellowtail flounder in relation to production of its prey. *Mar. Ecol. Prog. Ser.*, **36**, 205–213.

Crowder, L. B., Drenner, R. W., Kerfoot, W. C., McQueen, D. J. Mills, E. L., Sommer, U., Spencer, C. N. and Vanni, M. J., 1988. Food web interaction in lakes, in Carpenter, S. R. (Ed.), *Complex Interactions in Lake Communities*, Springer-Verlag, New York, 141–160.

Downing, J. A., Plante, C. and Lalonde, S., 1990. Fish production correlated with primary productivity, not the morphoedaphic index, *Can. J. Fish. Aquat. Sci.*, **47**, 1929–1936.

Evans, S. 1984. Energy budgets and predation impact of dominant epibenthic carnivores on a shallow soft-bottom community on the Swedish west coast. *Estuar. Coast. Shelf. Sci.*, **18**, 651–672.

Gliwicz, Z. M. and Prejs, A. 1977. Can planktivorous fish keep in check planktonic crustacean populations? A test of size-efficiency hypothesis in typical Polish lakes. *Ekol. Pol.*, **25**, 567–591.

Hakkari, L. 1978. Eläinplanktonlajien elintavoista ja merkityksestä kalanravintona. *Suomen kalastuslehti*, **85**, 80–83.

Hardy, J. and Gucinski, H., 1989. Stratopheric ozone depletion: implications for marine ecosystems, *Ocenography Magazine*, **2**, 18–21.

Helminen, H. and Sarvala, J., 1993. Changes in zooplanktovory by vendace (*Coregonus albula*) in Lake Pyhäjärvi due to variable recruitment, *Verh. Int. Ver. Limnol.*, **25**.

Helminen, H., Sarvala, J. and Hirvonen, A. 1990. Growth and food consumption of vendace (*Coregonus albula* (L.)) in Lake Pyhäjärvi, SW Finland: a bioenergetics modelling analysis. *Hydrobiologia*, **200/201**, 511–522.

Hewett, S. W. and Stewart, D. J. 1989. Zooplanktivory by alewives in Lake Michigan: ontogenetic seasonal, and historical patterns. *Trans. Am. Fish. Soc.*, **118**, 581–596.

Horppila, J., 1994. Interactions between roach (*Rutilus rutilus* (L.)) stock and water quality in Lake Vesijärvi (southern Finland), University of Helsinki (Lahti Research and Training Centre) Scientific Monographs No. 5.

Kitchell, J. F., Eby, L. A., He, X., Schindler, D. E. and Wright, R. A., 1994. Predator–prey dynamics in an ecosystem context. *J. Fish. Biol.* **45** (Suppl. A), 209–226.

Latja, R. 1978. Suomunjärven muikun ravintolajien biologiasta vuonna 1976. In. Meriläinen, J. (ed.). *Kalantuotantoon vaikuttavat tekijät luonnonvaraisessa dysoligotrofisessa järvessä* (Suomunjärvi, Lieksa). pp. 129–225.

McQueen, D. J., Johanne, M. R. S., Post, J. R. Stewart, T. J. & Lean, D. R. S. 1989. Bottom-up and top-down impacts on freshwater pelagic community structure. *Ecol. Monogr.*, **59**, 289–309.

McQueen, D. J., Post, J. R. and Mills, E. L., 1986. Trophic relationships in freshwater pelagic ecosystems, *Can. J. Fish. Aquat. Sci.*, **43**, 1571–1581.

Mattila, J. and Bonsdorff, E. 1988. A quantitative estimation of fish predation on shallow soft bottom benthos in SW Finland. Kieler Meeresforsch., *Sonderheft*, **6**, 111–125.

Meijer, M-L., de Haan, M. V., Brukelaar, A. W. and Buiteveld, H., 1990. Is reduction of the benthivorous fish an important cause of high transparency following biomanipulation in shallow lakes?, *Hydrobiologia*, **200/201**, 303–315.

Meisner, J. D., Goodier, J. L., Regier, H. A., Shuter, B. J. and Christie, W. J., 1987. An assessment of the effects of climate warming on Great Lakes basin fishes, *J. Great Lakes Res.*, **13**, 340–352.

Mills, E. L. and Forney, J. L., 1988. Trophic dynamics and development of freshwater pelagicfood webs, in Carpenter, S. R. (Ed.), *Complex Interactions in Lake Communities*, Springer-Verlag, New York, 11–30.

Moiseev, P. A., 1994. Present fish productivity and bioproduction potential of the world aquatic habitats, in Voigtlander, C. W. (Ed.), *The State of the World's Fisheries Resources*, Proceedings of the World Fisheries Congress, Oxford; IBH Publishing Company Ltd., New Delhi, India, 70–75.

Morgan, N., Backiel, T., Bretschko, G., Duncan, A., Hillbricht-Ilkowska, A., Kajak, Z., Kitchell, J., Larsson, P., Leveque, C., Nauwerck, A., Schmierer, F. and Thorpe, J. E., 1980. Secondary production, in LeCren, D. and Lowe-McDonnell, R. H. (Eds), *The Functioning of Freshwater Ecosystems*, IBM 22, Cambridge University Press, Cambridge, 247–340.

Nixon S. W., 1988. Physical energy inputs and the comparative ecology of lakes and marine ecosystems, *Limnol. Oceanogr.*, **33**, 1005–1025.

Persson, L., Andersson, G., Hamrin, S. F. and Johansson, L., 1988. Predator regulation and primary production along the productivity gradient of temperate lake ecosystems, in Carpenter, S. R. (Ed.), *Complex Interactions in Lake Communities*, Springer-Verlag, New York, 45–65.

Persson, L., Diehl, S., Johansson, L., Andersson, G. and Hamrin, S. F., 1992. Trophic interactions in temperate lake ecosystems: a test of food chain theory, *Am. Nat.*, **140**, 59–84.

Pihl, L. 1985. Food selection and consumption of mobile epibenthic fauna in shallow marine areas. *Mar. Ecol. Prog. Ser.*, **22**, 169–179.

Polis, G. A. and Holt, R. D., 1992. Intraguild predation – the dynamics of complex trophic interactions, *Trend. Ecol. Evolut.*, **7**, 151–154.

Radke, R. J. and Eckmann, R. 1996. Piscivorous eel in lake Constance: can they influence year-class strength of perch? *Ann. Zool. Fennici*, **33**, 489–494.

Regier, H. A., Magnuson, J. J. and Coutant, C. C., 1990. Introduction to Proceedings of the Symposium on Effects of Climate Change on Fish, *Trans. Am. Fish. Soc.*, **119**, 173–175.

Riemann, B., Søndergaard, M., Persson, L. and Johansson, L., 1986. Carbon metabolism and community regulation in eutrophic, temperate lakes, in Riemann, B. and Søndergaard, M. (Eds), *Carbon Dynamics of Eutrophic, Temperate Lakes*, Elsevier, Amsterdam, 267–280.

Rudstam, L. G., Aneer, G. and Hildén, M., 1994. Top-down control in the Baltic ecosystem, *Dana*, **10**, 105–129.

Sarvala, J., 1992. Food webs and energy flows: fish components of the aquatic ecosystem, *Suomen Kalatalous*, **60**, 91–109 (in Finnish).

Sarvala, J., Aulio, K., Mölsä, H., Rajasilta, M., Salo, J. and Vuorinen, I., 1984. Factors behind the exceptional high fish yield in Lake Pyhäjärvi, south-western Finland – hypotheses on the biological regulation of fish production, *Aqua Fenn.*, **14**, 49–57.

Sissenwine, M. P., Cohen, E. B. and Grosslein, M. D. 1984. Structure of the Georges Bank ecosystem. *Rapp.P.-v. Reun. Cons. int. Explor. Mer.*, **183**, 243–254.

Tátrai, I. and Istvánovics, V., 1986. The role of fish in the regulation of nutrient cycling in Lake Balaton, Hungary, *Freshwater Biol.*, **16**, 417–424.

Urho, L., 1994. Removal of fish by predators – theoretical aspects, in Cowx, I. G. (Ed.), *Rehabilitation of Freshwater Fisheries*, Blackwell Scientific, London, 93–101.

Van Densen, W. L. T., 1994. Predator enhancement in freshwater fish communities, in Cowx, I. G. (Ed.), *Rehabilitation of Freshwater Fisheries*, Blackwell Scientific, London, 102–119.

Chapter 2.7

Monitoring of Faecal Pollution in Finnish Surface Waters

R. MAARIT NIEMI AND JORMA S. NIEMI

Hydrological and Limnological Aspects of Lake Monitoring
Edited by Pertti Heinonen, Giuliano Ziglio and André Van der Beken
©2000 John Wiley & Sons, Ltd, ISBN 0 471 89988 7

2.7.1   INTRODUCTION

'Infectious diseases caused by pathogenic bacteria, viruses, and protozoa or by parasites are the most common and widespread health risk associated with drinking water. Infectious diseases are transmitted primarily through human and animal excreta, particularly faeces'. This is a quotation from the WHO Guidelines for drinking water quality (World Health Organization, 1993), which considers primarily the infections transmitted by the faecal–oral route, but also deals with opportunistic and other water-associated pathogens and cyanobacterial toxins. It emphasizes that frequent examination of faecal indicator bacteria is the most sensitive and specific way of assessing the hygienic quality of waters. Although the Guidelines concentrate mainly on drinking water, many of the aspects presented are also relevant in the monitoring of water bodies.

Because of the large numbers of different water-borne organisms that are known to be pathogenic to humans, including bacteria, viruses and protozoa, it is impossible to monitor them all. Bacteria that normally occur in the faeces of humans and homoiothermic animals are therefore used to indicate faecal pollution and the associated health risks. Of the faecal indicators, the most common group is coliform bacteria, especially *Escherichia coli*, and faecal enterococci.

In 1885, Th. Escherich detected *E. coli* in the faeces of babies, which started the history of using coliform bacteria and *E. coli* as faecal indicators (Escherich, 1885). Thereafter, coliform bacteria replaced the use of heterotrophic bacteria in the analysis of faecal contamination of waters. Early in this century, direct specific enumeration of *E. coli* from water samples was impossible, and quality estimates were therefore based on the enumeration of all coliform bacteria. The determination of coliform bacteria is still considered to be one of the most efficient means of assessing water hygiene, particularly that of drinking water. This method has helped to prevent and limit the transmission of water-borne infectious diseases, thus considerably improving human health.

Introduction of the membrane filtration method in the middle of this century promoted the use of faecal streptococci as faecal indicators. Their enumeration was an important contribution to the investigation of water hygiene, because their sources differ somewhat from those of coliform bacteria and also because they are generally more persistent than coliform bacteria in the aquatic environment.

The aim of this present chapter is twofold. First, it presents the principal methodological and ecological aspects of faecal indicator bacteria, namely coliform bacteria, *E. coli* and faecal enterococci, and discusses their enumeration methods. Secondly, it summarizes the results of recent Finnish case studies, based principally on monitoring data, that have focussed on the occurrence of faecal indicator bacteria in different aquatic environments, i.e. in different types of surface waters and wastewaters, in individual polluted rivers and in lakes, and in rivers throughout the country.

2.7.2 ECOLOGICAL AND METHODOLOGICAL ASPECTS OF COLIFORM BACTERIA AND *ESCHERICHIA COLI* AS FAECAL INDICATOR BACTERIA

E. coli belongs to the family of Enterobacteriaceae. This family includes a large number of species, some of which, such as several members of the genus *Salmonella*, are pathogens. *E. coli* species is part of the normal flora of the intestine of man and homoiothermic animals, although some pathogenic strains are known. This species is abundant in faeces and is therefore a suitable faecal indicator, particularly for indicating the presence of pathogenic bacteria of the Enterobacteriaceae family. It may, however, fail to indicate the presence of viruses and pathogenic protozoa, because the survival characteristics of *E. coli* differ from those of viruses and protozoa (environmental fate or treatment processes of activated sludge and removal by filtration).

Due to practical problems of selectivity encountered in the enumeration of *E. coli*, its use was replaced by the enumeration of wider groups, i.e. coliform bacteria or total coliform bacteria. Coliform bacteria is a group defined on the basis of its enumeration method rather than as a taxonomic unit of its own. Traditionally, coliform bacteria are defined as Gram-negative, rod-shaped, nonsporulating, oxidase-negative and lactose-fermenting bacteria. This definition is derived from the method in which lactose fermentation can be detected in a growth medium containing surfactant. The oxidase test, cell morphology and Gram reaction characteristics can be checked separately for confirmation. Different species of Enterobacteriaceae fulfil these criteria and will be identified as coliform bacteria. Within some species of Enterobacteriaceae, the majority of strains can ferment lactose, whereas within other species only a few strains have this ability. Of some species, only a few strains ferment lactose, e.g. *Salmonella* are lactose negative and are not identified as coliform bacteria. The composition of coliform bacterial flora found in the aquatic environment is a result of numerous factors such as the type of bacterial source, possible reproduction and survival characteristics.

Although the early methods used for analysing coliform bacteria were able to enumerate a wide spectrum of Enterobacteriaceae species, they often allowed the growth of interfering non-coliform bacteria, such as *Aeromonas*. In order to increase the specificity of the enumeration method, the incubation temperature during cultivation was first elevated, and later the membrane filtration technique was developed by Geldreich *et al.* (1965). In North America, coliform bacteria enumerated at elevated temperature are referred to as 'faecal coliform', whereas in Europe, within the ISO and CEN definitions the term 'thermotolerant coliform' is used.

A long research history and extensive use of coliform bacteria in the monitoring of faecal pollution has resulted in the development of a large number of different enumeration methods, which vary from country to country. Different sample types (e.g. turbidity, level of contamination, etc.) require different basic techniques and methods (sensitivity, selectivity, etc.). International standardization of methods is therefore necessary. Limited data available for the validation of methods has restricted standardization. However, revisions of the ISO 9803 set of standards for coliform bacteria and *E. coli* are under preparation. At present, methods based on lactose fermentation and tolerance to surfactants are widely used for the enumeration of coliform bacteria by membrane filtration techniques (e.g. LES endo, membrane lauryl sulfate broth and lactose TTC agar with Tergitol media). In the enumeration of thermotolerant coliform bacteria, the same media can be used if they are suitable for use at elevated incubation temperatures or, alternatively, a medium specifically designed for use at elevated incubation temperatures, e.g. mFC medium, may be used. In recent years, more specific methods have been developed and there is a tendency towards a direct determination of *E. coli*. These new methods are based on the indole test, or on chromogenic or fluorogenic metabolites. Commercially available identification test kits have encouraged studies on the species composition of coliform bacteria in different aquatic environments.

2.7.3   ECOLOGICAL AND METHODOLOGICAL ASPECTS OF FAECAL ENTEROCOCCI AS FAECAL INDICATOR BACTERIA

In recent years, new methods in molecular biology have rapidly advanced the development of microbial taxonomy. Streptococci have been divided into several genera, including the genus *Enterococcus*, which despite its name also includes non-faecal species. In the earlier usage, faecal streptococci included various *Enterococcus* species and the *Streptococcus bovis/equinus* group. Within ISO and CEN standardizations, the term 'faecal enterococci' has now been taken into use, thus replacing the term 'faecal streptococci' used earlier. Revision of the taxonomy of faecal enterococci is so recent, and their identification to species level so demanding, that information available on the ecology of different species is rather limited. Table 2.7.1 summarizes the published information on the occurrence of faecal enterococci and some related species as identified to species level.

The methods commonly used in the enumeration of faecal enterococci include liquid enrichment (e.g. in azide dextrose broth followed by

Table 2.7.1 Published information on the occurrence of *Enterococcus* and some *Streptococcus* species.

Species	Source	Reference
Enterococcus		
E. avium	Human, chicken, dog and cat faeces	Collins *et al.*, 1984a; Devriese *et al.*, 1983
E. casseliflavus	Plants; soil; silage	Collins *et al.*, 1984a Devriese *et al.*, 1987
E. cecorum	Chicken faeces	Devriese *et al.*, 1983
E. columbae	Pigeon faeces	Devriese *et al.*, 1990
E. dispar	Human clinical sources	Collins *et al.*, 1991
E. durans	Milk and milk products poultry and human faeces	Collins *et al.*, 1984a Devriese *et al.*, 1987, 1992
E. faecalis	Human, homeothermic and poikilothermic animal faeces; insects; plants; food products	Mundt, 1986; Devriese *et al.*, 1987
E. faecium	Human, homeothermic and poikilothermic animal faeces; insects; plants; poultry, cattle, pig, dog, horse, sheep, goat and rabbit faeces	Mundt, 1986; Devriese *et al.*, 1987, 1992
E. flavescens	Human clinical sources	Pompei *et al.*, 1992
E. gallinarum	Poultry faeces	Collins *et al.*, 1984a; Devriese *et al.*, 1987
E. hirae	Sick children, pig, poultry, cattle, dog, horse, sheep, goat and rabbit faeces	Farrow and Collins, 1985; Devriese *et al.*, 1987, 1992
E. malodoratus	Gouda cheese	Farrow *et al.*, 1983
E. mundtii	Cattle skin; milkers' hands; plants; soil	Collins *et al.*, 1986; Devriese *et al.*, 1987
E. pseudoavium	Mastitis	Collins *et al.*, 1989
E. raffinosus	Human clinical sources; cat tonsils	Collins *et al.*, 1989
E. saccharolyticus	Cow faeces and skin; straw bedding	Farrow *et al.*, 1984
E. solitarius	Human clinical sources	Collins *et al.*, 1989
E. sulfureus	Plants	Martinez-Murcia and Collins, 1991
Streptococcus		
S. alactolyticus	Chicken; pig	Farrow *et al.*, 1984
S. bovis	Cattle, sheep and other ruminants; horses; gastric and enteric environments; pig faeces; human faeces; milk and milk products	Krieg and Holt, 1984
S. canis	Dog; (maybe) cat; cattle	Devriese *et al.*, 1986
S. equinus group	cf. *S. bovis*	Krieg and Holt, 1984
S. hyointestinalis	Pig intestines	Devriese *et al.*, 1988
S. intestinalis	Pig intestines	Robinson *et al.*, 1988
S. porcinus	Sick pigs; milk	Collins *et al.*, 1984b

confirmation tests), colony counting (often by the membrane filtration method) (e.g. Levin *et al.*, 1975) and, as a new method, the use of a fluorogenic metabolite in the MPN technique. Within the ISO, the set of standards (ISO 7899) for faecal enterococci is under revision.

2.7.4   COLIFORM BACTERIA AND *ESCHERICHIA COLI* IN WATERS

As pointed out above, the species observed as coliform bacteria are not always of faecal origin. Gavini *et al.* (1988) classified the following as faecal bacteria, namely *E. coli, Citrobacter freundii, C. diversus, C. amalonaticus, Klebsiella pneumoniae, K. oxytoca, K. mobilis* and *Enterobacter cloacea*, whereas *Serratia fonticola, Rahnella aquatilis, Enterobacterium intermedium, E. amnigenus, E. gergoviae, E. sakazakii, Buttiauxella agrestis, Klebsiella trevisanii,* and *K. terrigena* were classified as saprophytic environmental species.

Several studies have been carried out with the aim of determining the species composition of coliform bacteria in waters receiving different types of contamination (see Figure 2.7.1). In domestic wastewater, *E. coli* dominated, but *Klebsiella* species were also common. This observation is in agreement with that of Dufour (1977), who found that in sewage the proportion of *Klebsiella* increases in comparison with that of *E. coli*. In forest industry wastewaters, *Klebsiella* and *Enterobacter* genera were the dominant coliform bacteria, but in the wastewater of one specific pulp and cardboard mill the growth of *E. coli* was observed (Niemi *et al.*, 1987). In food and textile industry wastewaters, the genus *Citrobacter* occurred in significant numbers in addition to *Klebsiella* and *Enterobacter* genera. In industrial wastewaters with high concentrations of organic matter, the importance of *Klebsiella* was pointed out by Vlassoff (1977). In fish farm waters, *Citrobacter* and the false positive coliform *Aeromonas hydrophila* were common isolates, together with the usually lactose-negative species *Hafnia alvei*. In pristine waters, the most common isolates in decreasing order were *S. fonticola, H. alvei, E. cloacae* and *E. coli*. The proportion of unidentified strains increased in the following order: domestic, food and textile, forest industry wastewater, fish farm effluents and pristine waters.

In pristine waters, about 100% of the thermotolerant coliform bacteria identified were *E. coli* (Niemelä and Niemi, 1989). In agricultural areas the corresponding percentage was 91% (Niemi and Niemi, 1991) and in fish farm waters 89% (Niemi, 1985). In all of these waters, thermotolerant coliform bacteria reliably indicated faecal pollution. However, it is possible that in the same type of waters in warmer climates, thermotolerant coliform bacteria are not equally reliable indicators. Irrespective of climate, thermotolerant coliform bacteria are not reliable indicators of faecal pollution in water bodies receiving

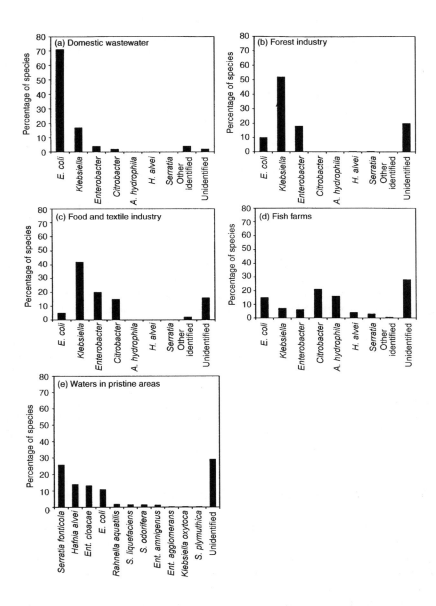

Figure 2.7.1 Species composition of coliform bacteria in different wastewaters: (a) domestic wastewater, $n = 674$; (b) forest industry, $n = 359$; (c) food and textile industry, $n = 331$; (d) fish farms, $n = 1244$; (e) pristine waters, $n = 372$ (Niemi, 1985; Niemelä and Niemi 1989; Niemi *et al.*, 1997c).

industrial wastewaters with a high content of organic matter, because in these conditions thermotolerant strains of the genera *Klebsiella* and *Enterobacter* can multiply.

There is ample evidence that several coliform and thermotolerant coliform species are able to grow outside their normal habitat, which is the intestine of man and homoiothermic animals. This fact naturally limits their suitability as faecal indicators. Consequently, in the monitoring of surface waters it is advisable to analyse *E. coli* instead of total coliform and thermotolerant coliform bacteria.

2.7.5   FAECAL ENTEROCOCCI IN WATERS

In treated wastewaters, 80% of faecal enterococci enumerated on primary cultivation media were confirmed as enterococci, whereas in watercourses of both agricultural and pristine areas the confirmation percentage was only 44% (Niemi and Niemi, 1991). The low proportion of confirmation is due to the high proportion of false positive colonies. In fish-farm waters, high numbers of bacteria grew on traditional primary enrichment broths and only 3.6% of the isolates were confirmed as faecal enterococci (Niemi, 1985). These observations indicate that confirmation tests should be carried out when using these primary cultivation media.

Faecal enterococci have been a more reliable indicator of water hygiene in waters contaminated by pulp and paper mills than thermotolerant coliform bacteria. Only limited information is available on the ecology of different *Enterococcus* species because the recent taxonomic changes make the use of old data questionable. However, growth of 'faecal species' of enterococci in the environment has been observed when using updated taxonomy (Anderson and Tyler, 1997). Faecal enterococci can be used as supplementary faecal indicators, when a confirmation step in their enumeration is included.

2.7.6   MONITORING OF FAECAL BACTERIA

Finnish inland waters have been monitored for faecal bacteria since the beginning of the 1960s. The first national water quality monitoring programmes, in rivers and lakes, were organized by the Environmental Authorities at the beginning of the 1960s. The objective was to provide a general picture of the quality of inland waters and a benchmark for further studies. At that time, the effects of industrial and domestic wastewaters on surface waters were of special concern. In the majority of programmes, mainly physical and chemical water quality variables were analysed, although faecal bacteria were also often included. Since then, the number of national and other monitoring programmes in progress has increased dramatically and the selection of measured water quality variables has become more extensive. The national monitoring programmes are carried out by the Finnish Environment Institute (FEI), together with 13 Regional Environment Centres (RECs).

In addition to national and regional monitoring programmes, there is a third class of monitoring programme which produces water quality data, including that of faecal bacteria. Some 1700 polluters, e.g. municipalities and factories discharging wastewaters, have an obligation to monitor the quality of their receiving waters in a total of about 4000 sites. This so-called statutory monitoring is carried out according to the Water Act by officially supervised laboratories coordinated by the FEI and RECs (OECD, 1988). It produces water quality data, particularly for waters under the influence of wastewater loading. In these waters, the enumeration of faecal indicator bacteria is often of particular importance.

Faecal bacteria are analysed by the laboratories of the RECs. In addition to this, some 20 other laboratories carry out statutory monitoring based on the decisions of Water Courts. All of the laboratories are co-ordinated and supervised by the FEI. Due to the standardization of the methods in operation since the mid-1970s, the methods used in the enumeration of faecal bacteria have been relatively uniform and the results are comparable. The data produced by national, regional and statutory monitoring programmes are stored in the information systems of the FEI and form a basis for studying faecal pollution and its historical development in the surface waters of the country.

2.7.7   BACTERIOLOGICAL QUALITY OF INDIVIDUAL RIVERS

Faecal bacteria are discharged to rivers from various sources, such as domestic and industrial wastewater, agriculture, diffuse loading and even in minor amounts from pristine areas due to wild animals. In river waters, faecal bacteria are outside their normal habitat and subject to the detrimental effects of environmental factors such as temperature, sedimentation, solar radiation, predation and toxic chemicals. All of these factors increase the die-off of bacteria. The concentrations of bacteria present in rivers are a result of a balance between the input of bacteria and their die-off rate.

Typical characteristics of polluted Finnish rivers are continuous input of faecal bacteria from wastewater treatment plants and other sources, good survival of bacteria during the long winter due to low temperatures under the ice cover, effective mixing and great fluctuations in discharge because of wide annual variations in climate and hydrology. These factors cause great temporal and spatial fluctuations in the concentrations of faecal bacteria.

The faecal pollution of Finnish rivers has been investigated in several studies, which were based mainly on the long-term monitoring data available since the 1960s. Examples of these case studies are summarized in the following.

The river Kymijoki situated in the southern coastal area is one of the largest rivers in the country. In the 1960s, the river was heavily loaded by untreated domestic and industrial wastewaters. However, loading considerably decreased due to the construction of wastewater treatment plants. In the forest industry situated along the river, mechanical wastewater treatment was brought into use in the 1970s and activated sludge treatment in the late 1980s. By 1992, all of the domestic and industrial wastewaters were treated by the activated sludge process. These treatment plants, as in the case of such plants in general, were designed to reduce nutrients, biological oxygen consumption and suspended solids, but not especially faecal bacteria.

Niemi *et al.* (1997a) investigated the bacteriological quality of this river by examining long-term monitoring data that included a total of 4804 enumerations of faecal enterococci from the period 1964–1992.

The results showed that bacteriological quality improved as the number of wastewater treatment plants increased and the treatment processes improved. The present faecal enterococci concentrations in the river (median 30–40 CFU, colony forming unit, per 100 ml) are probably the lowest that can be reached by applying normal wastewater treatment processes. Further decreases of faecal enterococci would probably require the application of water treatment processes specifically designed to reduce indicator bacteria.

The river Porvoonjoki is situated in the southern coastal area of the country. The river is polluted by treated domestic wastewaters and by agriculture, particularly manure application and cattle farming. Treated domestic waste-waters are discharged to the river from eleven wastewater treatment plants, the majority of which were constructed in the 1970s. Since 1976, the wastewaters of the town of Lahti, with a population of about 100 000, are discharged to the river Porvoonjoki.

Faecal pollution of the river was investigated by Niemi *et al.* (1996); the data consisted of about 4700 bacterial enumerations from 1963–1995. Median concentrations of both faecal enterococci and thermotolerant coliforms were high and gave identical pictures of the water quality of the river. There was a typical pattern in the occurrence of indicator bacteria between the headwaters and the mouth of the river.

During the period investigated, the medians of both bacterial groups decreased, especially at the beginning of the 1990s. The decrease was probably due to water management practices brought into use at that time, e.g. construction of wastewater plants, improvement of treatment processes and recently applied new agricultural practices. However, despite clear improve-ment, the hygienic quality of the river is still relatively poor.

The river Aurajoki flows through a densely populated area of the south-western part of the country and discharges to the northern Baltic Sea. It is a raw water supply for the city of Turku (about 170 000 individuals). The length of the river is about 70 km and the area of the total drainage basin is 885 km². The percentage of lakes of the drainage basin is only 0.2% and therefore the discharge varies rapidly and widely. Niemi *et al.* (1994) investigated the long-term temporal variation of thermotolerant coliforms and faecal enterococci in this river in one observation site upstream of the city of Turku.

Seasonal fluctuations in bacterial concentrations were similar each year, being high in spring, autumn and winter, but low in summer. Increasing or decreasing trends in concentrations were not observed. Regression analysis was applied to the material, and time-series analysis was tested for predicting bacterial concentrations. The correlation coefficient between thermotolerant coliforms and faecal enterococci was high ($r = 0.87$, $n = 411$). The correlation coefficient between discharge and faecal enterococci was typically higher than that between discharge and thermotolerant coliforms. Preliminary calculations showed that of the total bacterial flux of the river (concentration multiplied by discharge), the contribution of diffuse bacterial flux was on average 60–90% and that of treated wastewater 10–40%.

2.7.8　BACTERIOLOGICAL QUALITY OF FINNISH WATERS DURING 1963–1993

Because of low average population density (15 inhabitants per km²), abundant areal water resources (33 000 km²) and effective sewage treatment of municipal and industrial wastewaters, it might be assumed that the surface waters of Finland are generally of good quality. Unfortunately, this assumption is not always correct. Pronounced water quality problems exist, particularly in the more densely populated areas in the south and south-west, but locally also in other regions.

Niemi *et al.* (1997b) studied the general bacteriological quality of rivers and lakes using data from the years 1963–1993. The aim of the work was to determine the hygienic status of the inland waters of this country and to investigate whether the quality of waters had improved during this period. Special emphasis was placed on regional differences in the hygienic quality of lakes and rivers. Annual medians of thermotolerant coliforms and faecal enterococci were calculated separately for the lakes and rivers of ten large drainage basins that together cover the whole country. The data consisted of a total of about 210 000 bacterial enumerations, of which about 64 000 were thermotolerant coliform observations and about 147 000 faecal enterococci observations.

The results showed that lakes were clean and had good bacteriological quality. In the lakes of all ten basins the annual medians of faecal bacteria were low, typically less than 10 CFU per 100 ml, and often nondetectable, and thus trends were not observed. Lake waters clearly fulfilled the criterion of good swimming water (100 CFU per 100 ml).

Rivers were consistently more polluted than lakes. With respect to bacteriological quality, the rivers of the country could be divided into two groups, namely the coastal area and the rest of the country. The same division in the water quality of rivers was also observed in the regional investigation of other water quality variables of rivers (Niemi, 1998). Rivers of the northern, central and eastern parts of the country were clean. In these areas the annual medians of faecal bacteria never exceeded 100 CFU per 100 ml, being typically close to 10 CFU per 100 ml. By contrast, in coastal areas the rivers were more polluted and the medians oscillated around 100 CFU per 100 ml, typically exceeding the limit for good swimming water.

Both indicator bacteria gave a similar overall picture of pollution in lakes and rivers. These results indicate that water quality measures, where undertaken, have improved river water quality.

2.7.9   CONCLUSIONS

The most common faecal indicator bacteria, e.g. *E. coli* and faecal enterococci, are suitable for investigating the hygienic quality of surface waters. Total coliform bacteria have been useful in the past, but severe limitations to their applicability in surface waters have been observed. These bacteria, particularly direct enumeration of *E. coli*, are sensitive water quality variables which provide a reliable picture of the development of water quality. Long-term monitoring data of indicator bacteria, available from Finnish surface waters since the 1960s, showed that lakes were clean. The rivers of the coastal areas were more contaminated than other rivers.

In many rivers, particularly in certain districts of this country, the hygienic quality of surface waters has improved, particularly in the 1970s and 1980s.

It is probable that new, more specific methods based on specific substrates and chromogenic or fluorogenic metabolites will be introduced in this field in the near future. However, there is every reason to believe that until then the monitoring of faecal contamination of waters by using traditional indicators and enumeration techniques will continue to be useful in efforts to develop a cleaner and safer environment.

REFERENCES

Anderson, S. A. and Tyler, S. J., 1997. Enterococci in New Zealand environment: implications for water quality monitoring, in Morris, R., Grabow, O. K. and Jofre, J. (Eds.), *Health-Related Microbiology 1996, Water Sci. Technol.*, **35**, 325–332.

Collins, M. D., Jones, D., Farrow, J. A. E., Kilpper-Bälz, R. and Schleifer, K.-H., 1984a. *Enterococcus avium* nopm. Rev., comb. nov.; *E. casseliflavus* nom. Rev., comb. nov; *E. durans* nom Rev., comb. nov; *E. Gallinarum* comb. nov.; *E. gallinarum* cobm. nov; and *E. maldoratus* sp. nov., *Int. J. System. Bacteriol.*, **34**, 220–223.

Collins, M. D., Farrow, J. A. E., Katic, V. and Kandler, O., 1984b. Taxonomic studies of streptococci of serological groups E, P, U and V: Description of *Streptococcus porcinus* sp. nov., *System. Appl. Microbiol.*, **5**, 402–413.

Collins, M. D., Farrow, J. A. E. and Jones, D., 1986. *Enterococuccs mundtii* sp. nov., *Int. J. System. Bacteriol.*, **36**, 8–12.

Collins, M. D., Facklam, R. R., Farrow, J. A. E. and Williamson, R., 1989. *Enterococcus raffinosus* sp. nov., *Enterococcus solitarius* sp. nov. and *Enterococcus pseudoavium* sp. nov., *FEMS Microbiol. Lett.*, **57**, 283–288.

Collins, M. D., Rodrigues, U. M., Pigot, N. E. and Facklam, R. R., 1991. *Enterococcus dispar* sp. nov. A new *Enterococcus* species from human sources, *Lett. Appl. Microbiol.*, **12**, 95–98.

Devriese, L. A., Nutta, G. N., Farrow, J. A. E., van de Kerckhove, A. and Phillips, B. A., 1983. *Streptococcus cecorum*, a new species from chickens, *Int. J. System. Bact.*, **33**, 772–776.

Devriese, L. A., Hommez, J., Kilpper-Bälz, R. and Schleifer, K.-H., 1986. *Streptococcus canis* sp. nov.: a species of group G streptococci from animals. *Int. J. System. Bact.*, **36**, 422–425.

Devriese, L. A., van de Kerckhove, A., Kilpper-Bälz, R. and Schleifer, K.-H., 1987. Characterization and identification of *Enterococcus* species isolated from the intestines of animals, *Int. J. System. Bact.*, **37**, 257–259.

Devriese, L. A., Kilpper-Bälz, R. and Schleifer, K.-H., 1988. *Streptococcus hyointestinalis* sp. nov. from the gut of swine, *Int. J. System. Bact.*, **38**, 440–441.

Devriese, L. A., Ceyssens, K., Rodrigues, U. M. and Collins, M. D., 1990. *Enterococcus columbae*, a species from pigeon intestines, *FEMS Microbiol. Lett.*, **71**, 247–252.

Devriese, L. A., Collins, M. D. and Wirth, R., 1992. The genus *Enterococcus*, in Ballows, A., Trüber, H. G., Dworkin, M., Harder, W. and Schleifer, K.-H. (Eds), *The Prokaryotes*, 2nd Edn, Vol. II, Springer-Verlag, New York, 1465–1481.

Dufour, A. P., 1977. *Escherichia coli*: The faecal coliform, in Hoadley, A. W. and Dutka, B. J. (Eds), *Bacterial Indicators/Health Hazards Associated with Water*, STP 635, American Society of Testing and Materials, Philadelphia, PA, USA, 48–58.

Escherich, Th. 1885. Die Darmbakterien des Neugeboren und Sauglings, *Rotschr. Med.*, **3**, 515–522, 547–564 (in German); The intestinal bacteria of the neonate and breast-fed infant, *Rev. Infect. Dis.*, 1988, **10**, 1220–1225 (translated into English by K. A. Bettelheim).

Farrow, J. A. E., Jones, D., Phillips, B. A. and Collins, M. D., 1983. Taxonomic studies on some group D streptococci, *J. Gen. Microbiol.*, **129**, 1423–1432.

Farrow, J. A. E., Kruze, J., Phillips, B. A., Bramley, A. J. and Collins, M D., 1984. Taxonomic studies on *Streptococcus bovis* and *Streptococcus equinus*: Description of *Streptococcus alactolyticus* sp. nov. and *Streptococcus saccharolyticus* sp. nov., *System. Appl. Microbiol.*, **5**, 467–482.

Farrow, J. A. E. and Collins, M. D., 1985. *Enterococcus hirae*, a new species that includes amino acid assay strain NCDO 1258 as strain causing growth depression in young chickens, *Int. J. System. Bact.*, **35**, 73–75.

Gavini, F., Leclerc, M. and Mossel, D. A., 1988. Enterobacteriaceae of the 'Coliform Group' in drinking water: identification and worldwide distribution, *System. Appl. Microbiol.*, **6**, 312–318.

Geldreich, E. E., Clark, H. F., Huff, C. B. and Best, L. C., 1965. Faecal-coliform organism medium for the membrane filter technique, *J. Am. Water Works Assoc.*, **57**, 208–214.

Kreig, N. R. and Holt, J. G. (Eds), 1984. *Bergey's Manual of Systematic Bacteriology*, Vol. 1, Williams & Wilkins, Baltimore, MD, USA.

Levin, M. A., Fisher, J. R. and Cabelli, V. J. 1975. Membrane filter technique for enumeration of enterococci in marine waters, *Appl. Microbiol.*, **30**, 66–71.

Martinez-Murcia, A. J. and Collins, M. D., 1991. *Enterococcus sulfureus*, a new yellow-pigmented *Enterococcus* species. *FEMS Microbiol. Lett.*, **80**, 69–74.

Mundt, J. P., 1986. Enterococci, in Sneath, P. H. A., Mair, N. S., Sharpe, M. E. and Holt, J. G. (Eds), *Bergey's Manual of Systematic Bacteriology*, Vol. 2, Williams & Wilkins, Baltimore, MD, USA, 1063–1065.

Niemelä, S. I. and Niemi, R. M., 1989. Species distribution and temperature relations of coliform populations from uninhabited watershed areas, *Toxic. Assess.*, **4**, 271–280.

Niemi, J. S., 1998. The quality of river waters in Finland, *Eur. Water Manage.*, **1**, 36–40.

Niemi, J. S., Niemi, R. M. and Pajakko, P. M., 1994. Long-term temporal variation of hygienic indicator bacteria in a river, *Verh. Internat. Verein. Limnol.*, **25**, 1901–1909.

Niemi, J. S., Niemi, R. M. and Malin, V., 1996. Porvoonjoen hygieeninen laatu. (Hygienic quality of the river Porvoonjoki) *Vesitalous*, **6/1996**, 22–26 (In Finnish).

Niemi, J. S., Heitto, L. R., Niemi, M., Anttila-Huhtinen, M. and Malin, V., 1997a. Bacteriological purification of the Finnish river Kymi, *Environ. Monit. Assess.*, **46**, 241–253.

Niemi, J. S., Niemi, R. M., Malin, V. and Poikolainen, M-L., 1997b. Bacteriological quality of Finnish rivers and lakes, *Environ. Toxicol. Water Qual.*, **12**, 15–21.

Niemi, R. M., 1985. Faecal Indicator Bacteria at Freshwater Rainbow Trout (*Salmo gairdneri*) Farms, Publications of the Water Research Institue, No. 64. National Board of Waters, Helsinki, Finland.

Niemi, R. M. and Niemi, J. S., 1991. Bacterial pollution of waters in pristine and agricultural lands, *J. Environ. Qual.* **20**, 620–627.

Niemi, R. M., Niemelä, S. I., Mentu, J. and Siitonen, A., 1987. Growth of *Escherichia coli* in a pulp and cardboard mill, *Can. J. Microbiol.*, **33**, 541–545.

Niemi, R.M., Niemelä, S. I., Lahti, K. and Niemi, J. S., 1997c. Coliform and *E. coli* in Finnish surface waters, in Kay, D. and Fricker, C. (Eds), *Coliforms and E. coli – Problem or Solution*, The Royal Society of Chemistry, Cambridge, UK, 112–119.

OECD, 1988. Environmental Policies in Finland, Organization for Economic Co-operation and Development, Paris, France.

Pompei, R., Berlutti, F., Thaller, M. C., Ingianni, A., Cortis, G. and Dainelli, B., 1992. *Enterococcus flavescens* sp. nov., a new species of Enterococci of clinical origin, *Int. J. System. Bacteriol.*, **42**, 365–369.

Robinson, I. M., Stromley, J. M., Varel, V. H. and Cato, E. P., 1988. *Streptococcus intestinalis*, a new species from the colons and faeces of pigs, *Int. J. System Bacteriol.*, **38**, 245–248.

Vlassoff, L. T., 1977. *Klebsiella*, in Hoadley, A. W. and Dutka, B. J. (Eds), *Bacterial Indicators/Health Hazards Associated with Water*, STP 635, American Society for Testing and Materials, Philadelphia, PA, USA.

World Health Organization, 1993. *Guidelines for Drinking-Water Quality*, Vol. 1, *Recommendations*, 2nd Edn, World Health Organization, Geneva, Switzerland.

Part Three
Harmful Substances in the
Lake Environment

Chapter 3.1

Lake Sediments in Historical Monitoring of the Environment

HEIKKI SIMOLA

Hydrological and Limnological Aspects of Lake Monitoring
Edited by Pertti Heinonen, Giuliano Ziglio and André Van der Beken
©2000 John Wiley & Sons, Ltd, ISBN 0 471 89988 7

3.1.1 INTRODUCTION

Historical or retrospective environmental monitoring makes use of all kinds of natural archives, e.g. lake and marine sediments, peat deposits, snow and ice accretions, as well as various organism structures that contain a calendar reference: tree rings, corals, (sub)fossils, museum specimens etc. (MARC, 1985). Aquatic sediments are particularly rewarding objects for historical monitoring, where they also provide, besides an often detailed record of the external inputs, both terrestrial and atmospheric, a biological record of the lake ecosystem's response to the environmental change.

Palaeolimnological information, obtained by sediment stratigraphical analyses, is often an indispensable background for actual monitoring data, as most lakes have been impacted by various human activities for much longer than the time span of any systematic monitoring survey (Charles *et al.*, 1994). Furthermore, the sediment record may allow determination of the pre-disturbance ecosystem conditions, thus enabling prognostication of ecosystem recovery from an assault, or targeting of restoration measures to an attainable goal (Anderson and Rippey, 1994; Simola *et al.*, 1996).

A sediment deposit constitutes a continuous record of the ecosystem history in a chronological sequence, where the vertical dimension represents the passage of time, and the information contained in each horizontal stratum provides a more or less clearly defined image of a moment or a period in the ecosystem history. The temporal resolution of a sediment sequence, i.e. how sharply the past events can be distinguished, depends on the quantity of accumulating material and on eventual disturbance processes at the sediment surface. The most detailed temporal records are provided by annually laminated or varved sediments, in which even seasonal patterns may be discernible (Renberg, 1981; Simola, 1992). Most lake sediments are, however, affected by physical or biological mixing, so that a given layer represents a more or less homogenized sample of the sedimentation over several years or even decades (Håkanson and Jansson, 1983).

Lake sediments are composed of both autochthonous material, produced in the lake ecosystem, and allochthonous material, entering the lake from its drainage area and the atmosphere. In order to make full use of this information, a multidisciplinary approach, involving physical, chemical and biological analyses, is warranted. Palaeolimnology as an integrative discipline has developed rapidly since the 1970s (Haworth and Lund, 1984; Berglund, 1986; Smith *et al.*, 1991).

3.1.2 SEDIMENT CORING AND SAMPLING STRATEGY

A number of coring devices have been developed for obtaining stratigraphic sediment samples. Most of the ordinary models are either gravity or piston corers (Aaby and Digerfeldt, 1986); various special designs exist, e.g. for large sample volumes (Flower *et al.*, 1995), or for precise slicing of the sample core (Kansanen, *et al.*, 1991). *In situ* freezing is a special technique for studies of sediment microstructures, e.g. varves (Huttunen and Meriläinen, 1978). Sediment freezing is effected by the use of either dry ice or liquid nitrogen. Freeze-corer samples are well suited for high-resolution biostratigraphic analyses, but the freezing process can alter the chemical composition of the sample to some extent (Stephenson *et al.*, 1996); therefore, physical and chemical analyses should preferably be done from unfrozen sediment cores.

In most lakes, the sedimenting material is focused towards the deepest part of the basin, which therefore is the preferred site for taking the samples. In large or shallow lakes, in complex basins, and in wind-stressed conditions, however, sediment distribution mapping by systematic coring or echo-sounding may be necessary for locating the proper sedimentation areas (Whitmore *et al.*, 1996). Choice of sampling gear and the quantity and type of samples to be taken, depends on the monitoring problems to be assessed. Precise determination of sampling site location is important for possible future work on the same site. Because of local irregularities in sediment distribution, all samples for an investigation should preferably be taken from the selected site at the same time. For whole-lake inventories of total sedimentary fluxes of, e.g. nutrients or pollutants, multi-core sampling across the whole basin is warranted (Engstrom and Swain, 1986).

Dating of the sediment cores is crucial, while sedimentation rates may vary considerably between sites and within sequences. Each available method is subject to errors, so independent support provided by different dating methods is often required. For many practical aims, e.g. to identify sediment strata that are definitely pre-industrial, it may be sufficient to obtain a rough dating only.

3.1.3 ISOTOPE-DATING TECHNIQUES

Isotope dating using ^{210}Pb is a commonly used technique for recent (0–150 yr) sediment sequences (Olsson, 1986); ^{210}Pb is a member of the ^{238}U decay series, with a half-life of 22.26 yr. There are two sources of ^{210}Pb in the sediment, i.e. *supported* ^{210}Pb, which is formed by radioactive decay of catchment-derived ^{226}Ra within the sediment and constitutes a background activity in the sediment column, and *unsupported* ^{210}Pb, which is used for the dating, where the latter enters the lake from the atmosphere as a decay product of the inert gas ^{222}Rn. The unsupported ^{210}Pb is present in the youngest sediment layers only and decreases asymptotically downcore. The total inventory of unsupported ^{210}Pb can be calculated by subtracting the supported activity from the total in each analysed layer (usually 1 cm sediment slices). In routine analyses, the supported ^{210}Pb-level, as determined in the older (>150 yr) samples, is assumed to be constant within the datable sequence; more precise determinations require analysis of the parent ^{226}Ra stratigraphy (Oldfield and Appleby, 1984). The most commonly applied calculation model assumes a constant rate of supply of unsupported ^{210}Pb into the sediment. Error sources of ^{210}Pb-dating need also to be considered (von Gunten and Moser, 1993; Binford *et al.*, 1993).

Another commonly determined isotope, i.e. ^{137}Cs, records the history of atmospheric nuclear testing (peaking at around 1963) and the 1986 Chernobyl accident, and thus serves for dating the most recent sediments, as well as a control for ^{210}Pb-dating (e.g. Appleby, 1993). The use of ^{14}C-dating is not usually relevant for historical monitoring studies, while various anthropogenic emissions of both low and high ^{14}C activity have thoroughly disturbed the radiocarbon balance of the atmosphere during the industrial era. Other isotopes applicable to dating are described by Olsson (1986).

3.1.4   SEDIMENT VARVES AND MARKER LAYERS

Annually laminated or varved sediments are fairly common in deep stratifying lakes. Pronounced seasonality in sedimentation and hypolimnetic anoxia are common features that promote varve formation. Simola (1992) describes the structural features that are typically encountered in varved sediments and discusses problems in identifying the yearly units in structurally complex varves. Varved sediments are eminently suitable for historical monitoring studies, while the varves also provide a direct calendar dating for the sequence, and the retaining of the seasonal cycle of sedimentation indicates absence of bioturbation, and hence good qualitative preservation of the sedimenting material.

A number of different techniques are available for the microscopic analysis (Merkt, 1971; Simola *et al.*, 1986) and dissection (Schmidt *et al.*, 1995) of varved sediments for high-resolution stratigraphic studies.

3.1.5   CHEMICAL ANALYSES

The basic procedures for sediment chemical analyses have been presented by Engstrom and Wright (1984) and Bengtsson and Enell (1986). A wide range of analytical procedures, e.g. gas chromatography and mass spectrometry, developed for various environmental pollutants are also applicable to sediment studies.

A specific analysis relating to atmospheric industrial pollution is the microscopic counting of carbonaceous soot particles (Wik and Renberg, 1996). The universal features of soot particle stratigraphies, as well as those of atmospherically deposited heavy metals (Norton *et al.*, 1992) may also serve as dating tools, providing a sufficiently precise calendar framework for many practical aims and purposes in studies of recent environmental change.

Chemical analysis results are given as concentrations (usually mass per unit of dry weight of sediment). When dating is available for the investigated sediment sequence, the concentrations need to be transformed into accumulation rates (mass per unit area per year) or total inventories (mass per unit area within a period of time). Results of a multi-core, whole-lake investigation provide total lake inventories. All calculations involving dating, or within-core or across-lake integration, also introduce cumulative error terms into the results. For many purposes, it suffices to operate with the primary concentration data only.

3.1.6 BIOLOGICAL ANALYSES

The stratigraphies of biological remains (chitinous, siliceous or calcareous microfossils of different organisms, as well as different biochemical fossils, e.g. pigments) provide direct evidence of the past living conditions in the lake ecosystem. As regards biostratigraphic data, it is important to note that different organism groups each tell their own story: diatoms, other algae and cladocera mainly reflect the pelagial water quality and conditions in the littoral/littoriprofundal zone, while the profundal chironomid fauna directly indicates, often dramatic, changes in the hypolimnetic conditions.

The handbook of Berglund (1986) provides introductory chapters to the main groups of organisms commonly exploited in environmental palaeolimnology, including references to relevant taxonomic literature for each group.

It actually suffices to analyse the sedimentary assemblages as relative frequencies only, i.e. to present species numbers as percentages of the total in each analysed sample. However, quantitating the samples on to a volume or weight basis (Scherer, 1994) adds an important dimension to the data. Biological remains may occur in sediments over a very wide range of concentrations. Quantitative data provides indication of the degree of preservation of the remains, which may be crucial for proper interpretation of the results, and in favourable occasions may also allow a direct estimation of past production of the organisms (Anderson, 1994).

Differential preservation of the organism remains is a major problem concerning biostratigraphic analyses, particularly hampering comparisons between different lakes or even between different levels within a single sediment sequence. Diatoms are only occasionally found well preserved in calcareous sediments, and calcareous microfossils (e.g. ostracods and molluscs) tend to completely dissolve in soft-water sedimentary environments. Even when diatom preservation is generally good, the most thinly silicified forms (e.g. *Rhizosolenia*, *Attheya*) are rarely seen in the sediment assemblages. Species-specific dissolution may result in stratigraphic artefacts. This phenomenon is only rarely taken into account in the statistical analyses of sedimentary assemblages (Cameron, 1995).

3.1.7 DATA ANALYSIS FOR INTERPRETING ECOSYSTEM CHANGE

The classical approach to relate biological assemblages to environmental conditions involves the indicator-species concept. Empirical knowledge on the occurrence of different species in different environments allows the establishment of indicator-species groups that display a narrow or broader distribution along a particular environmental gradient. For the characterization of aquatic environments, indicator species for, e.g. nutrient levels, saprobicity, pH, salinity, temperature and different hydrographic conditions (depth, current and tidal regime), have been identified within various ecological (phyto- and zooplankton, benthos, etc.) and taxonomic groups. These groupings are readily applicable to fossil assemblages as well.

A number of biotic indices have been constructed on the basis of different indicator groups, thus providing numerical index values that relate to a specific environmental parameter or condition. The Benthic Quality Index (BQI) (Wiederholm, 1980), when applied to sedimentary chironomid assemblages, describes past conditions of the profundal habitat (Meriläinen and Hamina, 1993).

When a mathematical correlation between the biotic index and a particular environmental parameter can be established, it is possible to construct a transfer function by which a numerical value can be calculated for the parameter in question from the assemblage composition data. The development of the diatom-based pH-inference method (Battarbee *et al.*, 1986) is a classical example of reconstructed pH changes during the post-glacial development of a Danish lake. Meriläinen (1967) found a good logarithmic correlation between Nygaard's index α and the measured pH, while Renberg and Hellberg (1982) produced a somewhat refined transfer function for direct prediction of lakewater pH from diatom data.

The traditional indicator-species approach is always somewhat arbitrary, when it is based on subjective judgement of species preferences. In the modern approach to environment reconstruction, the transfer function is generated by computerized data processing by means of a calibration or training data set, which consists of modern species assemblages (usually surface sediment samples) and associated environmental data (water chemistry, catchment characteristics, climate etc.). The calibration data set should include at least

80–100 lakes that represent the range of environmental conditions that are potentially encountered in the reconstruction.

Canonical correspondence analysis (ter Braak, 1986) is a commonly used ordination technique for exploratory analysis of the calibration data set. It can be used to assess the importance of the desired variables (e.g. pH and nutrients) as contributors to the observed variance, to observe possible interdependence between different environmental variables, and to identify possible needs to modify the data set.

Weighted-averages regression and calibration (ter Braak and van Dam, 1989) is one of several computation procedures, and perhaps the one most commonly used, for quantitative reconstruction. At the first stage, the relationship between the species and the particular environmental variable in the training data set is established, and then this information is applied to provide an estimate for the parameter value from species composition in a fossil assemblage. Other procedures used for establishing the species–environment relationship include maximum likelihood or partial least squares estimations. Birks (1995) gives a detailed description of the different techniques available for constructing transfer functions, of the inherent assumptions involved and pitfalls to be avoided in this approach, and the rather complicated techniques used for assessing the statistical reliability of the reconstructions.

Another approach employed for past-environment reconstruction is the technique of modern analogue matching. This aims at a direct matching of fossil assemblages in a sediment core with those in modern lakes. Thus, the correspondence is directly sought between ecosystems, and no regression or calibration of the biota with environmental variables is needed. The matching is based on species-assemblage dissimilarity coefficients calculated for the modern and fossil assemblages (Flower *et al.*, 1997).

3.1.8 CONCLUSIONS

Quantitative palaeoecology, using numerical data analysis by multivariate statistical methods, has its roots in palaeolimnological studies of vegetation history. During the early 1980s, as lake acidification had emerged as a major environmental issue, quantitative methods became widely applied in palaeolimnological studies, primarily concerning sedimentary diatoms as indicators of pH history in lakes (Battarbee *et al.*, 1986; Charles *et al.*, 1989). Acidification research promoted considerable development of multi-variate statistics and appropriate computing programs as tools for quantitative analysis of palaeoecological data. This development is leading palaeoecology from an empirical-descriptive phase (detection of patterns), through a narrative phase (providing inductively based explanations), and so towards an analytical phase, when testable hypotheses for the observed patterns can be provided.

During the past decade, the transfer-function approach has been successfully applied to a wide range of environmental monitoring problems in different geographical areas and for different time-scales, usually providing retrospection for single environmental variables at a time relating to recent environmental or longer scale climatic changes.

REFERENCES

Aaby, B. and Digerfeldt, G., 1986. Sampling techniques for lakes and bogs, in Berglund, B. E. (Ed.), *Handbook of Holocene Palaeoecology and Palaeohydrology*, John Wiley & Sons, Chichester, UK, 181–194.

Anderson, N. J., 1994. Comparative planktonic diatom biomass responses to lake and catchment disturbance, *Journal of Plankton Research*, **16**, 133–150.

Anderson, N. J. and Rippey, B., 1994. Monitoring lake recovery from point-source eutrophication: the use of diatom-inferred epilimnetic total phosphorus and sediment chemistry, *Freshwater Biology*, **32**, 625–639.

Appleby, P. G., 1993. Forward to the lead-210 dating anniversary series, *Journal of Paleolimnology*, **9**, 155–160.

Battarbee, R., Smol, J. and Meriläinen, J., 1986. Diatoms as indicators of pH: a historical review, in: Smol, J., Battarbee, R., Davis, R. and Meriläinen, J. (Eds), *Diatoms and Lake Acidity*, Developments in Hydrobiology, Vol. 29, Dr W. Junte Publishers, Dordrecht, 5–14.

Bengtsson, L. and Enell, M., 1986. Chemical analysis, in Berglund, B. E. (Ed.), *Handbook of Holocene Palaeoecology and Palaeohydrology*. John Wiley & Sons, Chichester, UK, 423–451.

Berglund, B. E. (Ed.), 1986. *Handbook of Holocene Palaeoecology and Palaeohydrology*. John Wiley & Sons, Chichester, UK.

Binford, M., W., Kahl, J. S. and Norton, S. A., 1993. Interpretation of ^{210}Pb profiles and verification of the CRS dating model in PIRLA project lake sediment cores, *Journal of Paleolimnology*, **9**, 275–296.

Birks, H. J. B., 1995. Quantitative palaeoenvironmental reconstructions, in Maddy D. and Braw, S. (Eds), *Statistical Modelling of Quaternary Science Data*, Technical Guide No. 5, Quaternary Research Association, Cambridge, UK, 161–254.

Cameron, N., 1995. The representation of diatom communities by fossil assemblages in a small acid lake, *Journal of Paleolimnology*, **14**, 185–223.

Charles, D. F., Battarbee, R. W., Renberg, I., van Dam, H. and Smol. J. P., 1989. Paleoecological analysis of lake acidification trends in North America and Europe using diatoms and chrysophytes, in Norton, S. A., Lindberg, S. E. and Page, A. L. (Eds), *Acid Precipitation*, Vol. 4, *Soils, Aquatic Processes and Lake Acidification*, Springer-Verlag, Berlin, 207–276.

Charles, D., Smol, J. P. and Engstrom, D., 1994. Paleolimnological approaches to biological monitoring, in Loeb, S. L. and Spacie, A. (Eds), *Biological Monitoring of Aquatic Systems*, CRC Press, Boca Raton, FL, USA, 233–293.

Engstrom, D. and Swain E., 1986. The chemistry of lake sediments in time and space, *Hydrobiologia*, **143**, 37–44.

Engstrom, D. and Wright, H. E., 1984. Chemical stratigraphy of lake sediments as a record of environmental change, in Haworth, E. and Lund, J. W. G. (Eds), *Lake Sediments and Environmental History*, Leicester University Press, Leicester, UK, 1–67.

Flower, R. J., Monteith, D. T., Mackay, A. W., Chambers, J. M. and Appleby, P. G., 1995. The design and performance of a new box corer for collecting undisturbed samples of soft subaquatic sediments, *Journal of Paleolimnology* **14**, 101–111.

Flower, R. J., Juggins, S. and Battarbee, R.W., 1997. Matching diatom assemblages in lake sediment cores and modern surface sediment samples: the implications for lake conservation and restoration with special reference to acidified systems, *Hydrobiologia* **344**, 27–40.

Håkanson, L. and Jansson, M., 1983. *Principles of Lake Sedimentology*, Springer-Verlag, Berlin.

Haworth, E. and Lund, J. W. G. (Eds), 1984. *Lake Sediments and Environmental History*, Leicester University Press, Leicester, UK.

Huttunen, P. and Meriläinen, J., 1978. New freezing device providing large unmixed sediment samples from lakes, *Annales Botanici Fennici*, **15**, 128–130.

Kansanen, P., Jaakkola, T., Kulmala, T. and Suutarinen, R., 1991. Sedimentation and distribution of gamma-emitting radionuclides in bottom sediments of southern Lake Pijnne, Finland, after the Chernobyl accident, *Hydrobiologia*, **222**, 121–140.

MARC, 1985. Historical monitoring, MARC Reports No. 31, Monitoring and Assessment Research Centre, University of London, London, UK, 1–320.

Meriläinen, J., 1967. The diatom flora and the hydrogen-ion concentration of water, *Annales Botanici Fennici*, **4**, 51–57.

Meriläinen, J. J. and Hamina, V., 1993. Recent environmental history of a large, originally oligotrophic lake in Finland: A palaeolimnological study of chironomid remains, *Journal of Paleolimnology*, **9**, 129–140.

Merkt, J., 1971. Zuverlässige Auszählungen von Jahresschichten in Seesedimenten mit Hilfe von Gross-Dünnschliffen, *Archiv für Hydrobiologie* **69**, 145–154.

Norton, S., Bienert, R. W., Binford, M. W., and Kahl, J. S., 1992. Stratigraphy of total metals in PIRLA sediment cores, *Journal of Paleolimnology*, 7, 191–214.

Nygaard, G., 1956. Ancient and recent flora of diatoms and Chrysophyceae in Lake Gribsø, *Folia Limnologica Scandinavica*, **8**, 32–94.

Oldfield, F. and Appleby, P., 1984. Empirical testing of ^{210}Pb dating models for lake sediments, in Haworth, E. Y. and Lund, J. W. G. (Eds), *Lake Sediments and Environmental History*. Leicester University Press, Leicester, UK, 93–124.

Olsson, I. U., 1986. Radiometric dating, in Berglund, B. E. (Ed.), *Handbook of Holocene Palaeoecology and Palaeohydrology*, John Wiley & Sons, Chichester, 273–312.

Renberg, I., 1981. Improved methods of sampling, photographing and varve-counting of varved lake sediments, *Boreas* **10**, 255–258.

Renberg, I. and Hellberg, T., 1982. The pH history of lakes in south-western Sweden, as calculated from the subfossil diatom flora of the sediments, *Ambio*, **11**, 30–33.

Scherer, R., 1994. A new method for the determination of absolute abundance of diatoms and other silt-sized sedimentary particles, *Journal of Paleolimnology*, **12**, 171–179.

Schmidt, R., Höllerer, H. and Wallner, G., 1995. A vacuum sampler for subsampling freeze-dried laminated sediments with the application to *in situ* frozen varves of Mondsee, Austria, *Journal of Paleolimnology*, **14**, 93–96.

Simola, H., 1992. Structural elements in varved lake sediments, in Saarnisto, M. and Kahra, A. (Eds), INQUA Workshop on Laminated Sediments, Lammi, Finland, June 1990, Geological Survey of Finland, Helsinki, Finland, Special Paper **14**, 5–9.

Simola H., Huttunen, P. and Meriläinen, J., 1986. *Techniques for sediment freezing and treatment of frozen sediment samples*, Publications of Karelian Institute No. 79, University of Joensuu, Finland, 99–107.

Simola, H., Meriläinen, J., Sandman, O., Marttila, V., Karjalainen, H., Kukkonen, M., Julkunen-Tiitto, R. and Hakulinen, J., 1996: Palaeolimnological analyses as information source for large lake biomonitoring, in Simola, H., Viljanen, M., Slepukhina, T. and Murthy, R. (Eds), *Hydrobiologia*, **322** (Developments in Hydrobiology, No. 113), 283–292.

Smith, J., Appleby, P. G., Battarbee, R. W., Dearing, J. A., Flower, R., Haworth, E. Y., Oldfield, F. and O'Sullivan, P. E. (Eds), 1991. *Environmental History and Palaeolimnology*, Proceedings of the Vth International Symposium on Palaeolimnology, 1988, Cumbria, UK, *Hydrobiologia*, **214** (Developments in Hydrobiology No. 67), Kluwer, Dordrecht.

Stephenson, M., Klaverkamp, J., Motycka M., Baron C. and Schwartz, W., 1996. Coring artefacts and contaminant inventories in lake sediment, *Journal of Paleolimnology*, **15**, 99–106.

ter Braak, C. J. F., 1986. Canonical correspondence analysis: a new eigenvector technique for multivariate direct gradient analysis, *Ecology*, **67**, 1167–1179.

ter Braak, C. J. F. and van Dam, H., 1989. Inferring pH from diatoms: a comparison of old and new calibration methods, *Hydrobiologia*, **178**, 209–223.

von Gunten, H. R. and Moser, R. N., 1993. How reliable is the ^{210}Pb dating method? Old and new results from Switzerland, *Journal of Paleolimnology*, **9**, 161–178.

Whitmore, T., Brenner, M. and Schelske, C., 1996. Highly variable sediment distribution in shallow, wind-stressed lakes: a case for sediment-mapping surveys in paleolimnological studies, *Journal of Paleolimnology*, **15**, 207–221.

Wiederholm, T., 1980. Use of benthos in lake monitoring, *Journal of the Water Pollution Control Federation*, **52**, 537–547.

Wik, M. & Renberg, I., 1996. Environmental records of carbonaceous fly-ash particles from fossil fuel combustion, *Journal of Paleolimnology*, **15**, 193–206.

Chapter 3.2

Case Examples of Palaeolimnological Records of Lake Ecosystem Changes

HEIKKI SIMOLA

Hydrological and Limnological Aspects of Lake Monitoring
Edited by Pertti Heinonen, Giuliano Ziglio and André Van der Beken
©2000 John Wiley & Sons, Ltd, ISBN 0 471 89988 7

3.2.1   INTRODUCTION

This chapter presents some illustrative examples of sedimentary records of human impact on lakes. In all of these cases, significant past events are readily seen in the visual stratigraphy of the sediment. Four of the cases are from eastern Finland and one is from a former Finnish territory in NW Russia. The examples illuminate the importance and potential of the retrospective approach as a complement to lake monitoring surveys. Further examples have been given in the previous chapter (Chapter 3.1).

A common feature of all of the sediment sequences presented in this present chapter is that the sediments are at least partly varved or annually laminated. Sediment varves provide a detailed record of past changes, and in favourable cases even a direct calendar dating for these. *In situ* freezing of the sediment is often necessary to make full use of this information. The retaining of the seasonal pattern of sedimentation as varves implies a lack of bioturbation, which often reflects near-bottom oxygen deficiency. Lakes Ristijärvi and Heinälampi in Finland are examples of relatively deep, sheltered lakes in which this condition appears natural and the varves actually span thousands of years, while in the shallow Lake Varaslampi the onset of varve formation is related to recent eutrophication and subsequent anoxia.

3.2.2   LAKE RISTIJÄRVI IN VALTIMO: COMPLEX VARVES DETAIL LAND USE AND HISTORY

Lake Ristijärvi in Valtimo, North Karelia ($63°37'N$, $28°57'E$) is a 20 m deep lake surrounded by fine-grained waterlain silt and clay soils. For a pollen-analytical study of land use and cultivation history, the lake sediment was cored down to a 5.27 m depth in 1988. The entire sediment sequence proved to be regularly varved, with a total count of 5530 varves. The average sedimentation rate in this long sequence is thus slightly less than $1 \, mm \, yr^{-1}$. The earliest cereal pollen grain was encountered at a depth of 88 cm, or AD 1240. Evidence of continuous cultivation is present from AD 1500 onwards, with the period ca 1750–1950 being the most intensive agriculturally (Poutiainen *et al.*, 1994, Taavitsainen *et al.*, 1998).

Plate 1 shows the past century in the varve record of Lake Ristijärvi: the uppermost varve in this photograph represents 1987 (freeze-corer sample taken in March 1988), while AD 1890 is at 17 cm depth. Many of the varves in this sediment are rather complex; the diverse human activities around the lake have added plenty of structural details to the varves, resulting in fairly thick but variable annual deposits (see Simola, 1992a,b). In this type of sediment, one should look at the thin dark brown winter layers that appear the most consistent structural unit, and are regularly present in every varve – the more complicated the structure appears, then the thicker the yearly units. Many of the varves contain a light grey clay layer, presumably indicating ploughing of the near-shore arable fields. Several varves in the lower half of the sequence also contain thick green layers that consist of well-preserved planktonic diatoms (mainly *Aulacoseira* spp.). These diatom blooms show evidence of high inputs of dissolved silica into the lake, and their dwindling upcore may mark the end of slash-and-burn cultivation, which was extensively practised in this area until the early decades of this century. The overall decrease in varve thickness in the uppermost part of the core marks the gradual decline of agricultural activities in the area especially since the 1970s.

3.2.3 LAKE POHJALAMPI: STRATIGRAPHIC RECORD OF EUTROPHICATION AND ESCALATING INTERNAL LOADING

Lake Pohjalampi in Liperi, North Karelia (62°40′N, 29°33′E) is a small mesoeutrophic lake that over the past decades has been loaded by agricultural runoff, most notably by dairy farming effluents. The external loading has been effectively curbed, but internal phosphorus loading from the sediments still remains a problem. For several years, the lake has suffered from winter anoxia during the ice-cover period. The first massive blue-green algal bloom was observed in the lake in 1986, followed by similar blooms in 1990 and 1992. In 1993, an ecosystem restoration project was started on the lake by the mass removal of cyprinid fish, mainly roach and bream (Karjalainen and Leppä, 1995); the total removal of some 200 kg ha^{-1} of fish over five years appears to have favourably altered the food-web structure, while also decreasing the nutrient and chlorophyll levels of the water (Karjalainen *et al.*, 1999).

The escalating eutrophication development of Lake Pohjalampi is recorded in an *in situ* frozen sediment sample taken in 1995 Plate 2.

The olive-green algal gyttja of Lake Pohjalampi is less minerogenic than that of Lake Ristijärvi (see Plate 3.2.1), and thus the sediment is less compacted and the varved structures are more easily disturbed. Even though the lake is only 5.3 m deep, and despite of evident disturbances, the sediment still retains a laminated structure that provides at least a crude chronology for the past development. The alteration of black sulfide layers and greenish oxic layers in the sediment appears regular enough to be tentatively interpreted as the annual cycle of anoxic winters (black strata) and oxic summers. The 42 cm sample would thus include nearly 40 varves, or the period from the mid-1950s to the present, indicating a sediment accumulation rate of somewhat more than 1 cm yr^{-1}. The lower ones of these tentative varves appear horizontally continuous, but from the level of 32 cm upwards (late 1960s?) there is plenty of disturbance: one or a few successive varves are disrupted so that a crater-like gap appears, which is then filled by the subsequent layers. Most probably, these structures mark the escape of methane bubbles from the sediment. Methane is formed in organic sediments by anaerobic microbial decomposition, and methane ebullition is a major physical factor effecting the convection of sediment interstitial water with its dissolved nutrients into the water column, as previously observed by Ohle (1954). The ebullition craters grow progressively larger towards the top of the sample: according to the very tentative varve, dating the 12–15 cm level with heavy disturbance would represent the late 1980s, which tallies with the recorded history of algal blooming commencing at that time.

Other visually striking features in the sediment from Lake Pohjalampi are the vertically extended light-coloured blotches that are fairly abundant in the 15–30 cm interval (Plate 3.2.2). These appear to be chemical trace fossils of the conical dwelling tubes of large chironomid larvae that have been living in the sediment. Jónasson (1972) has described the structure of such tubes, and the formation of an oxic microenvironment (devoid of the sulfide colouration) around them as a consequence of the ventilation activity of the larva. The presence of these tube traces demonstrates that even large benthic animals are not necessarily disastrous for the preservation of sediment microstratigraphy. Quite similar traces are also present in the sediment obtained from Lake Varaslampi (Plate 3).

3.2.4 LAKE VARASLAMPI: URBAN LIMNOLOGY AND POLLUTION

Lake Varaslampi (62°36′N, 29°48′E) is a small pond within the urban area of the town of Joensuu, North Karelia. It is surrounded by a housing area, railway yard and roads. The lake is generally shallow, with the average depth being only ca 1 m, plus a rather small deeper area with a maximum depth of 3.5 m. Due to (nowadays diverted) sewage effluent pollution the lake became eutrophic and has suffered from oxygen depletion during most winters. Plate 3 shows an *in situ*-frozen sediment sample taken from the deepest part of the lake in 1986. The sediment above the 17 cm depth is variably laminated, with some conspicuous clay layers that obviously mark the effects of urban development and road-construction events on the lake's drainage area. By adopting very tentative varve counting, the 14–15 cm clay dates back to the late 1940s (post-war housing development?), while those at 10–11 cm date to the early 1960s (road construction?). As indicators of anoxia, black sulfide bands are present from the 12 cm level upwards, and the sulfide colouration appears almost continuous from 6 cm (ca 1968) upwards. New samples taken from the same site in February 1999 show that the deposition of black sapropel mud has continued up to the present day.

Plate 1   A varved sediment obtained from Lake Ristijärvi, Valtimo; the photographed sequence includes about 100 annual varves. The undisturbed, but structurally quite variable, varves record past ecosystem events in great detail.

Plate 2   Sediment sample taken from Lake Pohjalampi. Although the highly organic algal gyttja is easily disturbed, the stratigraphy provides a record of eutrophication and escalating internal loading. Crater-like structures disrupting the horizontal laminations indicate ebullition of methane bubbles from the sediment.

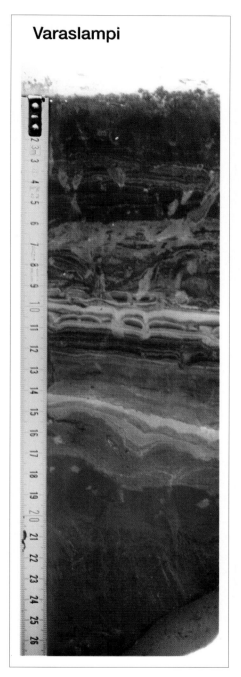

Plate 3 Sediment sample taken from Lake Varaslampi, Joensuu, which provides a history of urban development and pollution in the area.

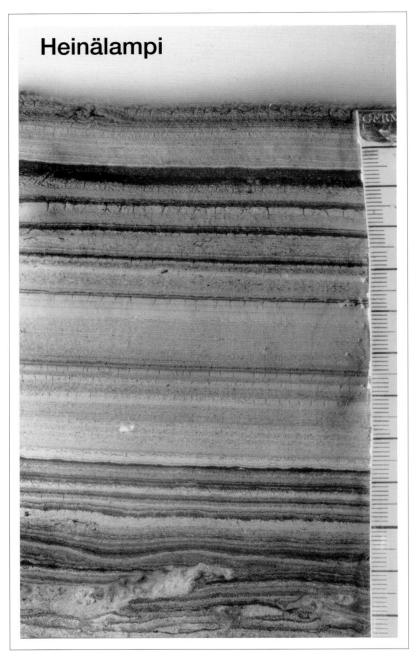

Plate 4 Sediment sample taken from Lake Heinälampi; the thick silt varves deposited during the 1980s provide evidence of great intensification of land use. In this case, forest drainage triggered massive erosion in the catchment area.

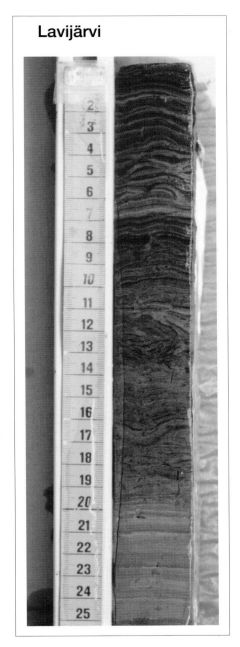

Plate 5 Sediment core sample obtained at a water depth of 22m from Lake Lavijärvi in ceded Karelia (Sortavala district, Karelian Republic, Russia). The sediment stratigraphy reveals the limnological consequences of dramatically weakened landuse in the post-war period.

Chemical and biological analyses of the sediment corroborate the visually striking record of pollution. Heavy metal concentrations (Zn, Cd, Cu, Pb, etc.) start increasing from the level of 14 cm upwards, reaching concentrations of 3–5 times the background levels in the uppermost sediment. Diatom stratigraphy gives evidence of eutrophication history of the lake, with *Aulacoseira distans* and *A. lirata* declining above 10 cm, *A. ambigua* and *A. subarctica* dominating the 8–3 cm sequence, and *Cyclotella pseudostelligera, Fragilaria construens* and *F. ulna* var. *acus* increasing in the uppermost 3 cm of sediment. A special feature is the occurrence of the very large heterotrophic diatom *Tryblionella plana* var. *fennica*, which in this sequence is clearly confined to the sulfide gyttja sequence only (Simola, 1990). There is also a profound change in the fossil chironomid stratigraphy: along with general upcore decrease in chironomid numbers, several species disappear completely and the large-sized *Chironomus anthracinus* gains dominance at the level of commencement of sulfide gyttja deposition. The stratigraphic record of the head-capsules of *C. anthracinus* corresponds rather well with the occurrence of the light-coloured blotches, similar to those in Lake Pohjalampi, which appear most abundant in the 11–3 cm interval (Häminen, 1997). Thus, it appears possible to interprete the blotches in this case as fossil dwelling tubes of this particular species.

3.2.5   LAKE HEINÄLAMPI: MASSIVE SILT EROSION FROM FIELD AND FOREST DRAINAGE

Lake Heinälampi in Siilinjärvi, North Savo (63°07′N, 27°39′E) is a small lake, 3.2 hr in area and 8 m deep, with a partly cultivated drainage area of 3.8 km². The drainage area is relatively shallow in relief and consists of fine-grained silty soils. Even though there are some small ponds in the upper tributaries, Lake Heinälampi serves as the first sedimentation basin for most of its catchment. Its sediment is fairly regularly varved down to 185 cm, spanning a period of some 2650 years, or its entire small-lake history since its isolation from the the large regressive Lake Kallavesi (Grönlund, 1991, 1995; Grönlund *et al.*, 1992).

The uppermost sediment of Heinälampi records in great detail the erosional history of the drainage area (Sandman *et al.*, 1990). From a 20 cm depth upwards, the varves contain conspicuous silt layers, reflecting intensified arable cultivation. Plate 4 shows a further abrupt increase in silt sedimentation that followed the ditching of some 50 hr of paludified forests which was undertaken in 1980. The two thickest silt varves, i.e. 8.7–6.5 and 6.5–4.8 cm, represent the years 1981–1982, with voluminous silt inflow continuing in the subsequent years – the pictured sample was taken in March 1988. The precise varve record of Lake Heinälampi, which was further confirmed by good correlation with ^{210}Pb and ^{137}Cs datings, made it possible to assess the sedimentary history of the lake in great detail on a yearly influx basis; (for details, see Sandman *et al.*, 1990). Overall, the sedimentary record of Lake Heinälampi is a representative, albeit rather extreme, example of the general intensification of land use throughout Finland over the past decades.

3.2.6 LAKE LAVIJÄRVI, SORTAVALA, RUSSIAN KARELIA: ECOSYSTEM RECOVERY FOLLOWING DEPOPULATION

About 40 000 km^2 of Finnish territory was ceded to Soviet Russia as a consequence of the 1939–1944 wars, and the entire Finnish population of some 400 000 was evacuated from the area. Landuse activities were slowly resumed after the war, and the present rural population in this area is now only about one third of the pre-war level. Weak or non-existent landuse during the post-war period in large parts of the ceded territory provides unique opportunities to study the ecosystem recovery following relaxation of human pressure.

Lake Lavijärvi (61°38′N, 30°30′E) in the vicinity of the town of Sortavala is the lowermost of three successive basins in a catchment area with abundant clayey soils, which were extensively used for arable cultivation before the war. Present-day meadows around the lake are former arable fields, which have been only used for cattle pasturing since the war. Forests in the area are mostly dense stands of mature mixed forest; logging activity has been minimal compared with other similar areas in Finland.

Plate 5 shows a sediment core from Lake Lavijärvi, taken at a 22 m water depth. The sediment stratigraphy reveals the limnological consequences of dramatically weakened landuse in the post-war period (Simola and Miettinen, 1997). An abrupt diminishing of mineral sedimentation is evident at a depth of 20 cm, seen as sediment colour and texture changes from silty clay to clay-gyttja, and obviously marking the cessation of arable field erosion during the war. Similar stratigraphy is observed in the two upstream lakes, Pitkjrvi and Kuokkalampi. These lakes seem to have been of the clay-turbid (argillotrophic) type, which is common in agricultural areas in southern Finland, but their water turbidity has considerably diminished since field ploughing ended.

The diatom stratigraphy of Lake Lavijärvi includes a marked floral change commencing at the level of sediment texture change (21 cm upwards): a species succession of about 10 yr duration transforms the sedimentary assemblage from a eutrophic one (*Aulacoseira ambigua* and epiphytic *Fragilaria*) towards one of mesotrophic character (*Cyclotella comta*, *Tabellaria flocculosa* var. *asterionelloides*, *Achnanthes minutissima*, etc.). In essential details, these changes are a mirror image of the eutrophication successions which have been observed, for example, in Lake Heinälampi and Lake Varaslampi. In the upper part of the Lake Lavijärvi core, there is again indication of some eutrophication, which has been assigned to increased cattle farming in the area from the late 1950s onwards. Of particular value for the interpretation of the ecosystem changes are the thorough limnological studies carried out in these lakes during the 1920s (Valle, 1927, 1928).

3.2.7 CONCLUSIONS

All of the example cases presented above deal with rather recent histories. For the practical aims of environment monitoring, this kind of short-core palaeolimnological investigation may indeed be sufficient, while the events of the past 200 years or so are often quite crucial for the assessment of the lake's present condition. However, one must bear in mind that the history of significant human impact in many lake ecosystems dates back much farther in time – this applies even for the present cases. For a proper understanding of an ecosystem state, it is often necessary to analyse long-term sedimentary records that definitely reach beyond the period of potential human interference. In particular, for ecosystem restoration projects it is important to establish the true pre-disturbance conditions.

Numerous palaeolimnological studies of the entire post-glacial history of lake ecosystems have demonstrated the significant effects of drainage basin evolution (pedological changes, paludification, effects of natural vegetation development, etc.) upon the limnological conditions in lakes. Such long-term studies may also reveal various natural or anthropogenic disturbance events (forest fires, land clearance etc.), which may help assess the sensitivity of the particular watershed–lake system to such disturbances. Pollen analysis, notably the identification of anthropogenic (pasturing and agricultural) indicators in the pollen flora, is an essential complement to the palaeolimnological analyses in interpreting whether human agency is involved in these instances.

REFERENCES

Grönlund, E., 1991. Sediment characteristics in relation to cultivation history in two varved lake sediments from east Finland, *Hydrobiologia*, **214**, 137–142.

Grönlund, E., 1995. A palaeoecological study of land-use history in East Finland, *University of Joensuu, Publications in Sciences*, **31**, 1–44 (plus appendices).

Grönlund, E., Kivinen, L. and Simola, H., 1992. Pollen-analytical evidence for Bronze-age cultivation in Eastern Finland, *Laborativ Arkeologi*, **6**, 37–42.

Hämäläinen, H., 1997. University of Joensu, unpublished observations.

Jónasson, P. M., 1972. Ecology and production of the profundal benthos, *Oikos Supplement*, **14**, 1–148.

Karjalainen, J. and Leppä, M., 1995. Liperin Pohjalammen ravintoketjukunnostus (Abstract: Biomanipulation research in Lake Pohjalampi), *Vesitalous*, **3**, 18–20, 39.

Karjalainen, J., Leppä, M., Rahkola, M. and Tolonen, K., 1999. The role of benthivorous and planktivorous fish in a mesotrophic lake ecosystem, *Hydrobiologia* (in press).

Ohle, W., 1954. Die Ursachen der rasanten Seeneutrophierung, *Verhandlungen der Internationalen Vereinigung für Limnologie*, **12**, 13–32.

Poutiainen, H., Grönlund, E., Koponen, M. and Kupiainen, R., 1994. Havaintoja Pohjois-Karjalan asutus- ja viljelyhistoriasta, *Kentltä poimittua*, **2**, 70–75.

Sandman, O., Liehu, A. and Simola, H., 1990. Drainage ditch erosion history as recorded in the varved sediment of a small lake in East Finland, *Journal of Paleolimnology*, **3**, 161–169.

Simola, H., 1990. 'Look at the big ones'. 11th International Symposium on Living and Fossil Diatoms, San Francisco, August 13–17, 1990, Abstracts, 106, California Academy of Sciences, San Franciso, CA, USA.

Simola, H., 1992a. Structural elements in varved lake sediments, in Saarnisto, M. and Kahra, A. (Eds), *INQUA Workshop on Laminated Sediments* (Lammi, Finland, June 1990), Geological Survey of Finland, Special Paper No. 14, 5–9.

Simola, H., 1992b. Seasonality and structure in sediment laminations, *Publications of Karelian Institute (University of Joensuu)*, **102**, 99–104.

Simola, H. and Miettinen, J., 1997. Mitä tapahtui kun väestö evakuoitiin? – luonnontilan palautumis – prosessi on tallentunut luovutetun Karjalan järvien sedimentteihin (Summary: Ecosystem recovery following depopulation: palaeolimnological records in former Finnish territory in Russian Karelia), *Geologi*, **49**, 86–89.

Taavitsainen, J.-P., Simola, H. and Groñlund, E., 1998. Cultivation history beyond the periphery – early agriculture in the north European boreal forest, *J. World Prehistory*, **12**, 199–253.

Valle, K. J., 1927. Ökologisch-limnologische Untersuchungen über die Boden- und Tiefenfauna in einigen Seen nördlich vom Ladoga: I, *Acta Zoologica Fennica*, **2**, 1–179.

Valle, K. J., 1928. Ökologisch-limnologische Untersuchungen über die Boden- und Tiefenfauna in einigen Seen nördlich vom Ladoga: II, *Acta Zoologica Fennica*, **4**, 1–231.

Chapter 3.3

Organochlorine Compounds in the Finnish Freshwaters

JAAKKO PAASIVIRTA

Hydrological and Limnological Aspects of Lake Monitoring
Edited by Pertti Heinonen, Giuliano Ziglio and André Van der Beken
©2000 John Wiley & Sons, Ltd, ISBN 0 471 89988 7

3.3.1   INTRODUCTION

Anthropogenic organochloro compounds (OCCs) in the freshwaters of Finland originate mainly from three type of sources, i.e. chlorination processes, especially pulp bleaching and water disinfection, wood preservation with chlorophenolics, including production of some preservants and incinerations for energy production, the metallurgy industry and waste treatments. During the last few decades, all of these OCC emissions have decreased greatly due to reduction in uses of chlorine chemicals and process improvements. Consequently, the relative role of atmospheric long-range transport of some semi-volatile and persistent OCCs has increased. More detailed information on these compounds and their environmental fate and ecotoxicity can be found in the book by Paasivirta (1991).

3.3.2   ORGANOCHLORINE COMPOUNDS FROM THE FOREST INDUSTRY

A great number of observations on the environmental fate of OCCs in Finnish freshwater ecosystems was carried out in studies of pulp-mill recipient watercourses, including background areas and some sites of wood preservation. OCCs were studied especially in sediments, fish and incubated mussels during the period 1978–1995. The main part (70–80%) of OCCs from pulp mills consists of persistent high molecular-weight, polydisperse 'chlorolignin' which has been deposited to recipient sediments. Analyses of total organically bound chlorine (TOCl) in dated sediment layers (Maatela *et al.*, 1990) show the history of chlorobleaching (Figure 3.3.1). Fifteen chlorophenolic compounds (chlorophenols, chlorocatechols, chloroguaiacols and a chlorovanillin) were commonly analysed in mill effluent. In recipient watercourses, chloroguaiacols in fish and mussels were the best indicators of exposure to bleaching mill effluent. The most common chlorophenols were non-specific to pulp mills, with combustion and especially wood preservation also leading to the production of such residues (Paasivirta *et al.*, 1998).

Free chlorophenolics occur in recipient sediments in much lower amounts than bound ones. Studies of mill samples showed that matrices containing chlorophenol were already formed in the bleaching process (Palm *et al.*, 1995). Annual depositions of specific groups of chlorophenolic compounds (Figure 3.3.2) analysed from recipient sediments are in good agreement with the history of bleaching.

In pulp-mill recipient watercourses, the depositions of OCCs to sediment layers and their concentrations in biota have decreased greatly due to process improvements. However, most of the mills which are currently using chlorine

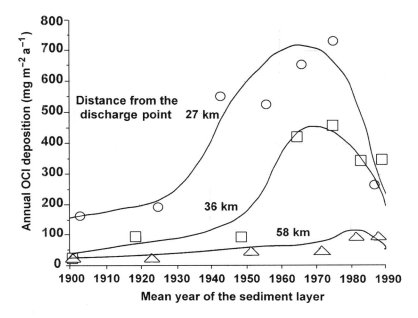

Figure 3.3.1 Annual deposition (sedimentation rate) of total organically bound chlorine (TOCl) in Jämsä watercourse over the period 1900–1990, where chlorobleaching of pulp had been stopped in 1981 (data of Maatela *et al.*, 1990).

dioxide as bleaching agent and secondary treatment, at present still discharge OCCs characteristic of mill effluents.

Chlorophenols are acute toxicants, and in particular some chloroguaiacols can strongly bioaccumulate. Their biomethylation products (chloroanisoles and chloroveratroles) are very effective tainting substances which have been shown to cause the bad taste of fish in bleaching pulp-mill recipient waters (Paasivirta *et al.*, 1990).

The major persistent chlorohydrocarbons from pulp bleaching were chlorinated cymenes, cymenenes alkyl bibenzyls, alkyl naphthalenes and alkyl phenanthrenes, which have all been found in mill samples, recipient sediments and biota (Koistinen *et al.*, 1992; Rantio and Paasivirta, 1996). Their levels indicated high bioaccumulation rates (Rantio *et al.*, 1996). Pulp chlorobleaching also caused formation of the notorious polychlorodibenzo-*p*-dioxins (PCDDs) and polychlorodibenzofurans (PCDFs), especially when an excess of elemental chlorine was used in the bleaching process (Rappe *et al.*, 1990). However, since 1985 the modern kraft mills in Finland have no longer emitted such amounts of the toxic polychlorodibenzo-*p*-dioxins, which could be detected in recipient sediment or biota (Paasivirta *et al.*, 1998). In recipient surface sediments of the Äänekoski watercourse, only two non-toxic isomers of

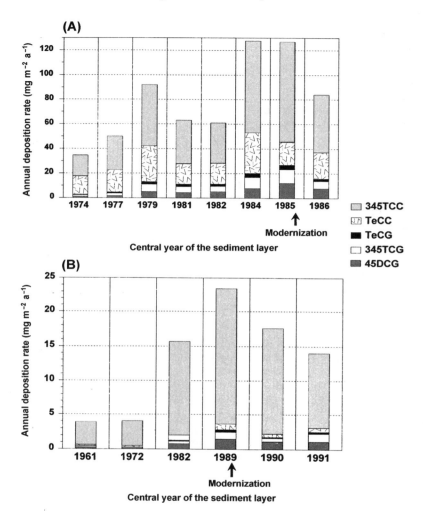

Figure 3.3.2 Annual deposition (sedimentation rate) of bound chlorophenolics in the sediments dowstream of two kraft mills, where modernization of the processes included increased bleached pulp production, extended cooking, substitution of elemental chlorine with chlorine dioxide and biological treatment (activated sludge) plant: (A) Äänekoski, central Finland, 2 km from the discharge point (data of Paasivirta *et al.*, 1990); (B) Pietarsaari (Bothnian Bay coast), 1 km from the discharge point (data of Palm *et al.*, 1995).

the tetrachlorodibenzo-*p*-dioxin were found, at levels lower than $2\,\mu g\,kg^{-1}$ of dry-weight sediment.

The bioaccumulation capacity of the persistent chlorohydrocarbons was estimated by comparison of their levels in effluent, spent biosludge, recipient fish and in mussels incubated in recipient watercourses. Chlororetenes (chlorinated methyl isopropyl phenanthrenes) appeared to have the highest accumulation potential (Rantio *et al.*, 1996).

Persistent chlorohydrocarbon emissions from Finnish pulp mills decreased rapidly after the pulp mills changed their processes (Rantio *et al.*, 1996). However, their traces in sediments are very persistent and are useful for identification of the impact of chlorobleaching and chlorodisinfection in freshwater lakes, as has been demonstrated in sediments of Lake Ladoga (Särkkä *et al.*, 1993).

3.3.3 WOOD-PRESERVATIVE WASTES

The chemical manufacture of tetrachlorophenolic wood preservatives in Kuusankoski during the period 1945–1984 has caused high levels of toxic OCCs to remain present in the Kymijoki river-bottom sediments. These sediments contain traces of the wood preservative and its impurities, especially polychlorophenoxyphenol congeners (predioxins), hexa-, hepta- and octa-chlorodibenzofurans, and polychlorodiphenylethers.

Some OCC pollutants may have caused the developmental abnormalities and population changes which have been observed in bottom biota of the River Kymijoki (Vuori and Parkko, 1996). Both of the chlorobleaching and wood-preservative industries produce OCCs that can redissolve from disturbed bottom solids and expose the biota. Despite non-detectable discharges since 1993, the Kymijoki river fish and incubated mussels still contain marked concentrations of these impurity compounds and biomethylated derivatives (Koistinen *et al.*, 1997).

Polychlorodiphenylethers in the River Kymijoki sediment and biota originate mainly from leakages of the earlier production. Some data are shown in Figure 3.3.3, where the current contents of these congeners in fishes are compared with the contents in the discharges from the wood-preservative industry (Koistinen *et al.*, 1995).

3.3.4 ORGANOCHLORINE COMPOUNDS FROM OTHER SOURCES

In Finland toxaphene congener residues were proven not to originate from pulp mills ((Rantio *et al.*, 1996), but from global atmospheric long-range transport, as are also the majority of hexachlorocyclohexane (especially α- and γ-isomers) and chlordane residues in Finnish biota (Paasivirta *et al.*, 1990, 1998). Hexachlorobenzene originates mainly from combustions and long-range transport. The bioaccumulation rate of hexachlorobenzene, however, is high. It

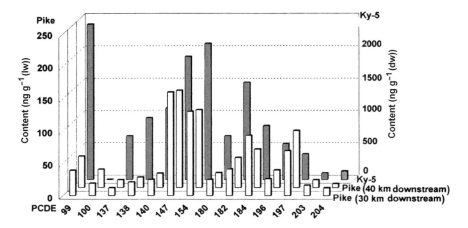

Figure 3.3.3   Contents of individual polychlorodiphenylether (PCDE) congeners found in the muscles of two samples of pike gaucht, compared with those found in a wood preservative sample (Ky-5), from the River Kymijoki, in spring 1993 (data of Koistinen *et al.*, 1995).

is possible that part of the bioaccumulated hexachlorobenzene could metabolize to octachlorodibenzodioxin, which explains the quite dominating presence of the latter compound in tissues of the background human population. Residues of the pesticide dichlorodiphenyltrichloroethane (DDT) (as its isomeric mixture) and technical mixtures of polychlorobiphenyls generally occur in Finnish freshwater biota at very low levels. Previous

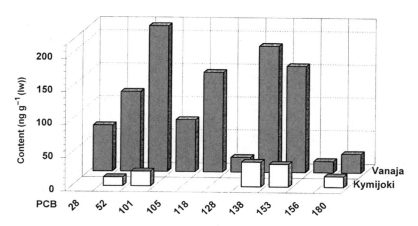

Figure 3.3.4   Contents of individual polychlorobiphenyl (PCB) congeners found in incubated musssels from the River Kymijoki and near Hämeenlinna in the Vanajavesi watercourse, in August 1995 (data of Koistinen *et al.*, 1997).

dumpings of extremely persistent polychlorobiphenyl (PCB) mixtures have caused some point-source types of freshwater ecosystem contaminations in Finland. The most serious one is still causing elevated PCB concentrations in fish and incubated mussels in the Vanajavesi watercourse in south Finland (Figure 3.3.4).

Waste combustion, energy production and traffic have been important thermal sources of dioxin emissions in industrialized countries (Paasivirta, 1991). More recent observations indicate that metallurgical processes are also essential sources of such emissions due to the revived catalytic (Cu, Fe- or Al-promoted) synthesis of chloroaromatics at temperatures between 300–700 °C (Fiedler, 1994). In Finland, aluminium reclamation processes were shown to emit significant amounts of chlorobenzenes, chlorophenols, PCBs, chloronaphthalenes, PCDDs, PCDFs and other polychlorinated aromatic compounds. These emissions could be controlled by process improvements and by filteration of the stack particles and gases (Aittola *et al.*, 1993, 1996).

REFERENCES

Aittola, J-P., Paasivirta, J. and Vattulainen, A., 1993. Measurements of organochloro compounds at a metal reclamation plant, *Chemosphere* **27**: 65–72.

Aittola, J-P., Paasivirta, J., Vattulainen, A., Sinkkonen, S. and Tarhanen, J., 1996. Formation of chloroaromatics at a metal reclamation plant and efficiency of stack filter in their removal from emission, *Chemosphere*, **32**: 99–108.

Fiedler, H., 1994. Sources of PCDD/PCDF and impact on the environment, *Organochloro Compounds* **20**, 229–236.

Koistinen, J., Nevalainen, T. and Tarhanen, J., 1992. Identification and level estimation of aromatic coeluates of polychlorinated dibenzo-*p*-dioxins and dibenzofurans in pulp mill products and wastes, *Envir. Sci. Technol.* **26**, 2499–2507.

Koistinen, J., Paasivirta, J., Suonperä, M. and Hyvärinen, H., 1995. Contamination of pike and sediment from the Kymijoki river by PCDEs, PCDDs, and PCDFs: Contents and patterns compared to pike and sediment from the Bothnian Bay and seals from Lake Saimaa, *Environ. Sci. Technol.*, **29**, 2541–2547.

Koistinen, J., Herve, S., Paukku, R., Lahtiperä, M. and Paasivirta, J., 1997. Chloroaromatic pollutants in mussels incubated in two watercourses polluted by industry, *Chemosphere*, **34**: 2553–2569.

Maatela, P., Paasivirta, J., Särkkä, J. and Paukku, R., 1990. Organic chlorine compounds in lake sediments. II. Organically bound chlorine, *Chemosphere*, **21**, 1343–1354.

Paasivirta, J., 1991. *Chemical Ecotoxicology*, Lewis, Chelsea MI, USA.

Paasivirta, J., Hakala, H., Knuutinen, J., Otollinen, T., Särkkä, J., Welling, L., Paukku, R. and Lammi, R., 1990. Organic chlorine compounds in lake sediments, III. Chlorohydrocarbons, free and chemically bound chlorophenols. *Chemosphere*, **21**, 1355–1370.

Paasivirta, J., Palm, H., Rantio, T., Koistinen, J., Maatela, P. and Lammi, R., 1998. Environmental fate of originated organochloro compounds from pulp mills, in

Turoski, V. (Ed.), *Chlorine and Chlorine Compounds in the Paper Industry.* Ann Arbor Press, Chelsea, MI, USA, Ch. 8, 105–117.

Palm, H., Paasivirta, J. and Lammi, R., 1995. Behavior of chlorinated phenolic compounds in bleach-plant, treatment-system and archipelago area, *Chemosphere* **31**, 2839–2852.

Rantio, T. and Paasivirta, J., 1996. Modeled and observed fate of chlorocymenes, *Chemosphere* **33**, 453–456.

Rantio, T., Koistinen, J. and Paasivirta, J., 1996. Bioaccumulation of pulp chlorobleaching-originated aromatic chlorohydrocarbons in recipient watercourses, in Servos, M. R., Munkittrick, K. R., Carey, J. H. and Van der Kraak, G. J. (Eds), *Environmental Fate and Effects of Pulp and Paper Mill Effluents.* St Lucie Press, Delray Beach, FL, USA, 341–345.

Rappe, C., Glas, B., and Wiberg, K., 1990. Solved and remaining PCDD and PCDF problems in the pulp industry, *Organohalogen Compounds*, **3**, 287–290.

Särkkä, J., Paasivirta, J., Häsänen, E., Koistinen, J., Manninen, P., Mäntykoski, K., Rantio, T. and Welling, L., 1993. Organic chlorine compounds in lake sediments. VI. Two bottom sites of Lake Ladoga near pulp mills. Chemosphere **26**, 2147–2160.

Vuori, K.-M. and Parkko, M., 1996. Assessing pollution of the River Kymijoki via hydropsychid caddis flies – population age structure, microdistribution and gill abnormalities in the *Cheumatopsyche lepida* and *Hydropsyche pellucidula* larvae, *Archiv für Hydrobiologie*, **136**, 171–190.

Chapter 3.4

Fate of Organic Xenobiotics in Sediments: Bioavailability and Toxicity

JUSSI V. K. KUKKONEN

Hydrological and Limnological Aspects of Lake Monitoring
Edited by Pertti Heinonen, Giuliano Ziglio and André Van der Beken
©2000 John Wiley & Sons, Ltd, ISBN 0 471 89988 7

3.4.1   INTRODUCTION

The bioaccumulation of compounds by organisms is the net exposure of the organism to a contaminant from various source compartments over time. This represents the balance between the flux into the organism and the loss through protective processes such as biotransformation and elimination. The importance of bioaccumulation is its direct link between the external contaminant concentrations in the sources and the potential effect of contaminants at various levels of biological structure and function. Bioaccumulated contaminants that attain sufficient concentration at a receptor site of a living organism for sufficient duration are responsible for exerting the pharmacological and/or toxicological effect of the compound on the same organism. Thus, the extent of bioaccumulation can be employed as a surrogate for the concentration at the receptor.

Accurate prediction and evaluation of xenobiotic contaminant exposure and accumulation from sediments remains difficult because of the complex interactions between the contaminant, the sediment, and the organism. These interactions depend on the following:

(a) chemical characteristics and concentration of the contaminant;
(b) physical and chemical characteristics of sediments;
(c) the presence of complex mixtures that can confound the contaminant interactions with both the sediment constituents and the biota;
(d) organism behaviour and physiology, influenced by such environmental factors as temperature, nutrient availability and habitat, which can modify the exposure both between species and temporally within a species;
(e) the length of sediment/contaminant contact time that can change bioavailability (Landrum *et al.*, 1996).

This present chapter will briefly review the accumulation of contaminants by aquatic organisms from sediments and the factors that influence the bioavailability and toxicity of organic compounds in such sediments.

3.4.2   PARTICLE–XENOBIOTIC INTERACTION AND BIOAVAILABILITY

Hydrophobic organic contaminants, such as polycyclic aromatic hydrocarbons and polychlorinated biphenyls, are bound by particles in the water and tend to accumulate in sediments. This binding to the particles is an important phenomenon and needs to be understood in order to evaluate the bioavailability of the compounds. The partitioning of organic xenobiotics between water and sediment or soil has been found to be linearly related to the organic content in both phases. Partitioning of organic contaminants between the two phases has also been correlated with the clay content of the sediment, where the highest partitioning has been reported to occur in the clay fraction. In most cases, the partition coefficient has been demonstrated to increase with the increasing hydrophobicity of the xenobiotics and the amount of organic matter in the sediment (Karickhoff *et al.* 1979; Schwarzenbach and Westall 1981). However, the role of the structural properties of the organic matter associated with the sediment particles has not been throughly evaluated. In the case of dissolved organic matter, changes in the structure can alter the observed partition coefficient, even on a carbon-normalized basis (Chiou *et al.*, 1986; Kukkonen and Oikari, 1991). In soil systems, some studies have demonstrated that partitioning varies with the soil organic matter composition. For instance, the organic-carbon-normalized partition coefficient decreases for a particular compound with increases in the soil organic matter polarity. The same effect is also observed for organic material extracted from sediments.

Although basic relationships between particle size and xenobiotic sorption have been established in laboratory studies (Barber *et al.*, 1992), the relationship between sorption, particle size, and bioavailability is not as clear. One reason for this is that most of the sorption studies using different particle-size fractions have used particles < 63 µm as the smallest size fraction. This fractionation is too coarse to relate the particle-size distribution to the bioavailability of compounds to benthic organisms.

In addition to the size, the physical-chemical characteristics of inorganic particles dictates the adsorption of natural organic material as well as xenobiotics on to the particles. A sediment particle can be viewed as an inorganic base that contains one or more minerals and which is coated with the natural organic molecules, for example, humic substances (Figure 3.4.1). Each of the mineral types has a characteristic surface charge (Sposito, 1992), and this charge influences the nature and extent of the interaction between natural dissolved organic matter and mineral surfaces (Tipping, 1981; Davis, 1982). Finally, this particle-associated organic matter largely controls the sorption of non-polar organic xenobiotics on to the sediment particles.

If different types of natural organic matter are bound by particles of differing size and composition, it is possible that the differing organic matter would have different affinities for different classes of xenobiotics. This, in fact, has been observed with dissolved organic matter in natural waters (Kukkonen *et al.*, 1990; Kukkonen and Oikari, 1991).

The differential bioavailability seen among sediments possessing the same organic carbon content but obtained from different sources (DeWitt *et al.*, 1992; Suedel *et al.*, 1993), points heavily to the influence of the sediment composition on exposure. Not only does the partitioning apparently vary with the organic carbon content and composition, but compounds of different chemical classes apparently bind to different portions of the organic matter.

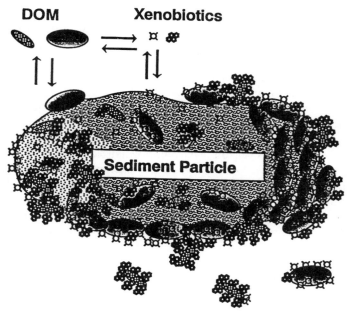

Figure 3.4.1 A conceptual model of a sediment particle where different 'molecules' of dissolved organic matter (DOM) form a coating on the inorganic particle with the subsequent binding of organic xenobiotics, mainly to the organic coating (modified from Kukkonen and Landrum, 1996).

Some studies also suggest that bacterial flora on the particles may affect the binding of the xenobiotics (Rao *et al.*, 1993), and this may also affect the bioavailability of the compounds, especially in the case of selective feeders.

3.4.3 BIOAVAILABILITY AND TOXICITY IN SEDIMENT EXPOSURES

Laboratory-contaminated sediment samples have been widely used to determine and to estimate the environmental fate, bioavailability and possible biological effects of sediment-associated pollutants (Heim *et al.*, 1994). The time gap between dosing the sediment and performing the bioassay varies normally from hours to a few months at a maximum. However, the bioavailability of some sediment-associated compounds has been observed to decrease with increased contact time between the sediment and the xenobiotic. For example, compounds such as fluorene, phenanthrene, and pyrene were more available to organisms in dosed sediments aged for less than one week than in those dosed and aged for 60 to 150 days (Harkey *et al.*, 1994). Even though reductions in accumulation have been observed, the potential impact of this process has not generally been recognized and taken account of in testing procedures.

The bioavailability of some compounds has also been studied in sediment cores taken from the field. The results obtained, however, are somewhat confusing. Ferraro *et al.*, (1990) reported a significant increase in calculated accumulation factors for benzo(*a*)pyrene as a measure of the bioavailability in surficial sediments (0 to 2 cm layer, i.e. recently contaminated sediment) versus material taken at 4–8 cm depths from the same sediment core. However, little change in the inaccumulation factor was seen for pyrene and chrysene. On the other hand, Harkey *et al.*, (1995) reported that the highest bioavailability of polyaromatic hydrocarbons (PAHs) was measured either at 4–8 cm or at 12–16 cm depths, but not at the surface layer (0–4 cm depths). These results could be explained by compositional differences of the natural organic matter associated with the particles among the sediment depths, but it is certain that more experimental data are needed in order to accurately explain the effect of contact time on the bioavailability of contaminants in the sediments.

Sorption was a strong function of organic carbon content when toxicity was determined for a mixture of chlorinated ethers in various sediments (Meyer *et al.*, 1993). In this case, the toxicity to *Hyalella azteca*, *Chironomus tentans*, and *Daphnia magna* depended on the sediment organic carbon. However, the interstitial water concentrations did not correlate with the observed effects for the two sediment-dwelling species, *C. tentans* and *H. azteca*. The bioavailability of fluoranthene to these same species was not similar in three laboratory-dosed sediments possessing the same physical and chemical characteristics (i.e. similar organic carbon, particle size fractions, etc.) because significant differences in species response were observed (Suedel *et al.*, 1993).

In a study where several different sediments were used, the bioavailability of polychlorinated biphenyls (PCBs), measured as the uptake clearance (amount of source compartment cleared of contaminant per mass of organism per hour), was controlled by the amount of organic carbon (Landrum *et al.*, 1997). However, for PAHs, bioavailability was better controlled by the polarity of the organic carbon.

In bioaccumulation studies, there was a tenfold difference in accumulated body residues when three benthic species (*Lumbriculus variegatus*, *C. riparius* and *Sphaerium corneum*) were exposed in the same contaminated sediment (Penttinen *et al.*, 1996). This implies that the behaviour of the organisms and their feeding habits affect the bioaccumulation of organic compounds in sediments. Similarly, the variation in bioavailability for three sediment-ingesting species was sevenfold when based on accumulation values normalized to contaminant concentration and organic carbon (Harkey, 1993). These reports suggest that factors other than total carbon in sediments and lipid content of organisms are responsible for the wide range of contaminant accumulation among sediment-dwelling species.

Epifaunal and infaunal sediment dwellers represent a myriad of feeding behaviours and life histories. Each of these behaviours can affect the relative contaminant exposure via manipulation of the environment surrounding the organisms. For example, infaunal oligochaetes burrow through sediment and obtain food from ingested sediment particles. These organisms, appropriately named 'conveyor belt deposit-feeders', ingest sediment over a range of depths, while they deposit gut contents on the sediment surface from posterior ends that protrude at the sediment–water interface (Robbins, 1986). This behavioural aspect has been used in advance when studying the feeding rate and effect of feeding on the bioaccumulation of compounds into the oligochaetes. Bioturbation produced by these organisms disrupts any equilibrium established among sediment-associated contaminants, thus affecting bioavailability not only to the oligochaetes, but to all biota in the reworking zone. This behaviour can also redistribute contaminants from buried deposits back into the feeding zone for shallower-feeding organisms (Keilty *et al.*, 1988). Both the enhancement and reduction in bioaccumulation that occurs with dose are thought to be strongly related to the feeding rate for sediment-associated contaminants, because an enhanced feeding rate can result in increased accumulation (Harkey *et al.*, 1994).

Selective feeding behaviour is one of a number of biological characteristics affecting contaminant availability. In amphipods, this selectivity is suggested as a major reason for the differential accumulation of chlorinated hydrocarbons as opposed to polycyclic aromatic hydrocarbons from sediments (Landrum, 1989). When the relative distribution among sediment particles is large, such selectivity is readily assumed. However, when the chemically measured differences in distribution are small, the picture is much more uncertain.

Once ingested by an organism, uptake of sorbed contaminants can be modified by gut processes that cause contaminant fugacity and concentration to increase as the volume of food decreases and lipids are hydrolysed (Lee, 1991, Gobas *et al.*, 1993). Weston and Mayer (1998) have shown that the digestive fluid from the gut of benthic organisms efficiently extracts PAHs from sediment particles.

3.4.4   MEASURING TOXICITY IN SEDIMENTS

Evaluating the effect of contaminants on various levels of the aquatic food chain has traditionally used concentrations in the external environment. When mixtures of chemicals or multiple sources are involved, and the exposure becomes complicated due to significant bioavailability limitations such as exposures in sediments, then assessing effects based on the external environment may not be very predictable. Rather, there is a body of knowledge that is developing to evaluate the effect of chemicals based on the internal concentration in organisms (McCarty and Mackay, 1993; Fitzgerald *et al.*, 1996, 1997). This is analogous to utilizing blood levels in mammals to predict drug effects and behaviour. The complication in using this approach is that most researchers who have measured toxicity have not measured the internal dose to the organisms.

The range of concentrations that produce effects (mortality or other waste response) vary with both the mechanism of action and the duration of exposure. The use of internal concentrations and the resultant effects is now beginning to develop, however, and the utility of this approach is obvious. When organisms are exposed to multiple sources, where none are dominant or where simple equilibrium models do not effectively reflect the concentration, then the prediction of body burdens through kinetic models and assessing effects based on internal dose should provide better estimates of environmental hazard.

REFERENCES

Barber, L. B., Thurman, E. M. and Runnells, D. D., 1992. Geochemical heterogeneity in a sand and gravel aquifer: Effects of sediment mineralogy and particle-size on the sorption of chlorobenzenes, *J. Contam. Hydrology*, **9**, 35–54.

Chiou, C. T., Malcolm, R. L., Brinton, T. I. and Kile, D. E., 1986. Water solubility enhancement of some organic pollutants and pesticides by dissolved humic and fulvic acids, *Environ. Sci. Technol.*, **20**, 502–508.

Davis, J. A., 1982. Adsorption of natural dissolved orgnic matter at the oxide/water interface, *Geochim. Cosmochim. Acta*, **46**, 2381–2393.

DeWitt, T. H., Ozretich, R. J., Swartz, R. C., Lamberson, J. O., Schults, D. W., Ditsworth, G. R., Jones, J. K. P., Hoselton, L. and Smith, L. M., 1992. The influence of organic matter quality on the toxicity and partitioning of sediment-associated fluoranthene, *Environ. Toxicol. Chem.*, **11**, 197–208.

Ferraro, S. P., Lee, H., Ozretich, R. J. and Specht, D. T., 1990. Predicting bioaccumulation potential: A test of a fugasity-based model, *Arch. Environ. Contam. Toxicol.*, **19**, 386–394.

Fitzgerald, D. G., Warner, K. A., Lanno, R. P. and Dixon, D. G., 1996. Assessing the effects of modifying factors on pentachlorophenol toxicity to earthworms: applications of body residues, *Environ. Toxicol. Chem.* **15**, 2299–2304.

Fitzgerald, D. G., Lanno, R. P., Klee, U., Farwell, A. and Dixon, D. G., 1997. Critical body residues (CBRs): Applications in the assessment of pentachlorophenol toxicity to *Eisenia fetida* in artificial soil, *Soil Biol. Biochem.*, **29**, 685–688.

Gobas, F. A. P. C., McCorquodale, J. R. and Haffner, G. D., 1993. Intestinal absorption and biomagnification of organochlorines. Environ, *Toxicol. Chem.* **12**, 567–576.

Harkey, G. A., 1993. Investigation of the Bioavailability of Sediment-Associated Hydrophobic Organic Contaminants via Laboratory Bioassays, PhD. Thesis, Clemson University, USA.

Harkey, G. A., Landrum, P. F. and Klaine, S. J. 1994. Comparison of whole sediment, elutriate and pore-water exposures for use in assessing sediment-associated organic contaminants in bioassays, *Environ. Toxicol. Chem.* **13**, 1315–1329.

Harkey, G. A., Van Hoof, P. L. and Landrum, P. F., 1995. Bioavailability of polycyclic aromatic hydrocarbons from a historically contaminated sediment core, *Environ. Toxicol. Chem.*, **14**, 1551–1560.

Heim, K., Schuphan, I. and Schmidt, B., 1994. Behavior of [^{14}C]-4-nitrophenol and [^{14}C]-3,4-dichloroaniline in laboratory sediment-water systems. 1. Metabolic fate and partitioning of radioactivity, *Environ.Toxicol.Chem.*, **13**, 879–888.

Karickhoff, S. W., Brown, D. S. and Scott, T. A., 1979. Sorption of hydrophobic pollutants on natural sediments, *Water Res.*, **13**, 241–248.

Keilty, T. J., White, D. S. and Landrum, P. F., 1988. Sublethal responses to endrin in sediment by *Lumbriculus hoffmeisteri* (Tubificidae) and in mixed culture with *Stylodrilius heringianus* (Lumbriculidae), *Aquat. Toxicol.*, **13**, 227–250.

Kukkonen, J. and Landrum, P. F., 1996. Distribution of organic carbon and organic xenobiotics among different particle size fractions in sediments, *Chemosphere* **32**, 1063–1076.

Kukkonen, J. and Oikari, A., 1991. Bioavailability of organic pollutants in boreal waters with varying levels of dissolved organic material, *Water Research*, **25**, 455–463.

Kukkonen, J., McCarthy, J. F. and Oikari, A., 1990. Effects of XAD-8 fractions of dissolved organic carbon on the sorption and bioavailability of organic micropollutants, *Arch. Environ. Contam. Toxicol.*, **19**, 551–557.

Landrum, P. F., 1989. Bioavailability and toxicokinetics of polycyclic aromatic hydrocarbons sorbed to sediments for the amphipod, *Pontoporeia hoyi.*, *Environ. Sci. Tech.*, **23**, 588–595.

Landrum, P. F., Harkey, G. A. and Kukkonen, J., 1996. Evaluation of organic contaminants exposure to aquatic organisms: The significance of bioconcentration and bioaccumulation, in Newman, M. C. and Jagoe, C. H. (Eds), *Ecotoxicology: A Hierarchical Treatment*. Lewis Publishers, Ann Arbor, MI, USA, 85–131.

Landrum, P. F., Gossiaux, D. C. and Kukkonen, J., 1997. Sediment characteristics influencing the bioavailability of nonpolar organic contaminants to *Diporeia* spp., *Chem. Speciat. Bioavail.*, **9**, 43–55.

Lee, H., 1991. A clam's eye view of the bioavailability of sediment-associated pollutants, in, Baker, R. A. (Ed.), *Organic Substances and Sediments in Water: Volume 3 Biological*. Lewis Publishers, Ann Arbor, MI, USA, 73–93.

McCarty, L. S. and Mackay, D., 1993. Enhancing ecotoxicological modeling and assessment, *Environ. Sci. Technol.*, **27**, 1719–1728.

Meyer, C. L., Suedel, B. C., Rodgers, J. H. and Dorn, P. B., 1993. Bioavailability of sediment-sorbed chlorinated ethers, *Environ. Toxicol. Chem.*, **12**, 493–505.

Penttinen, O.-P., Kukkonen, J. and Pellinen, J. 1996. Preliminary study to compare body residues and sublethal energetic responses in benthic invertebrates exposed to sediment-bound 2,4,5-trichlorophenol, *Environ. Toxicol. Chem.*, **15**, 160–166.

Penttinen, O.-P. and Kukkonen, J. 1998: Chemical stress and metabolic rate in aquatic invertebrates: Threshold, dose–response relatioships and mode of toxic action, *Environ. Toxicol. Chem.*, **17**, 883–890.

Rao, S. S., Dutka, B. J. and Taylor, C. M., 1993. Ecotoxicological implications of fluvial suspended particulates, in Rao, S. S. (Ed.), *Particulate Matter and Aquatic Contaminants*. Lewis Publishers, Boca Raton, FL, USA, 157–167.

Robbins, J. A., 1986. A model for particle-selective transport of tracers in sediments with conveyor belt deposit feeders, *J. Geophy. Res.*, **91**, 8542–8558.

Schwarzenbach, R. P. and Westall, J., 1981. Transport of nonpolar organic compounds from surface water to ground water. Laboratory sorption studies, *Environ. Sci. Technol.*, **15**, 1360–1367.

Sposito, G., 1992. Characterization of particle surface charge, in Buffle, J. and van Leeuwen, H. P. (Eds), *Environmental Particles*, Vol. 1. Lewis Publishers, Chelsea, MI, USA, 291–314.

Suedel, B. C., Rodgers, J. H. and Clifford, P. A., 1993. Bioavailability of fluoranthene in freshwater sediment toxicity tests, *Environ. Toxicol. Chem.*, **12**, 155–165.

Tipping, E., 1981. The adsorption of aquatic humic substances by iron oxides, *Geochim. Cosmochim. Acta*, **45**, 191–199.

Weston, D. P. and Mayer, L. M., 1998. *In vitro* digestive fluid extraction as a measure of the bioavailability of sediment-associated polycyclic aromatic hydrocarbons: Sources of variation and implications for partitioning models, *Environ. Toxicol. Chem.*, **17**, 820–829.

Chapter 3.5
Mercury: A Challenge for Lake Monitoring

MATTI VERTA

Hydrological and Limnological Aspects of Lake Monitoring
Edited by Pertti Heinonen, Giuliano Ziglio and André Van der Beken
©2000 John Wiley & Sons, Ltd, ISBN 0 471 89988 7

3.5.1 INTRODUCTION

The varied chemical species and forms of mercury promote its transport and cycling in the environment between bedrock, soil, water and air. Many sources – both natural and anthropogenic – contribute to the global mercury cycle, while several factors modify the leaching, evaporation and deposition, and the subsequent transformation, distribution, and bioaccumulation of mercury. Mercury sources and modifying factors vary extensively across regions, continents, and hemispheres, thus complicating the assessment of mercury's role as a global pollutant. From a biogeochemical perspective, researchers and decision makers need to know if the global pattern of mercury cycling has changed or is changing due to human activities. From a regional or site-specific perspective, monitoring will give detailed spatial and temporal information on mercury concentrations (distribution) in the environment, which may help to quantify the importance of local, regional or global mercury sources, assess the effect of environmental characteristics on the mercury cycle, estimate the anthropogenic effect of mercury releases on wildlife and human mercury exposure, etc.

Human exposure to methylmercury (MeHg) through the consumption of freshwater and marine fish is the principal public health concern with mercury in the environment. Elevated MeHg concentrations in fish are frequently found not only in heavily contaminated areas, but also in terrestrial waters distant from point sources in Scandinavia, Canada and the USA, and even in the Arctic environment. It has become evident that atmospheric, terrestrial and aquatic cycling of Hg, the synthesis of MeHg (methylation), and the bioaccumulation of MeHg is driven by complex chemical and biological reactions involving exceedingly small quantities of Hg at the bottom of aquatic food chains, but also leading to harmfully high concentrations at the top of some more vulnerable food chains. Accordingly, environmental investigations on mercury require highly sophisticated analytical approaches, as well as a detailed understanding of the biochemical and biological processes, and the ecosystem dynamics.

3.5.2 GLOBAL MERCURY CYCLE

Mercury has gaseous, aqueous, and particulate phases, as well as different chemical forms that strongly affect its transport and cycling in the environment, thus allowing it to cycle between water, soil and air. Because of its gaseous phase, mercury is widely distributed around the globe. Mercury is emitted through human activities and from natural sources such as volcanic eruptions and degassing or vaporization and re-emittence (Fitzgerald *et al.*,

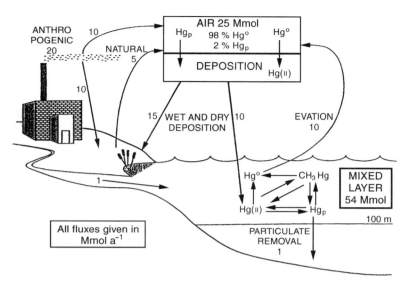

Figure 3.5.1 Current mercury budgets (Redrawn from Mason *et al.*, 1994).

1998). A recent review of mercury contamination arising from natural and anthropogenic sources indicates, despite uncertainties, that across large regions of the world, anthropogenic emissions have increased relative to natural sources since the onset of the industrial period (Fitzgerald *et al.*, 1998). Present estimates of anthropogenic emissions range between 50 and 75% of the yearly input to the atmosphere (Wilken *et al.*, 1996). Figure 3.5.1 shows a simplified conceptualization of the current global mercury cycle (Mason *et al.*, 1994). Important fluxes include anthropogenic and natural emissions, marine and terrestrial wet deposition and evasion from the oceans.

Knowledge of where mercury settles in the environment is incomplete; its source attribution, and the deposited mercury's origin, has proven difficult to quantify. Moreover, a lack of reliable data about the speciation of mercury in source emissions contributes to assessment difficulties. Although a plausible relationship between emissions from industrial sources and deposition exists, there is a need for more quantitative data about the amounts of mercury that are locally or regionally deposited or globally dispersed.

Point-source atmospheric emissions of mercury for 1994–1995 have been estimated by the US Environmental Protection Agency (EPA, 1998), which indicates total emissions of about 158 tons of mercury per year in the USA. A European estimate for 1990 was from 600 to 700 tons per year (Pacyna, 1996; Pirrone *et al.*, 1996), but the amount has declined considerably during the 1990s. The main source of emission is energy production, with the distribution between the different quantified sources being shown in Table 3.5.1.

Table 3.5.1   Distribution of atmospheric mercury emissions between different sources.

Source	Total $(t\ a^{-1})$	Proportion (%)
USA (1995–98)[a]	158	
Coal-fired boilers		45
Municipal-waste combusters		19
Medical-waste incinerators		10
Oil-fired boilers		4.8
Hazardous-waste combusters		4.5
Chlor-alkali manufacturing process		4.5
Area sources		2.2
Geothermal power		0.9
Other sources		8.5
Western Europe (1992)[b]	352	
Coal combustion		27
Solid-waste incineration		28
Oil combustion		11
Pyrometallurgical processes		21
Miscellaneous		14
Eastern Europe (1992)[b]	282	
Coal combustion		41
Solid waste incineration		15
Oil combustion		11
Pyrometallurgical processes		20
Miscellaneous		14

[a] EPA, 1998.
[b] Pirrone *et al.*, 1996.

3.5.3   PROCESSES OF MERCURY CYCLES IN LAKES

The complete mercury cycle has not yet been unravelled. There may still be unidentified processes which are significant, while some processes known to occur still remain to be quantified. A simplistic model to describe the major processes of the mercury cycle in lakes has been presented, e.g. by Harris *et al.*, 1996 (Figure 3.5.2). These processes include inflows and outflows (surface and groundwater), adsorption/desorption, particulate settling, resuspension and burial, atmospheric deposition, air/water gaseous exchange, industrial mercury sources, *in situ* transformations (e.g. methylation, demethylation, Hg(II) reduction, etc.), mercury kinetics in plankton, and bioenergetics related to methylmercury fluxes in fish. There remains a need to assess the importance of processes involved in mercury cycling in various types of aquatic systems, for example drainage and seepage lakes, reservoirs, and water bodies with direct mercury effluents.

Until the late 1980s, studies on the mercury cycle were hampered by an inability to measure concentrations of total mercury and methylmercury in

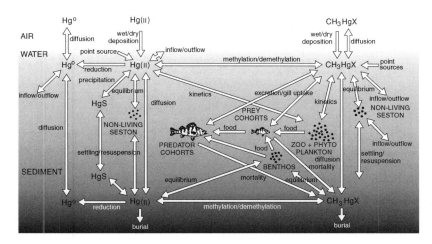

Figure 3.5.2   Schematic representation of the mercury cycle in lakes (from Harris *et al.*, 1996). Reproduced by permission of the author.

water, and by an inability to measure (natural) bacterial methylation rates in aquatic systems. Much progress have been made in recent years in terms of measuring different mercury species in the environment (Table 3.5.2), and eventually more and more studies on net methylation rates at natural level are in progress.

Table 3.5.2   General forms of mercury in the environment (Harris, 1991).

Form	Comments
Elemental Hg ($Hg°$)	This reduced mercury form is volatile and neutral; dominant form in air
Non-methylated Hg (II)	This category includes all non-methylated complexes of Hg^{++} in solution and adsorbed on solids; solid phase HgS precipitate is excluded; complexing and adsorption sites may be organic or inorganic; dominant Hg category found in nature
Monomethyl Hg (CH_3HgX)	This category includes dissolved complexes involving CH_3Hg^+, and CH_3Hg^+ bound to solids; complexing and adsorption sites may be organic or inorganic; dominant form in fish
Dimethyl Hg (($CH_3)_2Hg$)	This form is volatile and unstable in air; may be a product of biological methylation at high pH or degradation of methylmercury in anaerobic conditions
Solid mercuric sulfide	Solid precipitate (HgS) in the presence of sufficient free sulfide and Hg^{++}

(a) In water column and sediments

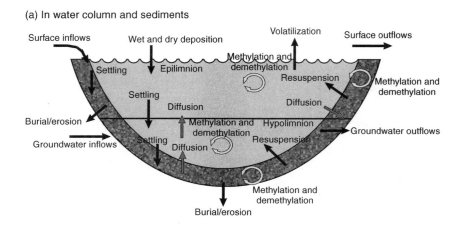

(b) In food chain

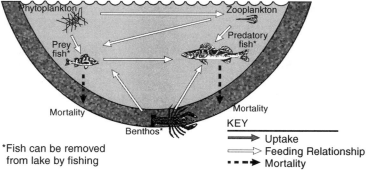

*Fish can be removed
from lake by fishing

Figure 3.5.3 Schematic representations of methylmercury cycle in lakes: (a) in the water column and sediments; (b) in the food chain (from Harris *et al.*, 1996). Reproduced by permission of the author.

3.5.4   METHYLMERCURY CYCLE IN AQUATIC SYSTEMS

When fish mercury levels are of interest, there is a need to consider MeHg cycling (Figure 3.5.3), since MeHg is the dominant mercury form in fish. The production of MeHg and its decomposition (demethylation) in the water column, sediments, catchment soil and wetlands is thought to be primarily a biological process, although chemical methylation by organics and chemical demethylation by sunlight have both been observed. In general terms, Hg(II) is methylated to monomethyl mercury by a variety of bacteria, while other bacteria demethylate mercury first to Hg(II) as an intermediate and finally to Hg^0 vapour. The focus of most studies have been on methylation. However, the

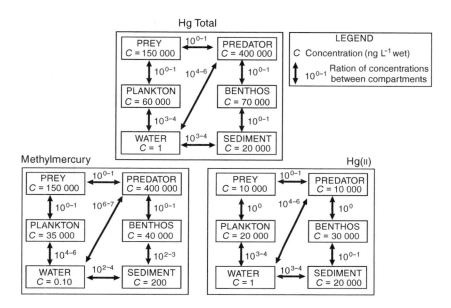

Figure 3.5.4 Generic estimates of mercury partitioning in freshwater systems (from Harris, 1991). Reproduced by permission of the author.

dominance of inorganic Hg in surface waters indicates that natural levels of MeHg are rather more controlled by demethylation, which seems to be a more efficient and variable process than methylation (Meili, 1997).

Other MeHg fluxes include watershed runoff and outflows from the waterbody, adsorption on solids in the water column and sediments, settling via suspended solids, air/water surface exchange, groundwater exchange, diffusion between porewater and the water column, wet and dry deposition from the atmosphere, uptake and excretion by biota, and sediment burial. Transformations of methylmercury to dimethylmercury, and vice versa, are also possible.

Figure 3.5.4 indicates that MeHg concentrations in fish tend to be 6–7 orders of magnitude greater than in their surrounding waters, but only 0–2 orders of magnitude higher than the MeHg concentrations in their diet. The greatest increase in mercury concentration between compartments tends to occur at the base of the food chain (eg. between plankton and water), particularly for MeHg. When considering the potential for bioaccumulation of MeHg in biota, it is important to know whether the mercury is taken in primarily from the water pathway (gills) or via the diet (biomagnification). In general, the higher in the food chain the organism is (eg. predatory fish), then the higher is the share of MeHg ingested through the diet. Uptake from food usually dominates even in primary consumers (Meili, 1997).

3.5.5   ENVIRONMENT CHANGE/METHYLMERCURY CONCENTRATION IN FISH

Many water quality and trophic variables have been considered to explain mercury cycling and concentrations in fish. These include pH, carbon, water-temperature bacterial activity, colour, oxygen, redox potential, sulfide, selenium, alkalinity, hardness, conductivity, lake productivity, and ratios of watershed area to lake area or volume. A selection of variables which may affect mercury cycling are presented in the following.

(a) pH associations with:

- watershed and atmospheric loading of Hg(II) and/or MeHg;
- available mercury for methylation (H^+ competition, dissolved organic carbon (DOC) aggregation/settling, etc.);
- cobalt availability of methyl-cobalamin-mediated methylation;
- biota biomass, metabolism, and diet;
- calcium effects on fish mercury uptake;
- microbial activity;
- hydrolysis of dimethyl mercury to monomethyl mercury.

(b) Carbon associations with:

- watershed loading of Hg(II) and MeHg;
- available mercury for methylation;
- microbial activity;
- biota biomass, metabolism and diet;
- water column reduction and volatilization of mercury.

(c) Productivity, dissolved oxygen and E_h associations with:

- microbial activity of methylators and demethylators;
- available mercury for methylation;
- biomass.

(d) Temperature associations with:

- microbial activity of methylators and demethylators.

(e) Microbial adaptation to elevated mercury concentrations:

  • microbial balance of methylators and demethylators.

Of special interest is pH (acidification), as increased acidity has often been associated with fish mercury content. Several thorough reviews of pH implications regarding MeHg bioavailability and/or content in fish have been presented (e.g. Richman *et al.*, 1988, Winfrey and Rudd, 1990). Generally, pH effects can be grouped into two categories, i.e. those affecting mercury loading to the waterbody and those affecting *in situ* cycling. Several possible pH-related interactions with mercury cycling and elevated fish MeHg levels are listed in the following (modified from Harris, 1991).

(a) Factors increasing the rate of *in situ* net methylation:

  • potential for increased available Hg(II) in solution due to pH-related shifts in complexation and adsorption at low pH;
  • potential for increased available Hg(II) in solution due to aggregation/ settling of humic acids in the waterbody at low pH;
  • potential for increased sediment Hg(II) due to aggregation/settling of humic acids from the water column at low pH;
  • lower rate of surface-water reduction and volatilization of Hg(II) by humic acids in acid conditions;
  • lower rate of surface-water reduction and volatilization of Hg(II) by H_2O_2 in acid conditions;
  • atmospheric oxidation of elemental mercury by H_2O_2 in acidic atmospheric conditions, which could lead to increased Hg(II) for methylation in surface waters, due to increased atmospheric deposition of Hg(II);
  • potential for increased methylation by sulfate-reducing bacteria in acidic conditions;
  • enhanced methyl-cobalamin-mediated methylation due to increased cobalt availability at low pH;
  • easier flow of Hg(II) across bacterial cell membranes at lower pH;
  • decreased rate of demethylation at low pH;
  • possible chemical hydrolysis of dimethyl mercury to monomethyl mercury at low pH.

(b) Factors increasing loading of methylmercury to a waterbody:

  • increased MeHg loading from acid-impacted watersheds due to increased abiotic methylation at low pH;
  • several of the above factors applicable to *in situ* methylation, which may also apply to the watershed.

(c) Impacts on biota:

- lower biomass in acid-stressed lakes, thus resulting in higher mercury available per fish;
- more metabolic activity for individual fish to reach given weights in acid-stressed waters, thus resulting in greater uptake of MeHg;
- increased MeHg in food (for a given food supply);
- increased MeHg in food (shift in prey species);
- shifts in the ability of fish to excrete mercury.

(d) Correlations between pH and other variables:

- correlations with hardness, alkalinity or calcium, which may reflect competition by anions for transport across membranes by fish;
- relationship between pH and humic content of water, which may reflect watershed Hg(II) or MeHg loading;
- correlations between anthropogenic acidic deposition and anthropogenic atmospheric mercury loading to a waterbody.

3.5.6 MERCURY CONCENTRATIONS IN AQUATIC SYSTEMS

Due to long-range atmospheric transport of mercury and other anthropogenic emissions, the current 'background' mercury cycling conditions probably reflect, even in remote areas, an anthropogenic influence. Within the scientific discussion on mercury concentrations and fluxes, therefore, terms such as 'pre-industrial background', 'present background', 'present', and 'elevated relative to present background' are typical. A wide variability in concentrations and fluxes may occur within each category. Table 3.5.3 provides estimates of present background concentrations for total mercury and methylmercury in water, sediments and fish, while Figure 3.5.5 provides a rough guide for

Table 3.5.3 Estimated background concentrations of total mercury and methylmercury in water, fish and sediments (Harris, 1991).

Water		Sediment		Predatory Fish (adult)		Prey Fish (adult)	
MeHg $(\mu g\,m^{-3})$	Total $(\mu g\,m^{-3})$	MeHg[a] $(\mu g\,g^{-1})$	Total[a] $(\mu g\,g^{-1})$	MeHg[b] $(\mu g\,g^{-1})$	Total[b] $(\mu g\,g^{-1})$	MeHg[b] $(\mu g\,g^{-1})$	Total[b] $(\mu g\,g^{-1})$
0.02–5.0[c]	5.0–10.0[d]	0.001–0.01	0.05–0.5	0.2–1.5	0.2–1.5	0.05–0.5	0.05–0.5

[a] Dry weight.
[b] Wet weight.
[c] Usually less than $0.5\,\mu g\,m^{-3}$ in aerated lake waters.
[d] Usually less than $2\,\mu g\,m^{-3}$ in aerated lake waters.

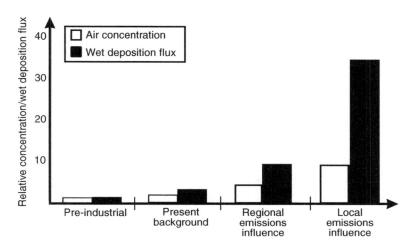

Figure 3.5.5 Illustration of the relative influences of emission source and geographic scale on air concentration and wet deposition of mercury (Redrawn from Wilken *et al.*, 1996).

discriminating between global, regional and local influences on air concentration and wet deposition (Wilken *et al.*, 1996).

Since direct measurements of mercury concentrations in water, sediments or biota from pre-industrial times do not exist, indirect means must be used to estimate historical trends. Arguments have been made on the basis of estimated anthropogenic emissions, soil cores, sediment cores and ice cores that concentrations in sediments and fish before the industrial revolution were significantly lower than today, e.g. lower than those indicated in Table 3.5.3.

3.5.7 OBJECTIVES FOR MERCURY MONITORING

Environmental monitoring performed on a national level must address policy issues of importance. The most important questions that provide the context for the development of monitoring network systems may be regarded as follows:

(a) To what extent are ecosystems (lakes) contaminated from anthropogenic pollutants (Hg, MeHg)?
(b) What effects are anthropogenic pollutants (Hg, MeHg) likely to have on food webs and on humans?

From these general policy questions, specific programme objectives can be developed, with some examples of these being presented in Table 3.5.4). These

Table 3.5.4   Examples of monitoring network design.

Feature	Comments
Example 1	
Objective	To obtain present Hg deposition load to lakes and lake catchments
Media	Bulk deposition (e.g. US National Atmospheric Deposition Programme (NADP))
Advantages	Direct deposition-rate estimate
Disadvantages	Variability in time and space; high costs; does not reveal anthropogenic share
Example 2	
Objective	To obtain recent and historic Hg loads to lakes
Media	Dated lake-sediment cores (e.g. Arctic Monitoring and Assessment Programme [a,b])
Advantages	Less short-term variation in Hg deposition; provides historic information on relative change (in some cases absolute); may be compared over large geographic areas; less costly than direct monitoring; anthropogenic share may be estimated; may provide estimate on total deposition
Uncertainties/ disadvantages	Generally only a relative measure of changing deposition rates; sensitive to selection of lake/lake site
Example 3	
Objective	To obtain data for lake biota exposure to Hg
Media	Fish Hg concentrations (e.g. Arctic Monitoring and Assessment Programme)
Advantages	Enables human exposure estimation
Uncertainties/ disadvantages	Large variability between different fish species, and between individual fish; does not provide estimate of anthropogenic impact; does not necessarily reflect lake Hg load changes
Example 4	
Objective	To measure human exposure to MeHg
Media	Fish Hg concentrations; estimate of the use of different fish species
Advantages	Direct estimate of human exposure; generally equals total MeHg exposure for humans
Uncertainties/ disadvantages	Large variability in fish Hg concentrations, and in fish use
Example 5	
Objective	To study effect of global warming on Hg concentration in fish
Media	Several key steps of Hg cycle
Uncertainties/ disadvantages	Climate-change effects on other key environmental characteristics have not been quantified (e.g. pH, sulfate cycle, temperature, sediment and hypolimnia oxygen, soil moisture, nutrients, fish population dynamics, etc.)

[a] e.g. Arctic Monitoring and Assessment Programme (AMAP, 1993).
[b] e.g. Porcella, 1996.

may be set out as shown in the following proposed scheme (modified from Landers *et al.*, (1992)):

(a) Document Hg and MeHg levels in the ecosystem;
(b) Evaluate the recent history of contamination and possible sources;
(c) Determine possible food web effects (including humans);
(d) Interpret results from international perspectives.

3.5.8 CONCLUSIONS

Monitoring is generally developed in direct response to policy questions considered to be important and of high priority. Regarding mercury's special characteristics, changes in environmental conditions may contribute to changes in Hg fluxes between different compartments of the Hg cycle, as well as concentrations and pools in environmental media. The answers to various questions, such as if the Hg concentration changes (as revealed by monitoring) are caused from changes in the anthropogenic load or from other environmental changes, have been proven to be difficult. In order to get answers to these specific questions, a wider knowledge, in addition to that achieved through monitoring, is required.

REFERENCES

AMAP, 1993. The Monitoring Programme for Arctic Monitoring and Assessment Programme, AMAP Report 93(3), AMAP, Oslo, Norway.

EPA, 1998. Mercury Study Report to Congress, EPA-452/R-97-003, US Environmental Protection Agency, Office of Air Quality Planning and Standards, Office of Research and Development, US Government Printing Office. Washington, DC, USA.

Fitzgerald, W. F., Engström, D. R., Mason, R. P. and Nater, E. A., 1998. The case for atmospheric mercury contamination in remote areas. *Environ. Sci. Technol.*, **32**, 1–7.

Harris, R., 1991. A Mechanistic Model to Examine Mercury in Aquatic Systems, PhD Thesis, McMaster University, Ontario, Canada.

Harris, R., Cherini, S. A. and Hudson, R., 1996. Regional Mercury Cycling Model (R-MCM): A Model for Mercury Cycling in Lakes, Draft User's Guide and Technical Reference, Tetra Tech Inc., Lafayette, CA, USA.

Landers, D. H., Ford, J., Gubala, C. P., Curtis, L., Urquhart, N. S. and Omernik, J. M., 1992. Arctic Contaminant Research Programme. Research Plan EPA/600/R-92/210, US Environmental Protection Agency, Environmental Research Laboratory, Corvallis, OR, USA.

Mason, R. P., Fitzgerald, W. F. and Morel, F. M. M., 1994. The biogeochemical cycling of elemental mercury: Anthropogenic influence, *Geochim. Cosmochim. Acta*, **58**, 3191–3198.

Meili, M., 1997. Mercury in lakes and rivers, in Sigel, A. and Sigel, H. (Eds), *Metal Ions in Biological Systems*, Vol. 34, *Mercury and its Effects on Environment and Biology*, Marcel Dekker, New York.

Pacyna, J. M., 1996. Emission inventories of almospheric mercury from anthropogenic sources, in Baeyens, W., Ebinghaus, R. and Vasilieu, O. (Eds), *Global and Regional Mercury Cycles: Sources, Fluxes and Mass Balances*, NATO ASI Series, 2. Environment, Vol. 21, Kluwer Academic, Dordrecht, Holland, 161–177.

Pirrone, N., Keeler, G. J. and Nriagu, J. O., 1996. Regional differences in woldwide emissions of mercury to the atmosphere, *Atmos. Environ.*, **30**, 2981–2987.

Porcella, D., 1996. Protocol for Estimating Historic Atmospheric Mercury Deposition, EPRI TR-106768, 3297, Electric Power Research Institute, Palo Alto, CA, USA.

Richman, L. A., Wren, C. D. and Stokes, P., 1988. Fact and fallacies concerning mercury uptake by fish in acid stressed lakes, *Water, Air, Soil Pollut.*, **37**, 465–273.

Wilken, R.-D., Lindberg, S. E., Horvat, M., Petersen, G., Porcella, D., Schroeder, B., Wisniewski, J. R., Wisniewski, J., Wheatley, B., Wheatley, M. and Wyzga, R., 1996. 4th International Conference on Mercury as a Global Pollutant, August 4–8, 1996, Hamburg, Germany, Conference Summary Report.

Winfrey, M. and Rudd, J. W. M., 1990. Environmental factors affecting the formation of methylmercury in low pH lakes: A review, *Environ. Toxicol. Chem.*, **9**, 853–869.

Part Four
New Approaches in
Lake Monitoring

Chapter 4.1
Integration of Different Approaches in Lake Monitoring

GUIDO PREMAZZI AND ANA-CRISTINA CARDOSO

Hydrological and Limnological Aspects of Lake Monitoring
Edited by Pertti Heinonen, Giuliano Ziglio and André Van der Beken
©2000 John Wiley & Sons, Ltd, ISBN 0 471 89988 7

4.1.1   INTRODUCTION

Humans have caused an enormous impact on the environment, exploiting it beyond what can be sustained in the long run. As an outcome of the abuses inflicted, the freshwater resources most easily available to us (i.e. from rivers and lakes), and which constitute already only 0.26% of all the freshwater on earth, are now threatened with many types of pollution and ecological disturbances (UNEP, 1994).

At present, European inland surface waters are mainly used for drinking water supplies, irrigation, waste disposal, industrial processes and cooling, transportation and hydroelectric power generation, but their use for recreational purposes is becoming increasingly important (Kristensen and Hansen, 1994). A growing number of both uses and users have increased the exploitative pressure on European freshwater resources. Consequently, surface freshwater is an increasing valuable natural resource with major impacts and benefits for Europe's population and environment, and a pan-European methodology for assessing inland surface water quality and pollution is urgently needed.

The exigency to define a methodology for the assessment of surface water quality has also been sustained by the European Community (EC) water policy since the end of the 1980s. In 1988, the 'Ministerial Seminar on Community Water Policy for the Nineties', held in Frankfurt, concluded of the need for an expansion and intensification of the EC policy on the protection and management of Community water resources within the context of the European Single Act. The Commission had then the task to prepare a proposal to fulfil this aim.

After a lengthy discussion period, an earlier proposal (*Council Directive on the Ecological Quality of Water*) (COM(93)680 Final, 1994) was dropped and a new proposal establishing a *Framework for Community Action in the Field of Water Policy* was published in February 1997 (COM(97)49 Final, 1997). This last proposal improves earlier water legislation by clarifying the relevant rules through a total or partial repeal of those directives that have became obsolete in practice. This proposal aims to maintain and improve the aquatic environment in the European Union (EU), with the overall objective being to ensure a '**Good Ecological/Chemical Status**' of surface waters, groundwaters and waters in protected areas. To this end, it proposes establishing the administrative structures and procedures needed to manage and protect the quality and quantity of water resources. Among the objectives of the Framework Directive is the collection and analysis of the information needed to check the state of the aquatic environment and to propose viable policies.

From the above, one can envisage the necessity to obtain reliable high quality information about the quality of inland surface waters and, thus to establish an adequate monitoring scheme for the European lakes.

Furthermore, there is the urgency to define an appropriate management plan of Lake Systems in order to assure the sustainability of their use. This means that not only the lake itself but also its watershed must be carefully managed. In this scenario, the SALMON project (SAtellite remote sensing for Lake MONitoring), a co-funded EC research project (1996–1999) within the topic "Space Techniques Applied to Environmental Monitoring and Research", is of major importance. The project involves eight participants representing three EC Member States (Finland, Italy and Sweden) and covers three different ecoregions; i.e. sub-alpine, boreal and sub-arctic.

The SALMON project was designed on the basis of a co-operative interaction between limnologists and remote-sensing scientists, with the goal of evaluating the capabilities and potentiality of current and forthcoming space-borne remote sensing for the monitoring of European lake water quality. It foresees as one of its main outcomes the production of guidelines and protocols for the definition of a suitable tool in water quality monitoring and management.

In the following text, we present some considerations about the current lake water quality problems and on the approach of limnologists to these problems. A brief discussion of the pros and cons of limnology and remote sensing in lake monitoring is also given.

4.1.2 MAJOR PROBLEMS AFFECTING LAKES

In Europe, there are more than 500 000 natural lakes larger than 0.01 km^2 in area, and more than 10 000 major reservoirs (man-made lakes) covering a total surface area of more than 100 000 km^2 (EEA, 1994). Unlike the rapidly moving waters in rivers, water remains in lakes for months or years, and therefore is more easily polluted. The water quality in lakes can be quickly degraded if human activities are intensified and the population increases in the drainage basin around them (UNEP, 1994; Matsui *et al.*, 1995).

The degradation of lake environments is now a world-wide issue. In most parts of the world, the finite supply of freshwater is put to heavy use. Industrial wastes, sewage and agricultural runoff can overload lakes with chemicals, wastes and nutrients, and can poison water supplies. Sediments from eroded land can silt up dams, rivers and hydroelectric schemes. Ill-conceived irrigation projects can suck dry irreplaceable groundwater reserves. Premazzi and Chiaudani (1992) pointed out six major cause categories that have been important in degrading the quality of the European lakes in recent decades. These are as follows:

(a) excessive inputs of nutrients and organic matter, leading to eutrophication;
(b) hydrologic and physical changes such as water-level stabilization;

(c) siltation from inadequate erosion control in agriculture activities;
(d) acidification from atmospheric sources;
(e) introduction of exotic species;
(f) contamination by metals and organic micropollutants.

4.1.3   DEFINITION OF LAKE WATER QUALITY

The quality state of a lake can be described by many variables, but the most often used indicators are as follows: physico-chemical, namely pH, dissolved oxygen (DO), conductivity, ammonia, nitrate, alkalinity, orthophosphate, chlorophyll, chemical oxygen demand (COD) and biological oxygen demand (BOD), microbiological, i.e. total plate count (TPC), coliforms (CF) and fecal coliforms (FC), and biological, i.e. phytoplankton and macrobenthos composition. The subject of monitoring and classifying standing waters is in constant progress and a comprehensive review of the theme, for lakes in temperate regions, was recently published (Moss *et al.*, 1996).

There are two main methods used for assessing water quality; namely the physico-chemical and biological approaches. The physico-chemical methods involve measurement variables such as suspended solids, biochemical oxygen demand, dissolved oxygen, nitrogen and phosphorus compounds. The biological methods rely on the fact that pollution of a water body will cause changes in the physical and chemical environment of the waters and that these changes will then disrupt the ecological balance of the system. Thus, by measuring the extent of the ecological upset, the severity of the pollution can be estimated.

Premazzi and Chiaudani (1992) made a proposal for a Pan-European quality classification system for freshwater lakes. The criteria chosen (redundancy/simplicity, fluctuation, integration, sensitivity and cost) for this classification system were carefully established and considered in order to derive a list of variables which can be practicably measured and quantified, and which have adequate sensitivity to provide meaningful assessments (Table 4.1.1). There is no single variable or group of variables, however, that will serve as a universal mechanism to detect changes in the trophic status of all of the diverse types of freshwater lentic environments.

In the search for variables (listed in Table 4.1.1) with which to measure the ecological state of lakes and to suggest measures for protecting, maintaining or improving such states, criteria such as redundancy/simplicity, fluctuation, integration, sensitivity and cost could all be used (as mentioned above) (Premazzi and Chiaudani, 1992):

(a) redundancy/simplicity – variables should not be selected that are closely related and/or correlated in such a manner that they provide similar information;

Table 4.1.1 Suggested variables for evaluating changes in the ecological state of freshwater standing waters.

Variable	Priority[a,b]
Physical	
temperature profile	2
mean depth	2
conductivity	1
Secchi disc transparency	1
Chemical	
DO profile	1
total P	1
inorganic N	2
organic matter[c]	2
nutrients[c]	2
inorganic and organic contaminants[c]	2
Biological	
chlorophyll[a]	1
chlorophyll derivate products[c]	2
phytoplankton diversity and abundance	1
algal assays	2
macrophytes – % coverage	1
zooplankton diversity and abundance	2
littoral macro-invertebrates' fauna	2
profundal macro-invertebrates' fauna	1
fish diversity and abundance	1

[a]Priority 1 – variables meet all of the criteria listed and have maximum utility in assessing changes in the ecological state
[b]Priority 2 – variables do not meet all of the criteria listed but have utility in certain situations, or may be useful in evaluating cause-and-effect relationships and/or management needs
[c]variables to be analysed in the sediments

(b) fluctuation – variables that are subject to severe hourly/daily fluctuations (i.e. lacking stability) should be avoided or subject to very careful interpretation;
(c) integration – variables whose levels are a function of the interacting effects of several physical, chemical and biological factors are highly desirable;
(d) sensitivity – variables should be sensitive to subtle perturbations of the system;
(e) cost – variables should be inexpensive and easy to measure.

4.1.4 LAKE STUDY APPROACH BY LIMNOLOGISTS: CURRENT TECHNOLOGY

The current technology available to measure the water quality variables cited above involves *in situ* measurements or collection of water samples for subsequent laboratory measurements (for more specific details, see Chapter 1.5, and Part 2 in this present text). The sampling of lakes and reservoirs for the purpose of assessing water quality is a complex process, as is the interpretation of the data obtained. The strategies employed for sampling and data interpretation are controlled by lake use, the problem being addressed, and the availability of resources. The structure of the lake must also be defined if a rational sampling design is to be undertaken.

Sampling should provide information of a representative portion of the waterbody. The most critical factors to define, when designing the sampling strategy, are points of sampling, frequency of sampling and maintenance of integrity of the samples for analysis.

At each fixed station, chosen with respect to the information desired and in conformity to local conditions, three depths at least, need to be considered, i.e. 1 m depth from the lake surface, 1m above the lake bottom and midway between these previous depths. For shallow lakes and reservoirs, the total number of samples within the water column can be reduced.

In order to establish the sampling frequency, one should consider the economic and ecological importance of the water bodies, the quality conditions, the morphological and hydrological characteristics, the year-to-year variations in the water quality, the quality and quantity of the wastes discharged into the recipient and the technical–economical possibilities. The frequency of sampling should be, at least, twice a year for natural lakes and similar waterbodies, i.e. one at the beginning of spring, during the complete overturn of the water column, and the other at the end of the summer stratification, just before the autumn overturn (Premazzi and Chiaudani, 1992).

An adequate monitoring programme, which provides enough data to lake management, has to consider the lake as an integral part of the watershed. However, while the above technology may give accurate measurements for a limited number of points in time and space, for larger areas it is expensive and time consuming. More importantly, it does not give either the spatial or temporal view of the water quality that is needed for accurate assessment and monitoring of surface water quality problems in a large lake, and even less in multiple lakes across the landscape. Moreover, problems still exist regarding the techniques to determine certain variables, namely the methods for pigment determination.

To summarize, modern limnology considers a lake not as an isolated body of water, but as a part of a complex ecosystem which also includes the lake's catchment area and the overlying atmosphere. The study of variable processes in such complex and dynamic natural systems requires observation methods capable of providing information over a broad range of spatial and temporal scales, often not achievable with conventional *in situ* observation methods. The incorporation of remote-sensing methods can enable, in principle, the gathering of information on processes with the desired temporal resolution, and a coverage of large areas or volumes of the systems to be resolved.

4.1.5 POSSIBLE USE OF REMOTE SENSING FOR LAKE QUALITY MONITORING

In the above, we have come to the conclusion that a technique is needed for monitoring changes in the surface water quality variables and to provide rapid assessments of both the spatial and temporal variabilities. Satellite imagery is a relatively new tool in aquatic sciences which can provide synoptic data on environmental variables that may affect the eutrophication processes of aquatic systems (for more details on the basic aspects of remote sensing and some of its application to lake monitoring, see also Chapter 4.3).

Today, remote-sensing technology can be used in many applications to provide spatial and temporal data about landscape features, the distribution of water quality components in ocean and coastal waters (Gordon *et al.*, 1980; Gordon and Morel, 1983), and to provide a cost-effective method for monitoring large segments of landscape. Practical applications of remote-sensing techniques are obtained through the combination of comprehensive field experiments with numerical modelling, controlled with simultaneous *in situ* limnological measurements (Ritchie *et al.,* 1990; Mayo *et al.*, 1995; Kondratyev *et al.*, 1996).

However, inland waters are still a problem for remote sensing because phytoplankton, detritus, non-organic suspended matter and dissolved organic matter (humic and fulvic acids, so-called yellow substance, etc.) vary in character and amount over very short distances and time intervals, and can modify the upwelling visible radiation that is measured by remote-sensing instruments (Bukata *et al.*, 1989; Kondratyev and Gitelson, 1989). Therefore, in the application of remote sensing to inland waters for the assessment of water quality, some questions are still not fully answered:

(a) In which conditions and geographical locations should spaceborne imagery be used?
(b) How accurately can we 'sense' inland water quality from space?
(c) What are the uncertainties in the remotely sensed quantities resulting from the trophic state of the water bodies?

These questions have been addressed in several studies by using a multi-sensor approach to spectral signature analysis of inland waters, and by using large data sets from different geographic areas, water bodies, and seasons (Dekker *et al.*, 1990; Gitelson *et al.*, 1993). However, the choice of adequate optical variables for each model is still a serious problem, particularly for highly productive and dynamic inland waters because the composition of the phytoplankton and suspended matter varies continually. Different species have different mean cell sizes and distributions. The optical proprieties of phytoplankton vary with species and their physiological state and are even seasonally dependent. Therefore, the ultimate answers should be obtained from the optical characteristics of water bodies of different trophic states, for different seasons, at different locations and under different irradiation conditions, when trying to find features of radiance spectra that can be used for the assessment of constituent concentrations in inland waters.

Thematic Mapper (TM) data for chlorophyll estimations in inland water have been used by several groups (Dekker *et al.*, 1992, Jaquet and Zang, 1989; Richie *et al.*, 1990; Mayo *et al.*, 1995). Though limited in their universal application, empirically derived algorithms can provide adequate estimation of chlorophyll concentration. For successful extrapolations to conditions other than those under which these algorithms are calibrated, adjustments of their variables are required. Table 4.1.2 gives some examples of the empirical models used to predict chlorophyll concentrations in inland waters based on spaceborne remote-sensing variables.

Some additional problems present themselves in the use of satellite images for monitoring. In Ireland, a lake remote-sensing programme has been in operation for many years (McGarrigle, 1997). Currently, it is in use as primarily an airborne platform (Cessna 172) with a PMS multispectral scanner. Initially, the concept was developed based on satellite remote-sensing data (Landsat MSS, TM) but it became less important when tape costs escalated and it became apparent that it would be very difficult to obtain sufficient ground-truth data corresponding to cloud-free days in order to develop chlorophyll detection algorithms.

Satellite imagery is a relative new tool in aquatic sciences, but it has already been used in many applications to provide spatial and temporal data about the distribution of water quality components in oceanic and coastal waters. In this context, it is a cost-effective technique for monitoring large segments of landscape. However, its application to inland waters is still limited, with these bodies of water posing particular problems in the use of satellite images.

High resolution is needed due to the often irregular distribution of the water quality variables, such as algae distribution, which can vary over very short distances and time intervals. Therefore, when trying to find features of the radiance spectra that can be used for the assessment of constituent concentrations in inland waters, answers should be obtained from studying the optical characteristics of water bodies of different trophic states, for different seasons, at different locations and under different irradiation conditions.

An additional universal problem in the use of satellite images for monitoring is the cloud coverage, with a clear sky being necessary for image acquisition. A

Table 4.1.2 Examples of algorithms used for chlorophyll determination in inland waters.

Algorithms	Water body	Chlorophyll (mg/m^3)	Estimated error (mg/m^{-3})	Reference
$Chl = 0.164[R^1 - R^2/R^3]$	Lake Kinneret	3–10	<0.85	Mayo et al., 1995
$Chl = a[C_{av.} Z^b(C_{av.})]$	Lake Balaton	49	18	Gitelson et al., 1993
$Chl = a[Cav. Z^b(C_{av.})]$	Lake Balaton	63	10	Gitelson et al., 1993
$Chl = a[C_{av.} Z^b(C_{av.})]$	Lake Balaton	10	2.5	Gitelson et al., 1993
$Chl = a[C_{av.} Z^b(C_{av.})]$	River Don	19	2.7	Gitelson et al., 1993
$Chl = a[C_{av.} Z^b(C_{av.})]$	River Donec	8	2.0	Gitelson et al., 1993
$Chl = -27.009 R^3/R^1 + 1.478 R^2 + (R^1)/2 - 13.385$	Lake Garda	0.7–4.0	–	SALMON Project, 1997
$Chl = 11.2R^1 - 8.96R^2 - 3.28$	Lake Iseo	5.5–7.0	0.054	SALMON Project, 1998

[a] Chl = chlorophyll concentration predicted by the algorithms; R^1 = reflectance in the blue band; R^2 = reflectance in the green band; R^3 = reflectance in the red band; $a = 1.46C_{av.} + 44.25$; $b = -0.013C_{av.} + 2.9$; Z = apparent chromaticity blue coordinate; $C_{av.}$ = measured average concentration of chlorophyll.

Table 4.1.3   Some features of the limnological and remote-sensing techniques used for lake water quality assessment.

Measure of ecological status	Limnology	Remote sensing
Discrimination (type of pollution)	Yes	No
Precision (pollutant concentration assessment)	Good	Poor
Monitoring costs	Relatively inexpensive	Limited; cost effective for large areas
Frequency of analysis	High	Limited; long time periods dependent on satellite orbit and period, and availability of cloud free images
Spatial coverage	Poor	High
Data interpretation	Proven capability	Good; dependant on ground-truth analysis and algorithm development
Data publication/communication	Poor/fair if not properly addressed	Appealing technique

comparison of some characteristics of limnological and remote-sensing techniques for lake water quality assessment is shown in Table 4.1.3.

4.1.6   FINAL CONSIDERATIONS

The "traditional" standing-water monitoring and classification methods are in constant progress of improvement. Efforts have been made to produce a common classification system for freshwater lakes, based on the determination of variables which can be practicably measured and quantified, and which have adequate sensitivities for providing meaningful assessments.

However, an adequate monitoring strategy, which provides enough data for lake management, is expensive and time consuming, and for large lakes or multiple lakes does not give the spatial or temporal views of the water quality that is needed for accurate assessment and monitoring of surface water. Moreover, problems still exist regarding the techniques for determining certain variables, namely the methods for pigment determination whose relative efficiencies vary with algae composition. Satellite imagery is a developing tool for lake quality monitoring and assessment, although some limitations still exist in the applications to inland waters.

To summarize, both of the approaches described above, i.e. classical limnology and remote sensing, have their pros and cons, but one may conceive a future where classical limnology is used mainly for the monitoring of small water bodies and to the calibration of remote-sensing data, while remote sensing will be used to acquire information on large water bodies or multiple lakes across the landscape.

REFERENCES

Bukata, R. P., Bruton, J. E. and Jerom, J. H., 1989. Particulate concentrations in Lake St Claire as recorded by a Shipborne Multisprectral Optical Monitoring System, *Remote Sensing of Environment*, **25**, 201–229.

COM(93)680 Final, 1994. Proposal for a Council Directive on the Ecological Quality of Water, *Official Journal of the European Communities*, **15**, 6.

COM(97)49 Final, 1997. Proposal for a Council Directive establishing a Framework for Community Action in the Field of Water Policy, *Official Journal of the European Communities*, **C184**(40), 17 February 1997.

Dekker, A. G., Malthus, T. J. and Seyhan, E., 1990. An inland water quality bandset for the CAESAR system based on spectral signature analysis, *Proceedings of International Symposium on Remote Sensing and Water Resources*, IAHS, Enschede, The Netherlands, 597–606.

Dekker, A. G., Malthus, T. J., Wijnen, M. M. and Seyhan, E., 1992. Remote sensing as a tool for assessing water quality in Loosdrecht lakes, *Hydrobiologia*, **233**, 137–159.

Gitelson, A. A., Garbuzov, G., Szilagyi, F., Mittenzwey, K.-H., Karnieli, A. and Kaiser, A., 1993. Quantitative remote sensing methods for real-time monitoring of inland waters quality, *International Journal of Remote Sensing*, **14**, 1269–1295.

Gordon, H. R., Clark, D. K., Muller, J. L. and Howis, W. A., 1980. Phytoplankton pigments from Nimbus 7 Coastal Zone Colar Scanner: comparisons with surface measurements, *Science*, **210**, 63–68.

Gordon, H. R. and Morel, A. Y., 1983. Water color measurements, in: *Oceanography from Space*, Gower, J. F. R. (Ed.), Plenum Press, New York.

Jaquet, J.-M. and Zang, B., 1989. Colour Analysis of Inland Waters using Landsat TM Data, European Coordinated Effort for Monitoring the Earth's Environment, ESA SP-1102, European Space Research and Technology Centre (ESTEC), Noordwijk, The Netherlands 57–70.

Kondratyrev, K. Y. A., Bobylev, L. P., Pozdnyakov, D. V., Melentyev, V. V., Naumenko, M. A., Mokiewsky, K. A., Korotkerich, O. E., Zaitsev, L. V., Karetnikov, S. G., Beletsky, D. V. and Litvinenko, A. V., 1996. Combined application of remote sensing and *in situ* measurements in monitoring environmental processes, *Hydrobiologia*, **322**, 227–232.

Kondragyev, K. Y. A. and Gitelson, A. A., 1989. Principles of remote monitoring of inland waters quality, *Transactions Docklady of the USSR Academy of Sciences: Earth Science Sections*, **299**, 39–44.

Kristensen, P. and Hanson, H. O. (Eds), 1994. *European Rivers and Lakes: Assessment of their Environmental State, EEA Environmental Monographs 1*, European Environment Agency, Silkeborg, Denmark.

Matsui, S., Ide, S. and Ando, M., 1995. Lakes and reservoirs: reflecting waters of sustainable use, *Water Science and Technology*, **32**, 221–224.

Mayo, M., Gitelson, A., Yacobi, Y. Z. and Ben-Avraham, Z., 1995. Chlorophyll distribution in lake Kinneret determined from Landsat Thematic Mapper data, *International Journal of Remote Sensing*, **16**, 175–182.

McGarrigle, M., 1997. Personal communication.

Moss, B., Johnes, P. and Phillips, G., 1996. The monitoring of ecological quality and the classification of standing waters in temperate regions: a review and proposal based on a worked scheme for British waters, *Biological Reviews*, **71**, 301–339.

Premazzi, G. and Chiaudani, G., 1992. Ecological Quality of Surface Waters. Quality Assessment Schemes for European Community Lakes, European Communities Commission, EUR 14563, Ecological Quality of Surface Quaters, Environmental Quality of Life Series, Environment Institute, University of Milan, Milan, Italy.

Ritchie, J. C., Cooper, C. M. and Schiebe, F. R., 1990. The relationship of MSS and TM digital data with suspended sediments, chlorophyll, and temperature in Moon Lake, Mississipi, *Remote Sensing of Environment*, **33**, 137–148.

SALMON Project, 1997. Satellite Remote Sensing for Lake Monitoring, First Year Technical Report, Uppsala, May 13–17, 35 p.

SALMON Project, 1998. Satellite Remote Sensing for Lake Monitoring, Second Year Technical Report, Venice, January 29–February 4, 98 p.

UNEP, 1994. The Pollution of Lakes and Reservoirs, UNEP Environment Library, No. 12.

Chapter 4.2

Design of the Freshwater Monitoring Network for the EEA Area

TIM LACK AND STEVE NIXON

Hydrological and Limnological Aspects of Lake Monitoring
Edited by Pertti Heinonen, Giuliano Ziglio and André Van der Beken
©2000 John Wiley & Sons, Ltd, ISBN 0 471 89988 7

4.2.1 INTRODUCTION

Member States have monitoring networks in place to assess inland water quality (essentially to determine the state and trends in the physico-chemical and biological quality of rivers, lakes and groundwaters) according to their national or international/European requirements.

Information provided by countries to the European Commission may seem to be an important source of data for the needs of the European Environment Agency (EEA) but inspection reveals a great disparity in its nature and comparability. The information required by the European Commission (EU) is primarily for assessing implementation of and compliance with directives, rather than for assessing the status of and temporal changes in water resources.

Therefore, in order to fulfil its legal mandate (provision of objective, reliable and comparable information enabling the Commission and Member States to assess, frame, implement, further develop or modify European environmental policy) the EEA has designed a European monitoring and observation network for inland waters, namely EUROWATERNET, which is based almost entirely on existing national monitoring networks.

In parallel, the proposed Water Resources Framework Directive (WFD) will require Member States to monitor the status of surface water and groundwater at the catchment and sub-catchment level. Technical specifications of this monitoring network are being discussed with the Council Presidency, the European Commission (Environment, Nuclear Safety and Civil Protection (DGXI)), Member States' experts and the EEA and its Topic Centre on Inland Waters (ETC/IW). Emphasis is placed upon the necessity to obtain comparable information on quality assessments between Member States and it is likely that EUROWATERNET will provide a suitable mechanism for this.

This chapter describes the EEA's data and information needs and the water monitoring requirements arising from them, with particular stress being placed on EUROWATERNET. During the preparation of this book, a 'Technical Report' was produced by the European Environment Agency, which gives technical guidelines for the implementation of EUROWATERNET (EEA, 1998b).

4.2.2 MANDATE AND INFORMATION NEEDS OF THE EEA

The EEA was set up by Order of the Council of Ministers in 1990 and has a number of statutory duties, the chief of which is to provide the European Union and Member States with:

> objective, reliable and comparable information at a European level enabling them to take the requisite measures to protect the environment, to assess the results of such measures and to ensure that the public is properly informed about the state of the environment.

In the field of water, the European Topic Centre on Inland Waters was appointed in December 1994 to act as a centre of expertise for use by the Agency and undertake part of the EEA's multi-annual work programme, in particular those relating to:

(a) the assessment of status and trends of surface and ground water quality and quantity;
(b) how that relates and responds to pressures on the environment (cause–effect relationships).

In 1995, the EEA produced the report 'Europe's Environment – the Dobris Assessment' (EEA, 1995), which was presented to the Conference of Environment Ministers in Sofia. The Sofia Ministerial Declaration requested that the EEA should build on this assessment, using the pan-European network for data collection, by reporting progress in main issues before the next conference in Aarhus, Denmark in June 1998.

At the end of 1994, the European Commission requested the EEA to contribute to the review of the 5th Environmental Action Programme and a report 'Environment in the European Union' was produced in 1995. The Commission have requested an update of this by the end of 1998.

4.2.3 INFORMATION NEEDS OF THE EEA

Information is therefore needed on the status of Europe's water resources in order to detect changes over time (trends), and how the status and trends are related to the pressures on the environment and to policies directed at the pressures or the societal forces creating those pressures. This information will only be credible if its component data are comparable and give a representative assessment of water types within a Member State and across Europe. A process of Integrated Environmental Assessment has been adopted by the EEA which operates within a framework of Driving forces, Pressures, State of environment, Impact on environment, and Responses in the form of policy and regulations (DPSIR).

Member Countries design networks and monitor water resources according to their national requirements (legal and operational) and international obligations (EU Directives and International Conventions).

Information obtained from the Member States' reports to the Commission might appear to be a rich source of information for the EEA's needs but inspection has revealed that, by and large, the reports are not suitable mainly because:

(a) The degree of comparability depends on the interpretation of the designation rules of directives and national differences of how these are implemented.

(b) Data requirements for those directives which require routine monitoring are generally site specific, e.g. at sites designated for a specific use and/or sites affected by a specific discharge.

(c) The Exchange of Information Decisions require data from agreed sites in main rivers. The choice of sampling location is, for most directives, related to areas designated in an *ad hoc* way by the Member States. It is not surprising therefore that a comparison of quality across Europe based on these designated waters gives an incomplete and heterogeneous picture.

The same observations have been made for information obtained from the International Convention which represent only the main transboundary catchments in Europe.

Consequently, the EEA has designed a European Water Monitoring Information and Reporting Network to provide comparable and representative information on the state and pressures on the environment through harmonized methodologies of selection and monitoring of all water bodies. EUROWA-TERNET is the process or system based upon the European Environmental Information and Observation Network (EIONET), which is a network of people, organizations, hardware and software, whereby data can be gathered from within Member Countries. This allows the EEA to obtain the information it requires in order to fulfil its legal tasks and allow the Commission and Member Countries to evaluate the effectiveness of policy. It should be emphasized that the information provided by the network will *not* be for the assessment of compliance of Member States with the requirements of European Commission Directives, this latter responsibility being held by the European Commission.

4.2.4 EUROWATERNET: DEFINITION AND GENERAL CONCEPT

EUROWATERNET is the **process** by which the EEA obtains the information on water resources (quality and quantity) it needs to answer **questions** raised by its customers. These questions may relate to statements on the general status (of rivers, lakes and groundwaters) or specific issues (e.g. water stress, nutrient status and acidification) at a **European** level.

The key concepts of EUROWATERNET are as follows:

- it samples existing national monitoring and information databases;
- it compares like-with-like;
- it has a statistically stratified design 'tailor-made' for specific issues and questions;
- it has a known power and precision.

The network is designed to give **a representative assessment** of water types and variations in human pressures within a Member Country and also across the

EEA area. It will ensure that similar types of water body are compared. The need to compare like-with-like is achieved with a stratified design with the identified and defined strata containing similar water bodies. The use of the same criteria for selecting strata and water types across Member Countries will help to ensure that valid status comparisons will be obtained.

A **basic network** of river stations and lakes based on the relative surface area of countries is the first step for Member Countries. However, it is likely that this will not answer all of the questions raised by the EEA's customers or perhaps not with the desired precision and confidence that is needed.

Therefore, a flexible approach will also be required for the selection of other monitoring stations included in national networks in order to be able to answer more specific questions, such as 'what is the extent of acidification in Europe?' or 'what is/will be the impact of the Urban Waste Water Treatment Directive on water quality?'. This is because the stations required for these questions may not always be located on the same water bodies/catchments as the basic network stations. In addition, more specific and detailed pressure information might be required. Thus, in order to meet some of the EEA's information needs, site selection within each country must be issue- or question driven. This, if necessary (in the light of experience), will form the **impact network** of EUROWATERNET.

A network which is **fully representative** of the differences in, and variability, of quality, quantity and pressures found in all water body types across Europe would be expected to answer most questions asked of the EEA. It is the **long-term aim** to make EUROWATERNET fully statistically representative. This will be achieved through the experience gained in implementing the basic and impact networks. This development will need to take into account the number of stations required to answer questions with defined, or at least known, levels of precision and confidence, and with knowledge of any inherent bias (for example, towards the most polluted water bodies) in the selected river stations, lakes or groundwater sampling wells.

Member Countries will be asked to provide aggregated data with supportive descriptive statistics which will enable an assessment to be made of the precision and confidence of the information. As part of the development of EUROWATERNET, the precision and confidence obtained from a different number of stations are being assessed. At present, a precision of 10% of the mean or percentile values (as appropriate) with a confidence of 90% would appear to be appropriate, or at least possible in some countries. Thus, the selection of the required number of stations or water bodies would also have a statistical basis such as:

> How many stations/rivers/lakes/sampling wells do you need to select in order to be within 10% of the true status of the total river, lake or groundwater body population within Europe with 90% confidence, and to be able to detect a 10% change between (six year) reporting periods?

4.2.5 FUNCTIONALITY OF EUROWATERNET

EUROWATERNET has to be able to answer questions about the state of the environment, and a number of 'status indicators' have been specified for the update of the Dobris report (EEA, 1995) and the State of the Environment 1998 report (EEA, 1998a). It is likely that these indicators will also form part of the national reporting process so there is now a real possibility of single sources of national data being used for multiple purposes (e.g. by the EEA, EC, Organization for European Co-operation and Development (OECD), Statistical Office of the European Committes (EUROSTAT), and so on).

The list of status indicators required for the rivers and lakes networks is shown in Table 4.2.1.

The status indicators for groundwater quality can be divided into seven groups (see Table 4.2.2).

EUROWATERNET also has to be able to answer questions regarding the pressures from society on the water environment to be able to relate to the changes in state. Examples of the physical characteristics and pressure indicators are shown in Table 4.2.3.

Pressure indicators are of greatest value if they are expressed on a catchment or a sub-catchment basis. This means that EUROWATERNET will be encouraging the use of geographic information systems (GIS) in displaying both state and pressure information. This will be in harmony with several Member Countries' use of river catchments as the basic unit of management and is also supportive of the Water Resources Framework Directive which takes a similar approach.

4.2.6 DETAILED TECHNICAL SPECIFICATIONS

It is inappropriate to provide here the detailed technical specifications of EUROWATERNET. They have been developed by the European Topic Centre on Inland Waters and thoroughly discussed and agreed by Member Countries at an EEA workshop held in Madrid in 1996. The technical specifications have been published by the European Environment Agency (EEA, 1998b).

4.2.7 RELATIONSHIP WITH THE PROPOSED WATER RESOURCES FRAMEWORK DIRECTIVE

The Water Resources Framework Directive (WFD) addresses all qualitative and quantitative aspects of surface and ground waters. It is stated in the

Table 4.2.1 Primary and secondary indicators required for rivers and lakes.

Indicator determinands	Problems/issues[a,b] (Examples of indicators)	EQ	AC	NS	TS	OP	WU	RA	PI	FL
Biological indicators	Macroinvertebrates; fish macrophytes; phytoplankton; chlorophyll	33	33	3	3	3	3	7	33	7
Descriptive determinands	Dissolved oxygen; pH; alkalinity; conductivity; temperature, suspended solids	3	33	3	3	33	33	7	3	3[c]
Flow	Flows, levels	33	3	3	3	3	33	7	33	33
Hydromorphology	Habitat features; structure of bed; sinuosity	3	33	3	7	33	3	7	7	7
Additional determinands	Biochemical oxygen demand; chemical oxygen demand; total organic carbon; Secchi disc; aluminium fractions	3	33	3	7	33	3	7	7	7
Nutrients	Total phosphorus; soluble reactive phosphorus; nitrate nitrite; ammonia; organic nitrogen; total nitrogen	3	7	33	7	3	7	7	7	33
Major ions	Calcium; sodium; potassium; magnesium; chloride; sulfate; bicarbonate	7	33	7	7	7	3	7	7	7
Heavy metals	Cadmium; mercury; based on catchment/land-use	7	7	7	33	7	3	7	7	33
Pesticides	Based on catchment/land-use	7	7	7	33	7	3	7	7	33
Other synthetic organic substances	PAH; PCBs; based on catchment/land-use	7	7	7	33	7	3	7	7	33
Microbes	Total and faecal coliforms, faecal streptococci; salmonella; enteroviruses	7	7	7	7	33	3	7	7	7
Radionuclides	Total alpha and beta activity; caesium-137	7	7	7	7	7	7	33	7	3

[a]EQ, ecological quality; AC, acidification; NS, nutrient status; TS, toxic substances; OP, organic pollution; WU, water use and availability; RA, radioactivity; PI, physical intervention; FL, fluxes
[b]33 represents key determinands – primary importance; 3 represents important, but not key, determinands – secondary importance; 7 indicates non-essential determinands
[c]Suspended solids

Table 4.2.2 Status indicators for the groundwater quality monitoring network.

Group	Indicator	Determinands
1	Descriptive determinands	Temperature; pH; DO; electrical conductivity
2	Major ions	Ca, Mg, Na, K, HCO_3, Cl, SO_4, P_{04}, NH_4, NO_3, NO_2; total organic carbon
3	Additional determinands	Choice depends partly on local pollution source as indicated by land-use framework
4	Heavy metals	Hg, Cd, Pb, Zn, Cu, Cr; choice depends partly on local pollution source as indicated by land-use framework
5	Organic substances	Aromatic hydrocarbons; halogenated hydrocarbons; phenols; chlorophenols; choice depends partly on local pollution sources as indicated by land-use framework
6	Pesticides	Choice depends in part on local usage, land-use framework and existing observed occurrences in groundwater
7	Microbes	Total coliforms; faecal coliforms

Table 4.2.3 Some examples of the physical characteristics and pressure information required for each river station and lake in the basic network, and the groundwater network.

Requirement	Rivers	Lakes	Groundwater
Physical characteristics			
Stream order at station	4		
Depth (mean)		4	
Surface area		4	
Catchment area upstream of station/lake	4	4	
Catchment area recharging/affecting groundwater body			
Station/lake altitude	4	4	
Longitude/latitude	4	4	4
Upstream river length to source	4		
Hydrogeology			4
Aquifer type			4
Aquifer area			4
Soil type/geology of catchment	4	4	4
Pressure information			
Population density in (upstream) catchment	4	4	4
Upstream catchment land use such as:			
% agricultural land	4	4	4
% arable	4	4	4
% pasture land	4	4	4
% forest	4	4	4
% urbanization	4	4	4
Point-source loads entering upstream	4	4	4
Fertilizer usage in catchment upstream	4	4	4

Explanatory Memorandum 'that all measures to achieve the environmental objectives for a sustainable protection and use of water are co-ordinated and their effect overseen and monitored within river basins'. The objective is to achieve for all waters a 'good' quality status and to prevent deterioration of current quality.

In this context, the WFD will require Member States to establish monitoring networks enabling them to carry out the following:

- assess and report on water status (chemical and ecological status for surface waters and chemical and quantitative status for groundwater) by using a harmonized classification scheme;
- diagnose problems, chart progress and develop appropriate action programmes for protection and improvement of waters.

When considering the importance and impact of the WFD on the Member States, and the need to ensure that Community policy is applied in a coherent and consistent way, it is essential that countries implement consistent monitoring programmes and harmonized interpretation methods to guarantee the implementation of coherent action programmes across Europe.

The technical specifications of the monitoring to be required under the WFD are currently being developed by the Commission, the EEA and Member States. To date, it is still difficult to clearly see the link between the likely requirements of the WFD and EUROWATERNET (in particular, articles 10 and 13). However, the Explanatory Memorandum of the Directive (COM(97)49 Final, 1997) states that 'data is collected largely for operational reasons to inform decision making within the individual river basins' and the articulation with EEA monitoring is also referred to by the comment 'the monitoring programmes required by this proposal will extend the range of stations which the Agency can draw upon in developing its network'.

4.2.8   BENEFITS TO MEMBER STATES

Additional monitoring is expensive and is unlikely to be undertaken purely for the 'European need'. This is why EUROWATERNET is firmly based on existing national programmes. By and large, most countries already have adequate national monitoring networks for rivers in order to meet the EEA's needs (in terms of number of stations, frequency of monitoring and determinands monitored). This is less true with regard to groundwater, and even less true for lakes/reservoirs, although there are some exceptions.

Where national networks can be shown to be insufficient in terms of providing data to allow a representative European picture to be obtained, it is also likely that a case can be made to the Member Country that it itself is not obtaining a representative picture of national water resources. An example of

Table 4.2.4 Progress to date on testing and implementing EUROWATERNET in Europe.

Country	Activity	Next steps
Austria	Pilot-testing groundwater criteria	Full implementation and institutionalization of WATERNET on EIONET
	Extended to surface water network	
Belgium (with France and The Netherlands)	Testing criteria on two international river basins – Scheldt and Meuse	Awaiting outcome of the tests
	Testing criteria on surface water network of Flanders	
Denmark	Pilot testing surface-water criteria (rivers and lakes)	Full implementation and institutionalisation of EUROWATERNET on EIONET
	Extended to groundwater networks	
	Comparison and regional aggregation with river data from England and Wales, and France	
	Comparison and regional aggregation of lakes with Ireland, Norway and Sweden	
Finland	Testing surface-water criteria on lakes and rivers	Institutionalize surface-water network on EIONET
	Testing criteria on groundwaters	Extend to groundwater
		With the support of Finland and funding from Nordic Council implement in Baltic Countries
France	Testing rivers criteria in two Agences de l'Eau	Apply to all surface waters in France
	Comparison and regional aggregation with river data from Denmark, and England and Wales	Consider extension to groundwater network
Germany	Testing rivers criteria on data from North Rhine Westfalia	Extend testing to other data with support of LAWA
		Extend testing to lakes and groundwater

Ireland	Testing criteria on lakes and rivers network	Extend to river network
Norway	Testing criteria on lakes network	Extend to river network
Spain	Pilot testing of river quantity aspects of proposed network	Establishment on a national basis and extension to groundwater
Sweden	Extending testing to river quality network Testing criteria on lakes network	Extend to river network
United Kingdom	England and Wales pilot test surface waters criteria (rivers) Comparison and regional aggregation with river data from Denmark and France Extension of comparisons and regional aggregations to include rivers from Austria, Ireland, North Rhine Westfalia and Spain, as appropriate	Scotland and Northern Ireland – pilot test surface waters (rivers) Extend to groundwater where appropriate Compare and regionally aggregate data from UK with other appropriate regions/countries of EEA area

this could be if a national programme has been designed to provide data on impacted lowland rivers. It would be safe to say therefore that this programme will not tell anything about unpolluted upland streams. If there was no other national programme on upland streams then, when the European picture on upland streams was painted, for that country the picture would have to be blank. However, this situation, should it arise, could be mitigated if there was, within that particular country, a priori knowledge about the state of upland streams and the stability of the pressures upon them such that a decision not to monitor them (or monitor them less frequently) could be justified. The decision whether or not to undertake additional monitoring therefore remains within national control. Member Countries also have the freedom to decide if and when to discontinue or reduce national programmes.

As stated elsewhere in this chapter, there is not enough comparable information to obtain a quantitative assessment of water resources across Europe at present. This can lead to unfair or incomplete comparisons being made and wrong conclusions drawn about the effectiveness or otherwise of policy instruments often implemented at great cost to the Member Countries.

Therefore, the Member Countries will benefit from the following:

- a better knowledge of policy impact, thus leading to improved European policy as a result of the better information;
- the ability to provide inputs which show their country in the best (correct) light, by making full use of national data;
- the ability to provide information on international environmental issues in a consistent and standardized way;
- the ability to inform on policy initiatives by providing comparative information on performance in other countries;
- the ability to provide information to their citizens on the state of the environment in their country, neighbouring countries, European regions and the whole of Europe.

4.2.9   PROGRESS TO DATE AND THE WAY FORWARD

The EEA's recommendations have been tested by several countries, namely Austria, Denmark, England and Wales, France, Norway and Spain. Further development and testing is continuing with the support of Belgium, Finland, Ireland and Sweden. Most recently, Germany (Landesumweltamt North Rhine Westfalia) has assessed the stratification criteria on its water resources database and is fully collaborating with the EEA on further implementation. Table 4.2.4 summarizes the progress to date.

It is hoped to complete many of the actions in the 'next steps' column in Table 4.2.4 by the summer of 1998 in order to allow a report to be prepared in advance of a workshop to be organized by the EEA in Budapest in October 1998. The purpose of the workshop will be to inform all Member Countries of progress to date, to demonstrate the benefits to be gained from implementing the EUROWATERNET system and to invite all of the remaining Member Countries (including the PHARE Programme countries[1]) to participate in the process.

The overall objective is to have EUROWATERNET fully operational on the EIONET by December 2000.

4.2.10   CONCLUSIONS

EUROWATERNET is firmly based on national programmes at the 'zero' or 'low cost' option. By and large, national networks are likely to be more than adequate (in terms of numbers of stations, frequency of monitoring and determinands monitored) to meet the EEA's need. EUROWATERNET will provide National Focal Points, National Reference Centres and other national experts with guidelines on how rivers and lakes should be selected for the EUROWATERNET programme. It will also provide guidelines for the design of a groundwater monitoring network. The approach for rivers and lakes is different from that for groundwaters. This is because river and lake national monitoring networks are generally more established than those for groundwater.

The immediate aim is for Member Countries to establish basic networks for rivers and lakes. This will lead to transfer of the requested monitoring information from the local, regional and national levels for the EEA by using the EIONET structure. In the short term, the ETC/IW will provide worked examples of the operation of EUROWATERNET. In the longer term, Member Countries will be asked to extend the basic network to all water types, and to test and develop a network which is fully representative of general status issues and to answer more specific questions.

The proposed EU Framework Water Directive (COM(97)49 Final, 1997) is a major piece of legislation affecting the water environment. It will require the integrated management of water quantity and water quality. EU Member States will need to collect information on the status of, and pressures on, rivers, lakes and groundwaters. EUROWATERNET incorporates water quality and quantity information and is fully compatible with the reporting needs of the Framework Water Directive.

---

[1]PHARE; Poland, Hungary-EU Assistance for the Reforms of the Economies (currently extended to 13 central and eastern European countries).

REFERENCES

COM(97)49 Final, 1997. Proposal for a Council Directive establishing a Framework for Community Action in the Field of Water Policy, *Official Journal of the European Communities*, **C184**(40), 17 February 1997.

EEA, 1995. *Europe's Environment, The Dobris Assessment*. Eds: D. Stanners and P. Bourdeau, European Environment Agency, Copenhagen.

EEA, 1998a. *Europe's Environment: The Second Assessment*. European Environment Agency, Copenhagen.

EEA, 1998b. EUROWATERNET, Technical Guidelines for Implementation, Technical Report No. 7, The European Agency's Monitoring and Information Network for Inland Water Resources, European Environment Agency, Copenhagen, Denmark, June 1998.

Chapter 4.3

Remote Sensing as a Tool for Monitoring Lake Water Quality

KARI KALLIO

Hydrological and Limnological Aspects of Lake Monitoring
Edited by Pertti Heinonen, Giuliano Ziglio and André Van der Beken
©2000 John Wiley & Sons, Ltd, ISBN 0 471 89988 7

4.3.1   INTRODUCTION

Monitoring of water quality is usually based on the concentrations of various substances at a few fixed stations obtained by water sampling and laboratory analyses. The effective management of surface waters also requires information on the spatial distribution and temporal variation of these concentrations. Remote sensing is in many cases the only technique available for detecting the spatial differences of water quality variables that can also be used to study the functioning of the water ecosystems.

Remote sensing can only be used to determine the optically active substances in the surface layer of water. Variables that cannot be monitored by remote sensing include the oxygen concentration of hypolimnion, toxic subtances and acidity. However, many of the water quality variables that it is possible to detect by remote sensing (chlorophyll *a*, suspended solids, turbidity, Secchi disc transparency, etc.) have a central role in lake monitoring.

4.3.2   BASICS OF REMOTE SENSING OF WATER QUALITY

The monitoring of water quality by remote sensing is based on the use of optical sensors, with usually the visible wavelengths (380–760 nm) of electromagnetic spectrum being used for detection. In addition, the surface water temperature can be monitored with thermal infrared wavelengths (3–30 μm). Temperature maps are used to monitor hydrodynamic phenomena important for water ecosystems, e.g. upwelling areas and thermal bars.

A remote-sensing sensor consists of a spectrometer that measures the upwelling radiance from the water body at various wavelength ranges (channels). In order to make the measurements made under different illumination conditions comparable, the remote-sensing measurement is usually converted into reflectance, i.e. the ratio of upwelling radiance to the downwelling irradiance at a certain wavelength.

4.3.2.1   Optically active substances

The direct and diffuse solar radiation that penetrates into the water column is absorbed and scattered by the water itself and also by other optically active substances. Interpretation of water quality is based on the differences in the magnitude and shape of the spectral reflectance, effected by the optical properties and concentrations of the substances in the water (Figure 4.3.1). The absorption of light by pure water increases with increasing wavelength. This property produces the blue colour of the water, if the concentrations of other optically active substances are very low. Aquatic humus (coloured dissolved

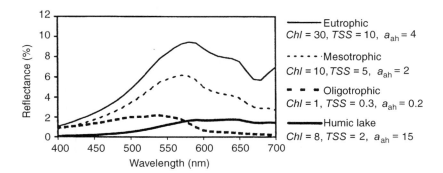

Figure 4.3.1 Underwater reflectance for four lakes of different type as simulated by a hyperspectral model described by Kutser (1997): Chl = chlorophyll a $(\mu g l^{-1})$; TSS = total suspended solids (mg l^{-1}); a_{ah}, absorption coefficient of aquatic humus at 400 nm (m^{-1}).

organic matter) absorbs light exponentially with decreasing wavelength, which results in the distinct brown colour of humic lakes. Suspended solids usually consist of mineral particles, phytoplankton and detritus; both mineral particles and phytoplankton scatter light. The optical properties of detritus are dominated by an absorption spectrum which is similar to aquatic humus. In addition, various pigments in phytoplankton absorb light. For example, chlorophyll a has an absorption spectrum with maxima at 430–450 and 660–680 nm.

4.3.3 REMOTE-SENSING SENSORS

These sensors can be divided in airborne and satellite sensors. In addition, portable spectroradiometers which can be located on board vessels are used to make point and transect measurements. Remote-sensing sensors are usually characterized by their spectral, spatial and temporal resolutions. The spectral resolution determines the width and wavelength of the channels and therefore defines which type of application the sensor can be used for. The spatial resolution of satellite sensors usually varies between 10 and 1000 m. The poor spatial resolution of some satellite sensors limits their use in small lakes and archipelago areas. A resolution of 100 m means that the mean water quality in a pixel of 100 m × 100 m can be detected. If a pixel includes land areas or islands, the remote-sensing signal can not be used for detection of water quality.

Temporal resolution is defined by the revisit time of the sensor and states how often an image can be obtained from the same area. Optical remote sensing by satellites is not possible in cloudy conditions and this means that proper images are available less often than that defined by the revisit time.

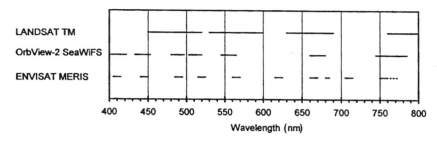

Figure 4.3.2   Spectral channels of selected satellite sensors.

Satellite sensors usually have revisit times between a few days (e.g. OrbView-2 SeaWiFS) and two weeks (e.g. LANDSAT TM).

The LANDSAT TM sensor, as for most of the satellite sensors that are available, was designed for land-use and vegetation mapping. It has a good spatial resolution (30 m), but the channels are wide. The OrbView-2 SeaWiFS sensor, launched in 1997, is the only existing sensor that has been specifically designed for water quality applications. However, its wavelengths are designed for clear-ocean-water applications and the pixel size of about 1100 m makes it less useful for lake monitoring. In the near future, sensors with good resolution and channel configurations which are suitable for lake and coastal water monitoring should become available. Such a sensor is, e.g. the ENVISAT MERIS, planned to be launched in 2001, with a pixel size of 300 m and with ten narrow channels in the visible range (Figure 4.3.2).

The satellite images cover wide areas and are less expensive than images produced by airborne remote sensing. Airborne surveys are suitable for local monitoring and they can be conducted whenever the weather is suitable, which makes airborne remote sensing less dependent on cloudiness. Airborne sensors also have a better spatial and spectral resolution than satellite sensors. Furthermore, the channels can be selected optimally for each survey, depending on the season and lake type in question.

4.3.4   INTERPRETATION OF WATER QUALITY FROM REMOTELY SENSED DATA

In lakes, inorganic suspended solids and allochthonous aquatic humus often dominate the optical properties of the water and they vary independently of the presence of phytoplankton. Therefore, the interpretation of water quality in lakes is more complex than in clear ocean waters. In oceans, phytoplankton and their associated materials (debris, autochthonous aquatic humus, etc.) have a dominant role in determining the optical properties and will all absorb light in the blue region of the visible range.

The two main methods used in the interpretation of water quality from remotely sensed data are empirical algorithms and analytical modelling.

4.3.4.1   Empirical algorithms

In this method, application is made of the statistical relationships between the measured spectral values (radiance or reflectance) and the water quality variables. The most common empirical method is the use of channel ratios (Dekker, 1993; Gitelson *et al.*, 1993; Bukata *et al.*, 1995; Kallio *et al.*, 1998; Kutser *et al.*, 1998). The wavelengths used in empirical algorithms should be based on the spectral characteristics of the water quality variables. Usually one wavelength is selected from the area of the spectrum where reflectance is strongly affected by the variable in question, while the other wavelength is obtained from the area where the effect on the reflectance is very small or opposite to that of the first wavelength.

Chlorophyll *a* in lakes is usually interpretated by the so-called absorption/ fluorescence algorithm, using the ratio $R(701\,\text{nm})/R(672\,\text{nm})$, where the dip in the reflectance spectrum at about 672 nm is due to chlorophyll absorption (Fig. 4.3.2 and 4.3.3). The peak at 685–715 nm could be due to fluorescence by chlorophyll *a*, a minimum in the combined absorption peaks of phytoplankton and water, or scattering by phytoplankton, as reviewed by Gitelson *et al.* (1993). The higher this ratio, then the higher is the concentration of chlorophyll *a* (for empirically derived algorithms for predicting chlorophyll concentrations in inland waters, see also Chapter 4.1).

Cyanobacteria have also been detected indirectly from hyperspectral reflectance by making use of the absorption maximum of cyanophycocyanin which occurs at about 620 nm (Decker, 1993). Suspended solids (and turbidity) usually increase the magnitute of the reflectance spectrum and therefore interpretation can be based on a single channel at 500–600 nm. In humic lakes, the absorption by aquatic humus clearly dominates the optical properties, and thus interpretation of chlorophyll *a* and suspended solids by using empirical methods is either impossible or requires lake-type-specific algorithms (Kutser, 1997, Kallio *et al.*, 1998) (see also Figure 4.3.3). The major advantage of using empirical algorithms is the fact that they are easy to apply, although the algorithms may vary from lake to lake or seasonally due to variations in the optical properties. Thus their use generally requires limnological sampling for algorithm testing and validation.

4.3.4.2   Analytical methods

The analytical methods used in this area (Fischer and Doerffer, 1987; Dekker, 1993; Kirk, 1994; Bukata *et al.*, 1995; Kutser, 1997) are based on the simulation of the effect of optically active substances on the water-leaving reflectance/radiance by taking into account their specific absorption and

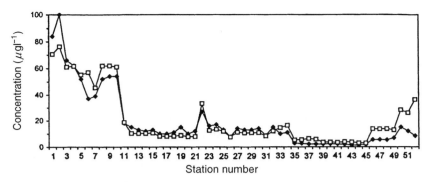

Figure 4.3.3 Calculated (□) and observed (◆) concentrations of chlorophyll *a* in August 1997 at 52 stations of the ten lakes located in southern Finland. The calculated values were obtained from the ratio of radiances measured at 702 and 674 nm using an AISA airborne spectrometer; stations 46–52 represent humic lakes. Redrawn after Kallio *et al.* (1998).

scattering coefficients at various wavelengths. The optical properties are obtained from site-specific measurements or from the literature. This method, sometimes known as hyperspectral modelling, makes it possible to demonstrate the effect of various optically active substances on the reflectance (see Figure 4.3.1) and to make sensitivity analyses. The hyperspectral model can also be used for the interpretation of remote-sensing data. In this inverse modelling technique, the concentrations are determined by minimizing the difference between the measured and simulated spectral reflectances.

4.3.5   LIMNOLOGICAL AND OPTICAL MEASUREMENTS FOR REMOTE SENSING

4.3.5.1   Sampling

Limnological sampling is needed to calibrate and validate the interpretation algorithms of the remote-sensing data. Therefore, sampling should be carefully planned in order to support the interpretation. The following aspects of sampling are particularly important (see also Van Stokkom *et al.*, 1993):

- the stations selected must cover a wide range of concentrations;
- sampling should take place on the same day and as close to the overpass of the sensor as possible, this being particularly important in water bodies with high temporal variability;
- sampling from depths of 0–0.5 m is usually sufficient;
- the recording of exact time and coordinates by a Differential Global Position System-receiver (DGPS);

- the intercalibration of analyses and instruments, if several water laboratories and field instruments are involved.

Regular monitoring programmes could be modified by taking into account the days of satellite overpasses. If the conditions during the overpass are cloudless, the field personnel could then change the sampling programme by taking samples only from the surface layer at several stations covering a wide range of concentrations. The analyses normally included in the monitoring programmes and which can usually be interpreted from the remote sensing data, are chlorophyll *a*, total suspended solids, turbidity, and Secchi disc transparency. Other analyses that support the interpretation are algae biomass and species composition, mineral suspended solids, dissolved organic carbon and the absorption coefficients from filtered samples at a fixed wavelength (400 nm) as a measure of aquatic humus. The detection of cyanobacteria by remote sensing requires the analysis of cyanophycocyanin concentration, where the latter is determined by using high performance liquid chromatography.

4.3.5.2 Optical measurements

The optical properties of aquatic humus, suspended solids and phytoplankton are needed in order to obtain data for understanding the effect of various substances on the reflectance and for algorithm development. It is also important to survey the lake-to-lake and seasonal variability of the optical properties. Apart from during the acquisition of remote-sensing data, such measurements can also be made in connection with normal monitoring programmes. These optical properties can be measured either in the laboratory or in the field. The laboratory methods include the measurement of the following:

- Absorption spectra (350–800 nm) of aquatic humus by a standard spectro-photometer using filtered water samples.
- Absorption by algae and detritus in a spectrophotometer equipped with a scattered-light transmission accessory. Before the measurements are carried out, the particle content is usually concentrated by filtration.

In addition to these laboratory measurements, several instruments have been specially developed for the *in situ* determination of optical properties in the field; these include boat-platform absorption/attenuation meters, back-scattering sensors and transmissiometers. Reflectances can also be measured by using portable spectroradiometers, both under and above the water surface.

4.3.6.   THE IMPLEMENTATION OF REMOTE SENSING TO LAKE MONITORING

The monitoring of lake water quality by remote sensing is not yet at the fully operational stage. However, in the near future satellite sensors such as ENVISAT MERIS, which is particularly suitable for monitoring lakes and coastal waters, will be available. More work is also needed in data processing and in the development of interpretation of algorithms. As remote-sensing techniques develop and interpreted images become easily available, remote sensing can then be utilized in lake monitoring in the following ways:

(A) In traditional monitoring based on point sampling, it is usually assumed that the water quality measured at one station represents a whole lake or part of a lake. Remote sensing is the only method that makes it possible to effectively monitor the spatial variation of water quality in a lake.

(B) The particular advantage of using remote-sensing techniques to monitor a large group of lakes is the fact that it provides simultaneous water quality estimates for all of the individual lakes. If the same large group of lakes is monitored on the basis of water sampling, the sampling itself would take several days or even weeks to complete. During this period, the weather conditions may vary considerably, thus changing the water quality. The results obtained from the water quality samples taken during or close to the overpass of a satellite can be used to calibrate the interpretation algorithms of the satellite data, thus making it possible to estimate water quality for all of the lakes included in the satellite image. In addition, if something exceptional or unusual is observed in the lake water quality on the basis of satellite image analysis, then traditional water sampling can subsequently be arranged.

(C) Since remote sensing provides information on the spatial distribution of water quality, the location of the sampling stations used in the regular monitoring programmes can be optimized. In this way, water samples from the most important, representative stations can be obtained. The remote-sensing data can also be used in planning the optimal location of sampling stations for each time a lake or a group of lakes is monitored. For such a use, the qualitative examination of remote-sensing data is often sufficient.

(D) Turbidity and high concentrations of suspended solids can be interpreted rather accurately from remote-sensing data. This can be utilized in monitoring the transport of inflowing waters which are rich in suspended solids and nutrients during high-flow periods. Another example of such an application is provided by water construction works (dredging), from which large loads of suspended solids can spread to the water bodies.

The effective use of remote-sensing images requires that the interpretated water quality maps are easily available to the end users. The images could be included in the environment data system, which will make it possible for end users to combine information in the images with other environment data such as laboratory water quality analyses, as well as hydrological and weather data. In addition, other geographical information systems (GIS) data which support the use of interpretated images, should be available. This includes the location of the monitoring stations and polluters, shore-line information and bathymetric maps.

REFERENCES

Bukata, R. P., Jerome, J. H., Kondratyev, K. Ya. and Pozdnyakov, D. V., 1995. *Optical Properties and Remote Sensing on Inland and Coastal Waters*. CRC Press. Boca Raton, FL, USA.

Dekker, A. G., 1993. Detection of Optical Water Quality Parameters for Eutrophic Waters by High Resolution Remote Sensing. PhD Thesis, Vrije Universiteit. Amsterdam, The Netherlands.

Fischer, J. and Doerffer, R., 1987. An inverse technique for remote detection of suspended matter, phytoplankton and yellow subtances from CZCS measurements, *Advances in Space Research* **7**, 21–26.

Gitelson, A., Garbuzov, G., Szilagyi, F., Mittenzwey, K.-H., Karniel, A. and Kaiser, A., 1993. Qualitative remote sensing methods for real-time monitoring of inland waters quality, *Int. J. Remote Sensing* **14**, 1269–1295.

Kallio, K., Kutser, T., Koponen, S., Hannonen, T. and Herlevi, A., 1998. Estimation of water quality in Finnish lakes by an airborne spectrometer. Proceedings of the Fifth International Conference on Remote Sensing for Marine and Coastal Environments, San Diego, CA, October 5–7, 1998, Volume II, 333–340.

Kirk, J. T. O., 1994. *Light and Photosynthesis in Aquatic Systems*, 2nd Edn, Cambridge University Press, New York.

Kutser, T., 1997. Estimation of Water Quality in Turbid Inland and Coastal Waters by Passive Optical Remote Sensing, PhD Thesis, University of Tartu, Estonia.

Kutser, T., Arst, H., Mäekivi, S. and Kallaste, K., 1998. Estimation of the water quality of the Baltic sea and lakes in Estonia and Finland by passive optical remote sensing measurements on board vessel, *Lakes and Reservoirs: Research and Management*, **3**, 53–66.

Van Stokkom, H. T. C., Stokman, G. N. M. and Hovenier, J. W., 1993. Quantative use of passive optical remote sensing over coastal and inland water bodies, *Int. J. Remote Sensing*, **14**, 541–563.

Chapter 4.4

Principles of Monitoring the Acidification of Lakes

JAAKKO MANNIO

Hydrological and Limnological Aspects of Lake Monitoring
Edited by Pertti Heinonen, Giuliano Ziglio and André Van der Beken
©2000 John Wiley & Sons, Ltd, ISBN 0 471 89988 7

4.4.1 INTRODUCTION

Acidification of freshwater systems has provided some of the earliest evidence of the damage caused by sulfur emissions. International agreements on the reduction of these emissions under the United Nations Economic Commission for Europe (UN/ECE) Convention on Long-Range Transboundary Air Pollution (LRTAP) have resulted in the general decline in sulfate deposition in Europe and North America since the mid-1980s. However, acidifying nitrogen and neutralizing base-cation deposition have more regional characteristics and have not shown large-scale monotonic trends. The decreased sulfate load from the atmosphere can already be related to the chemical recovery – decreasing sulfate concentration and increasing buffer capacity – of many aquatic systems in different parts of Europe and North America (Stoddard *et al.*, 1997).

It still remains unclear whether the current and planned emission reductions will be sufficient to meet the environmental objectives. Monitoring is needed to verify that the control programmes actually achieve the desired results (i.e. restoration of healthy biotic communities in areas affected by acidic precipitation), and to determine whether emission-reduction strategies require further adjustments.

The assessment of the causes and consequences of the changes in the environment requires complete and consistent water quality data sets which include all major cations and anions. The frequency of sampling is dependent on the system (lake/stream), scope (spatial/temporal/ long-term/short-term variation), expected magnitude of change, and resources available. However, in order to link the chemical changes to true environmental effects, information of tolerance limits of carefully selected, cosmopolitan indicator species are also needed.

If the data are intended to be used internationally, the compilation of different national data sets for common evaluation and the regionalization in data analyses set special demands on the harmonization of site selection, sampling and analysing. Regular intercomparison tests for chemical analyses, and preferably also for biological analyses, are extremely important. This chapter provides an overview of the basic principles followed in the monitoring activities under the co-operative programmes of the Convention on Long-Range Transboundary Air Pollution (CLRTAP), especially for the work of the International Cooperation Programme on Waters (ICP Waters), the manual for which is extensively referred to here (ICP Waters, 1996). The focus is on the monitoring of lakes, although most of the principles also apply to running waters.

4.4.2 MONITORING STRUCTURE

A monitoring programme to evaluate the environmental effect of acidic deposition on surface waters is best organized in a co-ordinated, and hierarchical or pyramidal manner. At the apex of the pyramid is a small

Table 4.4.1 Schematic overview of the lake-acidification 'monitoring pyramid'. Process-oriented, site-specific information is collected intensively at a few sites, whereas extensive regional information is collected less frequently, but spatially more representatively, in order to relate the problems to large lake populations (50 000 lakes).

Sampling	Sites per area	Main questions	Focus
Daily/ Weekly	< 10	How much? How fast? Why? Combined effects?	PROCESSES
Monthly/ Seasonally	20	How common? When?	"REALITY CHECK"
Annually	< 200	Which types?	
5–10 year interval	1000–2000	Where? How many?	EXTENT

number of intensively monitored sites where sufficient information is collected from relevant ecosystem components so that time-dependent models may be developed and applied to predict the future changes occurring under different deposition scenarios. Beneath the apex is a series of regional networks that employ progressively less comprehensive and frequent sampling but greater spatial coverage. Finally, the pyramid base is composed of a regional survey 'network' in which a large number of sites may be sampled as infrequently as once per decade (Table. 4.4.1).

The regional-scale assessment of the present and future statuses of the ecosystems depends on the continuing existence of all levels within the monitoring hierarchy. The regional monitoring/survey sites have generally been selected to be representative of the areas considered sensitive to acidic deposition. In lake-rich countries, statistical methods have been used in the monitoring/survey-site selection procedure, which is the most reliable way to assure the representativeness of these sites (Landers *et al.*, 1988; Brooksbank *et al.*, 1989; Forsius *et al.*, 1990; Henriksen *et al.*, 1998).

4.4.3 SITE-SELECTION CRITERIA

In regions where surface waters are universally sensitive to acidification, sites should be chosen to represent the diversity of the region (chemically, biologically and geographically). If surface waters exhibit a wide range of acid sensitivity, sites should be chosen from among the most susceptible to acidification. The aim of the site selection should be to focus primarily on sites that are likely to change in response to acid deposition, and secondarily to represent the region as a whole.

Selected sites should not have a pronounced influence from local sources of pollution in the catchment that may lead to misinterpretation of chemical and biological data (domestic sewage, industrial waste water, agriculture, road de-icing, etc). Valuable long-term records may also be lost due to significant local changes such as liming mitigation programmes and changes in forestry practices. Confidence in the future protection of the site from changes in local influences is important. National parks and nature reserves should be considered for sampling sites. Reference or background sites should be included if possible in order to provide information by which the analyses can be compared with respect to climatic or other influences.

Sufficient sites must be included to provide for statistical confidence in regional analysis of changes or trends. The number of sites should be balanced against the ability to support the monitoring on a sustainable and long-term basis. Sites with long time series are preferred, if the other main criteria are met.

Drainage lakes are best suited for monitoring. Lakes should be selected in the headwater part of the catchment, without a larger lake upstream. Lakes with very long residence times (seepage lakes) react slowly to changes in deposition of air pollutants and are therefore less suitable for detecting trends on decade time-scales. A sampling site that is not directly influenced by any inlet stream, preferably in the outlet area, is optimal. A lake outlet is also a possible option. For long-term monitoring, epilimnion samples or surface samples (0.1–1 m) are sufficient, with additional samples of the depth profile being optional. If a palaeolimnological investigation of sediments is planned, then the lake should be sufficiently deep to avoid resuspension.

4.4.4   SAMPLING FREQUENCY FOR WATER QUALITY

Sample collection is an essential link in the monitoring programme and the accuracy and reliability of the final results depend upon the representativeness of the sample of the actual site characteristics that are to be monitored. If the sample is not representative, then the data obtained and subsequent interpretation of trends may be incorrect or misleading.

Lakes exhibit a wide range of hydrologic characteristics, from very fast-flushing drainage lakes, to seepage lakes with long residence times. Sampling frequency should be designed to characterize well the lake's annual variability. Monthly samples are recommended for most fast-flushing lakes; more frequent sampling may be required occasionally in lakes that undergo short-lived acidic episodes or nitrate peaks. In addition, where flow data are available for calculations of yearly transport values of elements from catchments, increased sampling frequency in flood periods is recommended.

Quarterly or seasonal sampling is likely to be adequate in lakes with long residence times. In remote areas where frequent sampling is impossible for practical and economical reasons, even one sample per year may be useful for long-term monitoring. Such samples must be taken at the same time of the year, each year, preferably shortly after fall overturn. For yearly sampling, it is recommended that a group of lakes, rather than a single lake, is selected.

4.4.5 PHYSICO-CHEMICAL VARIABLES

Non-filtered samples are generally preferred for lake waters. The essential determinants are those which define the degree of and causes behind the acidification of surface waters such as the following:

- **anions** HCO_3^- or alkalinity, SO_4^{2-}, NO_3^-, Cl^- and (TOC)/(DOC) (to estimate the organic anion);
- **cations** Ca^{2+}, Mg^{2+}, K^+, Na^+, H^+ and Al^{n+} (labile Al);
- **other** total N, specific conductivity.

Desirable variables are those giving additional biogeochemical informations:

- $T(°C)$, NH_4^+, F^-, total P, soluble reactive PO_4, O_2, SiO_2, Fe, Mn, Cd, Zn, Cu, Ni, Pb, As, Cr, Mo, total Al

In many boreal regions with extensive amounts of brown-water lakes, particular attention must be paid to the role of the organic anion. In Finland, for example, the catchment derived organic acidity (A^-) commonly exceeds the anthropogenic acidity, i.e. $A^- > SO_4^{2-} + NO_3^-$ (Kortelainen and Mannio, 1990).

4.4.6 OTHER SITE INFORMATION

Interpretation of the data will require various types of information on the characteristics of the monitoring sites. The following data are needed for most analyses, i.e. lake area and average depth, catchment area, hydrologic type, elevation, precipitation and average runoff. Additional data for critical-load calculations include deposition of S and N, dominant bedrock type, soil type and depth, vegetation and land-use types. Details of process-oriented (Forsius *et al.*, 1996) or regional predictive modelling are beyond the scope of this present paper. A detailed analysis of the present knowledge of nitrogen processes in catchments can be found in Chapter 1.4.

4.4.7 QUALITY ASSURANCE

Laboratories participating in international programmes providing chemical analyses should be certified under one of the laboratory accreditation systems, e.g. EN 45001 and ISO/IEC Guide 25. Data from uncertified laboratories will be subject to more detailed scrutiny prior to acceptance. All laboratories that participate in any kind of co-operative programme (national or international) should provide documented evidence that in-laboratory quality control (QC) is maintained in order to assure the accuracy and uniformity of routine laboratory analyses. In-laboratory QC should include documentation of the methods of control (use of control samples and control charts), evidence of analytical performance (including analysis of external audit materials), accuracy of in-house standards and accuracy of methods employed, etc.

The quality of the sample-specific data, such as an adequate ionic balance or specific conductivity determination for individual samples is of great importance. The target accuracy for the ion balance should be that the difference between the sum of the cations and the sum of the anions should not exceed 10% of the total cations. Organic anions can be approximated from TOC/DOC measurements and pH (Oliver *et al.*, 1983), or with charge balance (cations minus inorganic anions). The calculated conductivity will indicate whether one or several analytical measurements are too low or too high.

Unless in-laboratory quality control is carried out as normal operating practice, there is little benefit of between-laboratory quality control programmes, which can be organized by obliging the laboratories to take part in ring tests under full identification (Hovind, 1996). Samples prepared by the control laboratory are targeted to test for any bias in analyses of the principal determinants of the programme. Participating laboratories are assessed on their reported determinations in relation to the other participants. If a particular variable lies outside an acceptable ($\pm 10\%$) deviation, no data related to that variable are included in the database. Only data from laboratories that have participated in ring tests are acceptable. When a laboratory's practices deviate from the recommended analytical method, the laboratory is required to demonstrate that the methods produce values that are similar ($\pm 10\%$) to the recommended method.

4.4.8   TREND DETECTION

In order to analyse long-term site-specific trends in water quality, the non-parametric Seasonal Kendall Test (SKT) (Hirsch *et al.*, 1982) has become a standard technique (Stoddard *et al.*, 1997). This test can accommodate the non-normality, missing data and seasonality that are common to these kinds of data. In a statistical sense, it is a powerful test for detecting a trend, that, however, must proceed in one direction only (decreasing or increasing) (Loftis and Taylor, 1989).

4.4.9   BIOLOGICAL MONITORING

In many cases, biota can integrate the average or worst conditions of the site more effectively than chemical monitoring (see also Chapters 1.5, 2.4 and 3.1).

At present, the most widely used indicator groups are invertebrates in running waters and fish and diatoms in lakes. A number of species are known to be sensitive to acid conditions and their presence/absence will indicate both current and recent past conditions. For example, molluscs are rarely found below a pH of 5.6, or gammarids below pH 6. Numerical relationships between the pH of water and species response have been developed for diatom assemblages, chrysophyte, chydorid and invertebrate communities where resolutions of 0.3 to 0.5 pH units are possible if calibrations for local conditions are developed (ICP Waters, 1996).

Species that are tolerant of acid conditions and favoured by the absence of predation should be monitored for relative abundance. For example, the abundance of Coleoptera, Corixids, Polycentropodidae, and the relative abundance of Ephemeroptera/Plecoptera, in running water will be indicative of the degree of acidity (Raddum and Fjellheim, 1994).

Monitoring of fish species with multimesh gill nets provides a figure for the fish community and the relative abundance (number and biomass) of catchable fish species in a lake (Appelberg *et al.*, 1995). For a single fish species, length frequency distribution, age distribution, back calculated growth and sex ratio can be determined. Electro-fishing can be applied in lake littorals in order to complete the information obtained by gill netting. Both methods should be applied at the end of the growth season but before the possible spawning migration.

A whole eco-system approach, including birds, for monitoring aquatic effects has been especially adopted in Canada. Changes in waterfowl distribution and production should reflect broad-scale changes in the habitats that they rely upon, with the predominant habitat change being available food resources. Hence, individual trends in species distribution and breeding success should mirror changes in populations and distributions of the fish and invertebrates that they rely upon, and thus should be suitable indicators for the aquatic ecosystems under consideraton (McNicol *et al.*, 1995).

4.4.10 FUTURE NEEDS

The reported lake monitoring records are relatively short, i.e. 5 to 15 years, even in countries well known to be affected by acidification (Driscoll *et al.*, 1995; Jeffries *et al.*, 1995; Mannio and Vuorenmaa, 1995; Skjelkvåle and Henriksen, 1995; Patrick *et al.*, 1996; Vuorenmaa, 1997; Wilander, 1997). The factors behind different responses could be understood better by combining and grouping already existing data. Both site-specific and regional analyses on the effects of short-term hydrological variation are needed, as well as soil and vegetation characteristics influencing sulfate retention and nitrogen saturation.

It is evident that in future the role of nitrogen will be more in focus than that of sulfur. This will set more data requirements, both on background information as well as on the intensity of observations, due to the more complex processes of nitrogen. Another task is the interaction of acidification/recovery processes with possible true trends in climate/hydrology resulting from global change.

In many countries, efforts during the past 10 years have been directed towards mapping and modelling future scenarios. As abiotic factors improve, the ecosystems will move towards recovery, but significant lags in biotic responses may be involved or irreversible changes may have occurred. The assessment of the rate and extent of biological recovery of acidified systems is a very logical but presently neglected topic.

REFERENCES

Appelberg, M., Berger, H. M., Hesthagen, T., Kleiven, E., Kurkilahti, M., Raitaniemi, J. and Rask, M., 1995. Development and intercalibration of methods in Nordic freshwater fish monitoring, *Water Air Soil Pollut*, **85**, 401–406.

Brooksbank, P., Haemmerli, J., Howell, G. and Johnston, L., 1989. Long Range Transport of Airborne Pollutants (LRTAP) Aquatic Effects Monitoring. Environment Canada Technical Bulletin No. 156.

Driscoll, C. T., Postek, K. M., Kretser, W. and Raynal, D. J., 1995. Long-term trends in the chemistry of precipitation and lake water in the Adirondack region of New York, USA, *Water Air Soil Pollut.*, **85**, 583–588.

Forsius, M., Malin, V., Mäkinen, I., Mannio, J., Kämäri, J., Kortelainen, P. and Verta, M., 1990. Finnish lake acidification survey: Survey design and random selection of lakes, *Environmetrics* **1**, 73–88.

Forsius, M., Alveteg, M., Jenkins, A., Johansson, M., Kleemola, S., Lükewille, A., Posch, M., Sverdrup, H., Syri, S. and Walse, C., 1996. Dynamic model applications at selected ICP IM sites, in: Kleemola, S. and Forsius, M. (Eds), 5th Annual Report 1996, UN/ECE ICP, Integrated Monitoring, The Finnish Environment, Vol. 27, Finnish Environment Institute, Helsinki, Finland, 10–24.

Henriksen, A., Skjelkvåle, B., Mannio, J., Wilander, A., Harriman, R., Curtis, C., Jensen, J. P., Fjeld, E. and Moiseenko, T., 1998. Northern European Lake Survey, 1995. Finland, Norway, Sweden, Denmark, Russian Kola, Russian Karelia, Scotland and Wales, *Ambio*, **27**, 80–91.

Hirsch, R. M., Slack, J. R. and Smith, R. A., 1982. Techniques of trend analysis for monthly water quality analysis, W*ater Resources Research* **18**, 107–121.

Hovind, H. 1996. Intercomparison 9610: pH, K_{25}, HCO3, $NO_3 + NO_2$,Cl, SO_4, Ca, Mg, Na, K, total Al, aluminium – reactive and non-labile, TOC and COD_{Mn}, NIVA Report 3550-96, Programme Centre, NIVA, Oslo, Norway.

ICP Waters, 1996. United Nations Economic Commission for Europe, Convention on Long-Range Transboundary Air Pollution (UN/ECE CLRTAP), NIVA Report Sno 3547-96, International Co-operative Programme on Assessment and Monitoring of Acidification of Rivers and Lakes, Programme Manual.

Jeffries, D., Clair, T. A., Dillon, P. J., Papineau, M. and Stainton, M. P., 1995. Trends in surface water acidification at ecological monitoring sites in southeastern Canada, *Water Air Soil Pollut.*, **85**, 577–582.

Kortelainen, P. and Mannio, J., 1990. Organic acidity in Finnish lakes, in *Acidification in Finland*, Kauppi, P., Anttila, P. and Kenttämies, K. (Eds), Springer-Verlag, Berlin, 849–863.

Landers, D. H., Eilers, J. M., Brakke, D. F. and Kellar, P. E., 1988. Characteristics of acidic lakes in the Eastern United States, *Verh. Internat. Verein. Limnol.*, **23**, 152–162.

Loftis, J. C. and Taylor, C. H., 1989. Detecting acid precipitation impacts on lake water quality, *Environmental Management*, **13**, 529–538.

Mannio, J. and Vuorenmaa, J., 1995. Regional monitoring of lake acidification in Finland, *Water Air Soil Pollut.*, **85**, 571–576.

McNicol D. K., Kerekes, J. J., Mallory, M. L., Ross, R. K. and Scheuhammer, A. M., 1995. The Canadian Wildife Service LRTAP Biomonitoring Program. Part 1. Strategy to monitor the biological recovery of aquatic ecosystems in Eastern Canada from the effects of acid rain. Technical Report Series, No. 245, Canadian Wildlife Service.

Oliver, B. G., Thurman, E. M. and Malcolm, R. L., 1983. The contribution of humic substances to the acidity of colored natural waters. *Geochim. Cosmochim. Acta*, **47**: 2031–2035.

Patrick, S., Battarbee, R. W. and Jenkins, A., 1996. Monitoring acid waters in the UK: An overview of the U.K. Acid Waters Monitoring Network and summary of the first interpretative excercise, *Freshwater Biol.*, **36**, 131–150.

Raddum, G. G. and Fjellheim, A., 1994. Invertebrate community changes caused by reduced acidification, in Steinberg, C. E. W. and Wright, R. F. (Eds), *Acidification of Freshwater Ecosystems: Implications for the Future*, John Wiley & Sons.

Skjelkvåle, B.-L. and Henriksen, A., 1995. Acidification in Norway – status and trends. Surface and ground water, *Water Air Soil Pollut.*, **85**, 629–634.

Stoddard, J. L., Jeffries, D. S., Lükewille, A., Clair, T., Dillon, P. J., Driscoll, C. T., Forsius, M., Johannessen, M., Kahl, J. S., Kellogg, J. H., Kemp, A., Mannio, J., Monteith, D., Murdoch, P., Patrick, S., Rebsdorf, A., Skjelkvåle, B.-L., Stainton, M. P., Traaen, T., van Dam, H., Webster, K., Wieting, J. and Wilander, A., 1999. Regional trends in aquatic recovery from acidification in North America and Europe. *Nature*, **401**, 575–578.

Vuorenmaa, J., 1997. Trend assessment of bulk and troughfall deposition and runoff water chemistry at IM sites, in Kleemola, S. and Forsius, M. (Eds), 6th Annual Report 1997, UN/ECE ICP, Integrated Monitoring, The Finnish Environment, Vol. 116, Finnish Environment Institute, Helsinki, Finland, 24–26.

Wilander, A., 1997. Referenssjöarnas vattenkemi under 12 år; tillstånd och trender, Naturvårdsverket Rapport 4652 (in Swedish).

Part Five
Quality Assessment

Chapter 5.1

Quality Assurance for Water Analysis

PENTTI MINKKINEN

Hydrological and Limnological Aspects of Lake Monitoring
Edited by Pertti Heinonen, Giuliano Ziglio and André Van der Beken
©2000 John Wiley & Sons, Ltd, ISBN 0 471 89988 7

5.1.1 INTRODUCTION

Analytical determination is usually a multi-step process which starts by sampling and ends with the actual measurement and interpretation of the results (see Figure 5.1.1). It is also an error-generating process where each step may generate its own error component; these errors may be either random or systematic. A good quality control system has to take both of these error types into account.

When the overall reliability of an analytical determination is estimated, the variance (squared errors) of the individual error-generating steps of the whole analytical process determines the overall quality of the result. Sampling and the probable error components which are generated by carrying this out, therefore, cannot be neglected when optimal analytical strategy for a given determination is being considered.

An international guide for quality control for water analysis has been produced (ISO, 1997). In this present paper, various methods will be discussed, which are useful for the following:

- estimating and controlling of errors related to sampling;
- method comparison;
- estimating the precision (total random error) of routine analyses;
- analysing interlaboratory comparisons.

5.1.2 ESTIMATION AND CONTROL OF ERRORS RELATED TO SAMPLING

A thorough study of the sampling error has been made by Gy (1992) and by Pitard (1989). Gy's theory (1992) is a general one, and also applies to water sampling and is useful when sampling procedures and equipment are designed or their perfomance is audited. The essential theory on sampling for analytical purposes can be found in a recent reference (Gy, 1998). In the sampling of flowing streams, the main source generating the sampling errors is the heterogeneity of the sampling target (lot) which is manifested as the fluctuation of the concentration of the determinant along the process stream and with respect to the location across the cross-section of the process stream. Figure 5.1.2 shows Gy's classification of the components of the overall determination error. The sampling error is especially important if the process stream or material to be analysed consists of an intrinsically heterogeneous material, i.e. it consists either of particulate solids or, in the case of process streams, of other type of multiphase flows such as, e.g. slurries, non-mixing liquids, emulsions, and dusty gas flows.

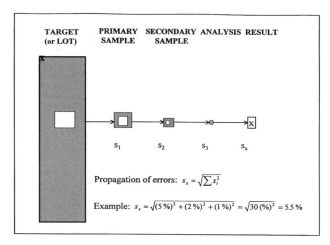

Figure 5.1.1 An analytical process is usually a chain of error-generating processes, where the weakest link often determines the overall reliability.

5.1.3 COMPONENTS OF THE SAMPLING ERROR

According to Gy's sampling theory, the total sampling error consists of two main components, namely the sample-selection error (caused by the heterogeneity of the sampling target, which makes the sampling error dependent upon the sampling time and location) and the preparation error (which can occur at any stage of the sampling process). Gy (1992) further divided the components of the sample-selection error into three different classes based on the sources of the errors. These are the integration error, the sample-materialization error and the weighting error (see below). The integration error may have three further components which are caused by *short-range quality fluctuations*, *long-range quality fluctuations*, and *periodic quality fluctuations*. The sample-materialization error may also have three different components; these are *sample-delimitation errors*, *sample-extraction errors* and *sample-preparation errors*. All of these components will be discussed further below.

5.1.4 WEIGHTING ERROR

The weighting error is caused by the possible fluctuation in the flow-rate of the process stream. Both the mean concentration over a given time domain and the total mass flow of the component to be analysed will be erroneous unless the individual results are weighted by the respective flow-rates. This error becomes

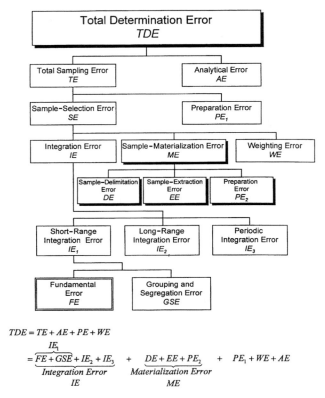

Figure 5.1.2 Components of an analytical determination error, according to Guy (1992).

smaller, however, when the number of results are increased and is completely removed if either the flow-rate is, or if it can be kept, constant, or if the material is homogeneous, having no time-dependent variation in its composition. A weighting error is also generated if the simple mean of the strata of different sizes is calculated without weighting the results with the relative sizes of the different strata.

5.1.5 INCREMENT DELIMITATION AND EXTRACTION ERRORS

If the critical component is not evenly distributed in the cross-section of the process stream from which the actual sample is drawn, the increment-delimitation error may be significant. This error can be cancelled, however, if the sample is made of a complete slice, with a uniform thickness, from the process stream. Another method used to avoid or to minimize this error is to homogenize the distribution by mixing the stream before sampling. With liquids having just one phase, complete mixing can be achieved but with, e.g. suspended solids, it is more difficult. Particulate materials have a high tendency to segregate during transportation. The increment-extraction error refers to the actual removal of a sample from the process stream. This error is likely to be significant if any material escapes from the intended sample, or any extra material from outside the intended sample profile comes into the sample. It is usually caused by improper design of the sampling device, and, consequently, can be avoided by proper design. These errors have to be considered whenever a heterogeneous stream is sampled by drawing a sample stream through a fixed probe inserted into the process stream.

5.1.6   SHORT-TERM INTEGRATION ERROR

Short-term integration error may have two components, namely the fundamental error and the grouping and segregation error. The fundamental error is the pure random sampling error, i.e. the error component which is left if all of the other errors have been eliminated, which, at least theoretically, can be achieved by ideal mixing of the lot prior to sampling and by using a correct sampling technique. It is the only error whose long-term mean necessarily approaches zero. All of the other components of the sampling error may cause biased results, i.e. their long-term mean does not necessarily tend to zero. The fundamental error depends solely on the sample size and on some basic properties of the material to be sampled. It is also the only error component that can be estimated theoretically, if the necessary material properties have been estimated. All of the other error components have to be estimated empirically. The estimation of the fundamental error, therefore, forms the basis of the evaluation of the reliability of all sampling procedures for intrinsically heterogeneous materials. When a reliable estimate of the fundamental sampling error is available, the significance of other error components can be tested against it. For intrinsically homogeneous materials (completely mixed gases and liquids), the fundamental error is practically zero, unless extremely small volumes and dilute mixtures are being analysed.

If this is the case, the standard deviation of the number of particles in the samples (s_n) and the relative standard deviation (s_r) of the fundamental error can be easily estimated as follows:

$$s_n = \sqrt{n} \tag{1}$$

$$s_r = \frac{100(\%)}{\sqrt{n}} \tag{2}$$

where n is the number of molecules, atoms or particles of the critical component in the sample. Equations (1) and (2) are applicable if the expectancy value of the number of critical particles in the sample can be estimated. These equations can be used for all methods in which the particles are counted, e.g. estimation of bacteria concentration. For particulate materials the relative standard deviation of the fundamental sampling error can be estimated from certain basic material properties.

The grouping and segregation error may be significant if the distribution of the critical particles cannot be completely homogenized before sampling, as is often the case with particulate process streams having high flow-rates. This error component will be minimized by (1), making the gross sample from as many increments as possible (taking care that an increment-delimitation error is not introduced) and (2), homogenizing the distribution of the critical particles by careful mixing before sampling. If the correct sampling practice is obeyed, it is safe to assume that the variance of the total short-range integration error is smaller than or equal to the fundamental-error variance.

5.1.7 LONG-RANGE AND PERIODIC COMPONENTS OF THE INTEGRATION ERRORS

The long-range integration error for a given sampling strategy can be estimated by using a variographic technique. This has been used in geochemistry to estimate spatial averages and to estimate concentrations in points within sampling grids by interpolation (a method known as kriging). Gy (1992) has modified this method to deal with sampling targets that can be handled as one-dimensional lots. Rivers and process streams can be regarded as being one-dimensional from the sampling point of view if the sample consists of a full cross-section of the stream or if a complete mixing is achieved before sampling so that the sample is representative of the cross-section of the stream at the time of sampling.

To characterize the variability (heterogeneity) of the target to be sampled a variographic experiment can be carried out. This involves of a series of samples taken at fixed intervals. From this experiment, the heterogeneity of the process can be calculated, together with the variance estimates for different sampling strategies in estimating the means.

If the measurements in a time series are auto-correlated, the points which are not far apart show a lower variability than the whole process, and where the variance and standard deviation of the mean of several (say n) samples depends on the strategy used to select these samples.

The three basic selection methods are as follows:

- Random selection (samples are taken at randomly selected intervals).

- Stratified (random) selection. When this method is used, the lot to be sampled is first divided into n sub-lots of equal sizes and then one sample is drawn randomly from each sub-lot. For an autocorrelated time series, this method always gives a better precision than the random selection.
- Systematic (stratified) selection. In this method, n samples are drawn from the lot at fixed intervals. This method usually gives the best precision, but there is one important exception. If the process has a periodic variation and the sampling frequency is a submultiple of the process frequency, then this method may have the poorest precision. When this method is used, therefore, it is important to check that there is no interfering periodicity in the process. If the process shows periodicity, either stratified sampling selection should be used or systematic sampling with a frequency of at least two samples per process period.

5.1.8 CONTROL OF RANDOM ERRORS

Random errors can always be estimated by repeating the measurements. However, reference materials cannot be used to evaluate the overall precision of the analytical results. The main reason for this is that they usually enter the analytical process at the level where the analytical sample is prepared and thus important error sources, especially the primary sampling steps, are left out. In addition, the matrix interferences in reference materials may differ from those in routine samples. Therefore, the estimates of the random errors should always be based on routine samples covering the concentration range and sample types for which the analytical method is used.

5.1.9 USE OF REPLICATES TO ESTIMATE RANDOM ERRORS

The absolute or relative standard deviations can be calculated from replicated determinations. One difficulty here is that usually both standard deviations depend on concentration. In a typical case, the absolute standard deviation increases when a concentration increases and the relative standard deviation (coefficient of variation) increases rapidly when the concentration level approaches the detection limit of the method. Thompson and Howarth (1973, 1976) used duplicate measurements to estimate the concentration dependence of the standard deviation. Their procedure can be simplified if the method is used at its optimal concentration range, because the relative standard deviation may then be regarded as being independent of concentration. In this case, the relative standard deviation may then be regarded as being independent of the concentration and the individual estimates may be pooled into a new estimate representing the whole concentration range used

Table 5.1.1 Estimation of the routine precision of ammonium determinations by using duplicate samples.

Sample no.	x_1 $(\mu g\,L)^{-1}$	x_2 $(\mu g\,L)^{-1}$	Mean $(\mu g\,L)^{-1}$	s $(\mu g\,L)^{-1}$	s_r (%)
1	23.3	22.1	22.70	0.85	3.74
2	41.2	41.1	41.15	0.07	0.17
3	42.0	42.0	42.00	0.00	0.00
4	51.0	51.0	51.00	0.00	0.00
5	52.9	49.6	51.25	2.33	4.55
6	88.5	85.9	87.24	1.84	2.10
7	111.6	112.1	111.85	0.36	0.32
8	377.7	387.0	382.35	6.58	1.72
9	409.8	402.4	406.10	5.24	1.29
10	447.0	455.0	451.00	5.66	1.25
11	508.5	479.0	493.75	20.9	4.22
12	2915	2555.0	2735.0	254.6	9.31
13	3200	3300.0	3250.0	70.7	2.18

(Minkkinen, 1986). The absolute standard deviation can usually be regarded as being constant only if a relatively narrow concentration range is in use. Table 5.1.1 shows an example of how duplicate wastewater samples were used to estimate the laboratory precision in ammonium determinations.

Duplicate determinations were made on randomly selected samples. The absolute standard deviation clearly increases with concentration whereas the relative standard deviation estimates show no clear concentration-dependence. Both of the standard deviation estimates from the duplicates have only one degree of freedom, and thus they alone are not very reliable estimators of the random error. If they can be assumed to be independent estimates of the same parent standard deviation, such as the relative standard deviation estimates in this case, they can then be pooled into a more reliable estimate having n (i.e. number of replicates) degrees of freedom, as follows:

$$s_r = \sqrt{\frac{1}{n}\sum s_{r_i}^2} = 3.45\% \qquad (3)$$

This value can be used for the relative standard deviation for the whole concentration range. With the help of s_r, the confidence interval for the random errors over this concentration range is also obtained by multiplying s_r with the appropriate value from Student's t-distribution. The 95% confidence interval for the relative random errors is then given as follows:

$$E_r = \pm 2.16 \times 3.45\% = \pm 7.5\% \qquad (4)$$

When planning a quality control programme, one should carefully think whether the purpose is to control the complete analytical process or just some

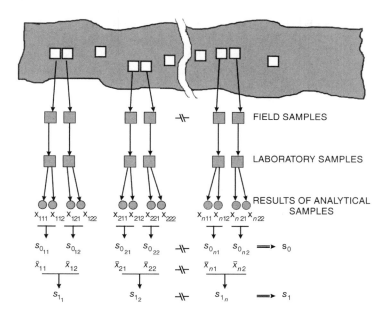

Figure 5.1.3 Components of the random-error variance can be estimated from replicate samples; field replicates are needed to control the overall variance, while the within-laboratory variance can be estimated from replicates made from the laboratory samples.

parts of it. Normally the replicates for control purposes should be selected from two levels, i.e. closely taken field samples and replicates from laboratory samples (Figure 5.1.3).

From this design, the most important error variances (and standard deviations) can be estimated and their level controlled. The overall variance of an analytical result can be modelled by using the following equation:

$$\sigma_T^2 = \sigma_b^2 + \sigma_s^2 + \sigma_0^2 \tag{5}$$

where σ_b^2 is the variance of the long-term and integration error of sampling, which for a systematic and stratified sampling plan should be estimated by using, e.g. Gy's variographic technique (Gy, 1992, 1998). In equation (5), σ_s^2 is the variance component which represents the sum of the variances of the short-term integration error and the preparation of the laboratory sample, while σ_0^2 is the variance component which represents the sum of the variances of the analytical measurement and the preparation of the analytical sample.

In order to estimate the two latter variances (σ_s^2 and σ_0^2), the data provided by the analytical replicate field samples and analytical replicate laboratory samples can be used.

5.1.10   CONTROL OF SYSTEMATIC ERRORS

Unlike random errors, which can always be estimated from repeated measurements, the systematic errors can be detected and estimated only if a reference value (a known, or assumed true, or consensus value) is available. The usual methods which are used to detect systematic errors (bias) are as follows:

• Analysis of reagent blanks. These are carried out especially to control the contamination of samples.
• Analysis of reference materials (synthetic or certified) with known concentrations. If possible, these should be run among routine samples so that they do not receive any special treatment.
• Recovery tests with spiked samples. This method can be used, e.g. when suitable reference materials are not available.
• Method comparisons. The methods tested should be independent, i.e. sample preparation and calibration should be made separately for the different methods to be tested.
• Interlaboratory comparisons. These are carried out as part of method development and standardization (all laboratories are advised to use the same method), or to test the performance of laboratories (proficiency tests, where laboratories can use their own methods). Interlaboratory comparisons are also often organized by producers of reference material to obtain the reference values (consensus values) for their materials.
• Plausibility checks based on the experience of the analyst.

5.1.11   COMPARISON TO REFERENCE VALUES

Systematic errors are also often dependent on concentration and, as in the case of random-error monitoring, bias monitoring tests should therefore be designed to cover the whole useful concentration range over which the methods are being used. Often the best method to detect concentration dependencies and to plan the subsequent data analysis is to plot the absolute deviations (equation (6)) and relative deviations (equation (7)) from the reference values as a function of concentration, as shown in Figure 5.1.5 below (from Bruce *et al.*, 1998):

$$d_i = x_i - x_{\text{ref}_i} \qquad\qquad (6)$$

$$d_{r_i} = \frac{x_i - x_{\text{ref}_i}}{x_{\text{ref}_i}} 100\% \qquad\qquad (7)$$

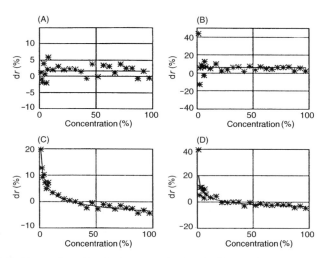

Figure 5.1.4 Examples of typical curves obtained when the relative deviations from the reference values are plotted as a function of concentration in analytical determinations.

In these two equations x_i is the test result on sample i, while x_{ref_i} is the corresponding reference value (result from the reference method carried out on sample i, the recommended or known value of reference material i used in tests, or the mean of the accepted values for sample i in interlaboratory comparisons). Figure 5.1.4 shows the four basic shapes of the curves obtained (A–D) when the deviations from the reference values are plotted as a function of concentration:

(A) Neither the random nor the systematic error have structure, i.e. the points are scattered around a line parallel to the concentration axis, which is the mean of the deviations. In this case, the t-test can be used to check if the mean of the relative deviations differs significantly from zero, i.e.

$$t = \frac{\bar{d}_r - 0}{s_{d_r}} \sqrt{n} \qquad (8)$$

where s_{d_r} is the standard deviation of the deviations and n is the number of test results.

If the test is significant, the mean estimate of the systematic relative random error always increases rapidly at the vicinity of the detection limit.

(B) The systematic error shows no trend, but the random error is heteroscedastic and is dependent on concentration (the absolute random error is usually an increasing function of concentration). The mean of the deviations is also in this case the estimate of the random error, but its significance cannot be tested by using the t-test. This can be done by using

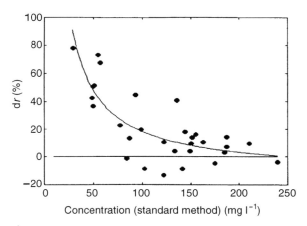

Figure 5.1.5 Illustration of the relative differences found between the results obtained by using the standard method and a modified (standard) method for BOD_7 determinations (see text for details). The data used includes results for both syntetic and 'real' samples.

    non-parametric tests, e.g. the Wilcoxon matched-pairs signed-rank test (Beyer, 1986).

(C) The random error can be regarded as being independent of concentration (homoscedastic), while the systematic error is clearly a function of the concentration. The systematic error can be modelled by using ordinary least-squares regression and, if necessary, the test results can be corrected for the systematic error by using this function.

(D) Both of the random and systematic error seem to be dependent on concentration. The systematic error can also be modelled in this case, but instead of ordinary least squares, a weighted regression calculation should be used.

Figure 5.1.5 shows another example, in which this method was used successfully to detect problems in BOD_7 determinations. In this case, a factory had changed the standard method into a simpler method after testing it with synthetic and wastewater samples against the standard method. In order to check the method synthetic samples were run regularly with actual samples, and the results showed that the method seemed to perform reliably. The factory also participated in interlaboratory testing programmes with other laboratories who used the standard methods, but the results were not consistent; in some samples, both methods agreed, while in other samples the non-standard method seemed to differ from the standard. When the results were plotted as shown in Figure 5.5.1, the problem was obvious. This figure shows that the method used gave much higher results than the standard method at lower concentrations, being approximately 25% too high at the factory's allowed level of discharge, which corresponded to about 100 mg BOD L^{-1}. Based on these erroneous results, the factory believed that they were exceeding their allowed BOD discharge levels and an emergency investment plan of 20 million FIM (3.3 million Euros) on a new wastewater treatment system had already been accepted by the management. This example emphasizes two important points. First, the control programme should cover the whole concentration range of interest and secondly, the data, after they have been obtained, should also be analysed properly and not just recorded. The basic problem here was that when the new method was adopted the operation level was at about 200 mg BOD L^{-1}. Over the following years when production increased and tighter regulations were imposed by the water authorities, the operation level was reduced to lower concentrations. The synthetic check samples, however, were prepared according to the original method specification and had a concentration equivalent of 200 mg BOD L^{-1}, where both methods agreed, and thus the problem was not recognized previously by using the synthetic control samples.

5.1.12 INTERLABORATORY COMPARISONS

Interlaboratory comparison are an effective method for testing the performance of analytical methods and laboratories. Guidance on how to organize interlaboratory exercises are given in various international standards (ISO, 1986) and guides (Youden and Steiner, 1975; Caulcutt and Boddy, 1983; Wernimont, 1985; Horwitz, 1994, 1995). The method presented here is especially designed for estimating the error components; further details can be found in the paper by Minkkinen (1995).

In interlaboratory comparisons, the total variance σ_t^2 of one determination (or reproducibility variance) can be broken up into three components, as follows:

$$\sigma_t^2 = \sigma_b^2 + \sigma_{ia}^2 + \sigma_0^2 \qquad (9)$$

where σ_0 is the random standard deviation of the method used (or repeatability standard deviation) and σ_{ia} and σ_b are the standard deviations of two different kind of systematic errors. Of these, σ_{ia} is the sample-laboratory interaction standard deviation (caused by systematic errors which may vary from sample to sample), and σ_b is the standard deviation of the bias (caused by constant systematic differences between the laboratories). All three variance components can be estimated by analysis of the variance from a properly designed experiment, if all laboratories make replicate determinations. If only single determinations are made, then only two variance components of the total variance can be estimated:

$$\sigma_t^2 = \sigma_b^2 + \sigma_1^2 \qquad (10)$$

where σ_1^2 is the within-laboratory variance (sum of the interaction and repeatability variances, i.e. $\sigma_{ia}^2 + \sigma_0^2$). Here the latter type of interlaboratory comparison data are analysed as an example. The comparison was carried out in order to check the total phosphorus determination from wastewater samples, and Table 5.1.2 gives the raw data and some statistics calculated from them, both with and without laboratories 7 and 8, which were regarded as outliers.

This table shows that the relative standard deviation seems to be independent of concentration over this concentration range. The data analysis was therefore based on relative differences, calculated for each result as the relative deviations from the sample means:

$$d_{r_{ij}} = \frac{x_{ij} - \bar{x}_j}{\bar{x}_j} 100\% \tag{11}$$

Table 5.1.3 gives these results for a set of laboratories. It can be seen that Laboratory 8 has produced 70% lower results than the overall average, while Laboratory 7 has a much higher within-laboratory standard deviation than the other laboratories, and therefore both of these sets of results can be rejected from the calculations. The data shown in Table 5.1.4, which contains the relative deviations for the accepted laboratories, can be used to estimate the variance and standard deviation components. Following various calculations

Table 5.1.2 Interlaboratory comparison: Determination of total phosphorus in wastewater samples.

| Laboratory | Results, x_{ij} (mg L^{-1}) | | | | |
	Sample 1	Sample 2	Sample 3	Sample 4	Sample 5
1	0.284	1.93	0.780	2.60	5.14
2	0.280	1.87	0.753	2.46	5.07
3	0.280	1.82	0.750	2.46	4.94
4	0.269	1.78	0.698	2.41	4.83
5	0.263	1.65	0.760	2.26	5.04
6	0.260	1.75	0.740	2.18	4.50
7	0.390	1.77	0.580	2.01	3.80
8	0.150	0.404	0.189	0.259	1.09
9	0.247	1.60	0.670	2.25	4.39
10	0.273	1.767	0.701	2.373	4.727
Mean (mg L^{-1})	0.270	1.634	0.662	2.126	4.353
	0.270[a]	1.771[a]	0.732[a]	2.374[a]	4.830[a]
s (mg L^{-1})	0.058	0.443	0.176	0.777	1.215
	0.012[a]	0.108[a]	0.038[a]	0.138[a]	0.273[a]
s_r (%)	21.4	27.1	26.6	31.8	27.9
	4.61[a]	6.11[a]	5.13[a]	5.81[a]	5.66[a]

[a]Values obtained after Laboratories 7 and 8 were rejected.

Table 5.1.3 Relative deviations, plus their means and standard deviations, as calculated from the overall sample means[a].

	Relative deviations from sample means						
Laboratory	Sample 1	Sample 2	Sample 3	Sample 4	Sample 5	Mean (%)	s^b (%)
1	5.34	18.11	17.81	22.29	18.09	16.33	6.42
2	3.86	14.44	13.73	15.70	16.50	12.84	5.13
3	3.86	11.38	13.28	15.70	13.50	11.54	4.56
4	−0.22	8.93	5.42	13.35	10.97	7.69	5.29
5	−2.45	0.97	14.79	6.30	15.79	7.09	8.12
6	−3.56	7.09	11.77	2.53	3.38	4.24	5.69
7	*44.66*	8.32	−12.40	−5.47	−12.70	4.48	*24.02*
8	*−44.36*	*−75.28*	*−71.45*	*−87.8*	*−74.96*	*−70.77*	16.02
9	−8.38	−2.09	1.19	5.82	0.86	−0.52	5.23
10	1.26	8.13	5.88	11.61	8.60	7.10	3.85

[a]Entries shown in italics indicate results differing signficantly from overall average values.
[b]Within-laboratory standard deviation.

Table 5.1.4 Relative deviations, plus their means and standard deviations, as calculated for the accepted laboratories, i.e. without laboratories 7 and 8.

	Relative deviations from sample means (%)						
Laboratory	Sample 1	Sample 2	Sample 3	Sample 4	Sample 5	Mean (%)	s^a (%)
1	5.38	8.99	6.63	9.51	6.43	7.39	1.78
2	3.90	5.60	2.94	3.62	4.98	4.21	1.07
3	3.90	2.77	2.53	3.62	2.29	3.02	0.70
4	−0.19	0.52	−4.58	1.51	0.01	−0.55	2.35
5	−2.41	−6.83	3.90	−4.81	4.36	−1.16	5.07
6	−3.53	−1.18	1.16	−8.18	−6.83	−3.71	3.87
9	−8.35	−9.65	−8.41	−5.23	−9.10	−8.15	1.72
10	1.30	−0.22	−4.17	−0.05	−2.12	−1.05	2.13

Standard deviation of the laboratory means (s_2)						4.84	
Pooled within-laboratory standard deviation (s_1)							2.70

[a]Within-laboratory standard deviation.

the bias (systematic error) and reproducibility standard deviation can be estimated, after first obtaining the means and standard deviations of the d_r values for each laboratory, as shown in Table 5.1.4.

5.1.13 CONCLUSIONS

A very large number of water analyses are carried out for different purposes, e.g. to control the quality of potable water and wastewater. Sea, lake and river waters are regularly analysed to monitor their quality and the impact of wastewater and pollutants on recipient water bodies. In water research, large data sets and long time series, covering some tens of years, are commonplace. In such cases, many laboratories using different methods may have produced the results. Strict quality control and harmonization of the methods of measurement is therefore essential, as otherwise the comparability of the results cannot be guaranteed. It is important to realize that the measurement is not under statistical control unless all of the independent steps of the error-generating process, and the different error types, are taken into account in the quality control programme. If the quality control fails and erroneous results are produced, the consequence at best is just the waste of money and effort, while at the worst the consequences may be disastrous. The optimum analytical strategy, and the error components to be considered, depend on the purpose of the determination, i.e. whether point estimates of the concentrations at different sampling locations or a general description (average) of the lot is what is required. It is also important when planning analytical programmes that the acceptable uncertainty levels for both the analytical determinations and the related quality control programmes are realistically set. Normally, the acceptable uncertainty level depends on how and where the results are used. If the target is too ambitious, then the programme will be costly. Provided that the plan is otherwise optimized, the cost–benefit ratios will not be favorable for uncertainty levels which are narrower than necessary, e.g. if an uncertainty level is set to half of what is necessary, then a four times bigger budget will be required for the analytical work.

REFERENCES

Beyer, W. H. (Ed.), 1986. *CRC Handbook of Tables for Probability and Statistics*, 2nd Edn, CRC Press, Boca Raton, FL, USA, 397–400.

Bruce, P., Minkkinen, P. and Riekkola, M.-L., 1998. Practical method validation: validation sufficient for an analysis method, *Mikrochimica Acta* **128**, 93–106.

Caulcutt, R. and Boddy, R., 1983. *Statistics for Analytical Chemists*, Chapman & Hall, New York.

Gy, P., 1992. *Sampling of Heterogeneous and Dynamic Material Systems. Theories on Heterogeneity, Sampling and Homogenizing*, Elsevier, Amsterdam.

Gy, P., 1998. *Sampling for Analytical Purposes*, John Wiley & Sons, New York.

ISO, 1986. Precision of Test Methods – Determination of Repeatability and Reproducibility for a Standard Test Method by Interlaboratory Tests, ISO 5725-1986, International Organization for Standardization, Geneva, Switzerland.

ISO, 1997. Water Quality – Guide to Analytical Quality Control for Water Analysis, ISO/TR 13530: 1997 (E), International Organization for Standardization, Geneva, Switzerland.

Horwitz, W., 1994. Nomenclature of interlaboratory analytical studies, *Pure Appl. Chem.*, **66**, 1903–1911.

Horwitz, W., 1995. Protocol for the Design, Conduct and Interpretation of Method-performance Studies: Revised 1994 (Technical Report), *Pure Appl. Chem.*, **67**, 331–343.

Minkkinen, P., 1986. Monitoring the precision of routine analyses by using duplicate determinations, *Anal. Chim. Acta*, **191**, 369–376.

Minkkinen, P., 1995. Estimation of variants components from the results of interlaboratory comparisons, *Chemom. Intell. Lab. Syst.*, **29**, 263–270.

Pitard F. F., 1989. *Pierre Gy's Sampling Theory and Sampling Practice*, Vols I and II, CRC Press, Boca Raton, FL, USA.

Thompson, M. and Howarth, R. J., 1973. The rapid estimation and control of precision by duplicate determinations, *Analyst*, **98**, 153–160.

Thompson, M. and Howarth, R. J., 1976. Duplicate analysis in geochemical practice, Part 1: Theoretical approach and estimation of analytical reproducibility, *Analyst*, **101**, 690–698.

Wernimont G. T., 1985. *Use of Statistics to Develop and Evaluate Analytical Methods*, Association of Official Analytical Chemists, Arlington, DE, USA.

Youden, W. J. and Steiner, E. H., 1975). *Statistical Manual of the Association of Official Analytical Chemists*, Association of Official Analytical Chemists, Washington, DC, USA.

Chapter 5.2

Performance Characteristics of Microbiological Water Analysis Methods

SEPPO I. NIEMELÄ

Hydrological and Limnological Aspects of Lake Monitoring
Edited by Pertti Heinonen, Giuliano Ziglio and André Van der Beken
©2000 John Wiley & Sons, Ltd, ISBN 0 471 89988 7

5.2.1   INTRODUCTION

In order to have confidence in their analytical results, laboratories need assurance that the methods they have chosen are not only generally valid but also work satisfactorily in their hands. Choosing a widely accepted standard method is the obvious first step but does not automatically guarantee analytical quality. Every analytical method has a reliability domain, i.e. a set of conditions under which it can be expected to work best. Awareness of the existence of such a domain in microbiological methods is implicit in statements about the lowest and highest reliable colony numbers per plate. Such recommendations have been given almost from the beginning of quantitative microbiology. Two aspects essentially determine the quality of analytical results, namely accuracy and precision. Accuracy means freedom from systematic errors, while precision means closeness of independent repeated measurements or determinations. The limits of the reliable domain are exceeded whenever systematic errors and/or imprecision become uncomfortably high. Due to the special nature of microbiological methods, these aspects are rather tightly bound to the number of colonies per plate.

5.2.2   SPECIAL CHARACTERISTICS OF MICROBIOLOGICAL METHODS

Microbiological methods can be described as the means of counting or estimating living particles. Typically, the detectors used in microbiological methods, which make use of the petri plate count and the most probable number (MPN) series, only function at low particle concentrations. With very few exceptions, the only necessary preparative procedure is homogenization and dilution of the sample. The most distinguishing feature of microbiological methods, however, is the interplay of two biological partners in the analysis. Microbes themselves are needed for amplification, through biological growth, of the signal (presence of a living particle), while a second biological partner, i.e. the analyst, is needed for interpreting and counting the signals.

5.2.2.1   Determination of the lower limit

In chemical metrology, two characteristics, namely the **limit of detection** and **limit of determination**, refer to the low end of the reliability scale. These are derived from precision data. The detection limit is reached when the signal given by the analyte is of the same magnitude as the 'noise' of blank solutions. This idea is not transferrable to microbiological methods because a 'blank' in microbiology is a sterile sample and can only give a signal through contamination. The idea of the limit of determination follows from the observation that with many methods the relative precision tends to be inversely related to the concentration of the analyte. It therefore makes sense to agree on a limit, i.e. the lowest concentration where precision is equal or better than a chosen value. In most cases, finding the limiting value requires specially designed method-performance tests. The precision of low microbiological colony counts is strictly related to the particle number in the analytical portion. The precision is determined by a statistical law, i.e. the Poisson distribution. There is no need to study the limit of determination empirically. The variance (s^2) of the Poisson distribution is numerically equal to the mean (m). The relative standard deviation (*RSD*) or coefficient of variation (*CV*) is in inverse relationship to the mean count, or more generally, to the total count (c) of the detection set (Figure 5.2.1), as follows:

$$RSD = \frac{s}{c} = \frac{\sqrt{c}}{c} = \sqrt{\frac{1}{c}} \tag{1}$$

For example, a single plate count of a sample taken from a perfectly mixed suspension (say 48 colonies), has the theoretical relative standard deviation (or relative precision) of $1/\sqrt{48} = \pm0.144$ (*CV*=14.4%).

The graph given in Figure 5.2.1 shows why certain colony numbers, e.g. 15, 20, 25 or 30, have traditionally been selected as the lowest statistically reliable counts. The rule is universal because it is based on properties of the suspension

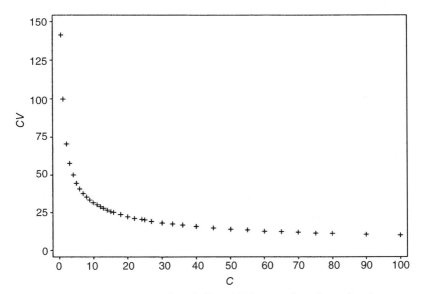

Figure 5.2.1 The coefficient of variation (*CV*) as a function of colony count (c), illustrating the precision of the Poisson distribution.

which are independent of the type of particle. It is not confined to water analysis or any particular cultivation technique.

Similar considerations are not valid for the MPN methods. The precision of MPN estimates depends more on the design (number of parallel tubes and the dilution step) than on the concentration of particles (Cochran, 1950). If the standard 3×5 design is chosen, then the estimate, even at its best, is less precise than is the count of five colonies.

In conclusion, the lower working limit of microbe detectors can be chosen on purely statistical grounds and can be expressed as the lowest reliable count per plate (or set of plates).

One method may detect more microbes in the same sample than the other. In other words, the probability of detecting a particular type of microbe is different for the two methods in the same situation. Both methods have, however, the same limit of detection according to the above definition. The difference is best brought out by comparing the relative recovery of the two methods.

5.2.2.2   Factors determining the highest reliable count

The upper working limit is the most distinctly unique performance characteristic of every microbiological method. It is also a most complex characteristic.

If particle statistics were the only determining factor, the precision of microbiological methods would improve indefinitely with increasing particle numbers (see Figure 5.2.1).

Microbial colonies occupy at least a millionfold space when compared to the original particles. Space becomes limiting as the number of colonies per plate increases; such crowding brings with it an increased overlap error, i.e. a systematic lowering of the count from the expectation. The trueness of the results thus decreases.

Different bacterial species may have very different colony sizes. The size also varies with the composition and even thickness of the nutrient medium. For these reasons, the extent of crowding varies with the bacterial species, nutrient medium, and external conditions. Every method has a characteristic, although rather vague **upper working limit** in terms of particles per test portion.

Dilution of a suspension is a simple and straightforward operation. Perfect **linearity**, i.e. the proportionality of colony counts with volume or dilution of the sample, should be observed. The theoretical linearity, however, breaks down when crowding becomes excessive. Testing for linearity is therefore a most effective way of finding the upper working limit.

5.2.2.3   Problems with selectivity

The presence and development of non-target growth on the plates also influences the upper limit. In this way, the **selectivity** of the detector and the sample microbial population are important determinants of the upper limit. Pure culture colonies can be counted reliably up to several hundred per plate, but when target colonies are outnumbered one to ten or one to a hundred by non-target colonies, the situation is obviously different. The upper limit is considerably lowered, but the amount of bias depends on the average size and metabolic type of the non-target colonies.

5.2.3   PRECISION CHARACTERISTICS

Working within the reliable domain of the method should produce unbiased estimates of acceptable or at least known precision. The foregoing discussion was focused on determining the limits where reasonably reliable estimates of microbial concentration should be possible. There remains the question of determining the precision numerically.

Precision, defined as *closeness of independent repeated observations*, is conceptually simple. It is usually expressed in terms of imprecision and computed as a standard deviation of the test results. However, practical and economical reasons preclude the possibility of repeating each determination enough times to enable computation of the precision of the results individually. One must trust the precision of the method in general or construct an individual case from other information.

Two measures of precision are considered necessary for a method-performance description. **Repeatability** means closeness of agreement under the same conditions of measurement, while **reproducibility** is the closeness of agreement of repeated measurements under changed conditions of measurement. Reproducibility testing normally involves the whole procedure. The basic aspects of these topics can also be found in Chapter 5.1. Due to the decisive role of the operator in counting target colonies, the **reproducibility of counting** is an important intermediate precision characteristic. This is tested and determined by repeating only the last stage of the procedure, i.e. the counting of colonies by different operators. The trust in (international) comparability of results made with the same methods means among other things that every person involved with colony counting would return exactly the same count on the same plate. Methods, however, appear to be very unequal in this respect. Reproducibility of counting is one of the best performance characteristics for comparing different methods.

Robustness (ruggedness) refers to the insensitivity of a method (or actually of the results) to small changes in procedure or in the testing environment. The latter aspect is particularly important in microbiology. The incubation time and temperature cannot be kept within very narrow limits in the long run. In addition, the moisture and the gas atmosphere will vary, and the microbes' ability to recover after stressful storage may depend on the method.

The permissible limits of the influential factors should be ascertained and reported as supplementary information among the method-performance characteristics.

5.2.4   AN ILLUSTRATIVE EXAMPLE

The best method protocols contain a considerable amount of performance data, usually scattered over different parts of the text. The standard total coliform membrane-filter procedure described in the American Public Health Association (APHA) publication, *Standard Methods for the Examination of Water and Wastewater* (APHA, 1992), can be taken as an example. The statement, 'All bacteria that produce a red colony with a metallic sheen within 24 h of incubation are considered members of the coliform group,' provides a qualitative definition of the target group in terms of colony appearance. This has the purpose of harmonizing the interpretations of different operators and provides a rough guide to incubation times. In other parts of the protocol, the following statements appear, 'incubate for 20 to 22 h at $35 \pm 0.5\,°C$', and 'samples of disinfected waters or wastewater effluent may include stressed organisms that grow relatively slowly and produce a maximum sheen in 22 to 24 h; organisms from undisinfected sources may produce a sheen after 16 to 18 h, and the sheen subsequently may fade after 24 to 30 h.' The latter comments give detailed information about the testing conditions under which results are expected to be robust. At the end of the protocol, there is also important information on the reliable quantitative domain, i.e. 'compute the count, using membrane filters with 20 to 80 coliform colonies and not more than 200 colonies of all types per membrane'. This passage gives the upper and lower working limits in terms of target colonies, and also an absolute limit of application in terms of total colonies. Taking both rules simultaneously into account means that the method should not be used whenever non-target colonies outnumber the target colonies by more than ten to one. The lower limit of 20 is obviously selected in order to ensure a tolerable precision, while the upper limits of 80 and 200 undoubtedly reflect the experience that above these limits, cases of overcrowding begin to be troublesome and affect the trueness or precision.

5.2.5   DERIVATION OF PRECISION ESTIMATES

The coliform protocol quoted in the foregoing section provides no information on the actual precision of the analytical result. There are basically three ways of obtaining precision values without actually repeating the particular analysis; these are as follows:

(A) Precision estimates are obtained by collaborative method-performance studies conducted by central organizations. The measured values are believed to be characteristic of the method in all situations resembling the conditions of the collaborative test.

(B) The assumption of a statistical distribution, such as the Poisson distribution, which is known to apply in perfectly mixed microbial suspensions.

(C) The construction of a combined uncertainty estimate from known uncertainties of the different phases of the procedure, according to principles laid out by Eurachem (1995).

5.2.6 CONCLUSION: PRECISION ESTIMATION IN A SIMPLE CASE

Suppose that a total colony count estimate of a sample had been obtained by counting 48 colonies from a single plate seeded with 1 mL of dilution 10^{-4}. The estimate is $Y = 48 \times 10^4 = 4.8 \times 10^5$ per mL, and it would be desirable to attach an uncertainty (precision) value to this estimate.

Lacking any repeated observations, the uncertainty estimate must be obtained by other means. No collaborative method-performance data seem to be available for microbiological water analysis methods based on colony counts. Some coliform MPN estimates have been studied (Edberg *et al.*, 1991). In a study of food analytical methods, Ginn *et al.* (1986) reported results for the repeatabilty and reproducibility of the total colony count of milk in the form of standard deviations at $\log_{10}$ scales. For the repeatability within one laboratory, these workers gave a standard deviation s_r of 0.05, which corresponds rougly to the percentage uncertainty (coefficient of variation) of 11.5% on an arithmetic scale. Milk and water are rather similar matrices. Attaching the repeatability value of 0.05 to the logarithm of the analytical result given above, 4.8×10^5, yields 5.68 ± 0.05. Converting back to an arithmetic scale, the upper limit of the estimate (with two significant digits) would be $N\log 5.73 = 540\,000$ per mL, while the lower limit would be $N\log 5.63 = 430\,000$ per mL.

If we assume a perfectly mixed suspension and no volumetric errors, the Poisson distribution should then be valid. Consequently, the count with standard deviation can be estimated as $48 \pm \sqrt{48} = 48 \pm 7$ ($\pm 14.4\%$). This method of estimation yields the upper limit of $550\,000$ and the lower limit of $410\,000$ per mL.

It is not possible to dilute a sample and measure the inoculum without some volumetric uncertainty. The uncertainties (standard deviations) of different phases of the microbiological process have been studied extensively, and real values can be found in the literature (e.g. Jarvis, 1989). Assuming that the determination was made by diluting the water sample twice by 1:100

(1 mL + 99 mL) and then inoculating 1 mL in the petri dish, the total volumetric uncertainty in the end is due to three consecutive 1 mL measurements. Jarvis (1989) suggests that 1 mL transfers can be made with an average relative standard deviation of about $\pm 2\%$. As the transfers are independent of each other, the combined uncertainties of the three consecutive transfers (ignoring the uncertainties of the 99 mL blanks) amounts to a *dilution error* of:

$$\sqrt{2^2 + 2^2 + 2^2} = \sqrt{12} = 3.46\% \tag{2}$$

The counting of colonies is not completely free from uncertainty either. However, the estimates found in the literature vary considerably. It can be assumed that in a very favourable situation the count is repeatable within the relative standard deviation of $\pm 3\%$. Lastly, the variation of particle numbers in 1 mL of the final suspension due to the Poisson distribution has a coefficient of variation of $100/\sqrt{48}\% = \pm 14.4\%$.

Therefore combining all three sources of uncertainty yields the *compound* relative uncertainty figure as follows:

$$\sqrt{3.46^2 + 3^2 + 14.4^2} = \sqrt{12 + 9 + 208} = 15.1\% \tag{3}$$

and thus $480\,000 \pm 15.1\%$ gives the upper and lower limit as 550 000 and 410 000, respectively.

In this example all three methods of estimation gave results which were very comparable to each other. Should methods (A) and (C) described above give very different results, it might mean that the assumptions about the magnitude of the error components in method (C) were unrealistic or that the collaborative study was made with microbial concentrations far different from the case at hand.

REFERENCES

APHA, 1992. Standard Methods for the Examination of Water and Wastewater, 18th Edn, American Public Health Association/American Water Works Association/Water Environment Federation.

Cochran, W. G., 1950. Estimation of bacterial densities by means of the 'Most Probable Number', *Biometrics* **6**, 39–52.

Edberg, S. C., Allen, M. J. and Smith, D. B., 1991. Defined substrate technology method for rapid and specific simultaneous enumeration of total coliforms and *Escherichia coli* from water: collaborative study. *J. Assoc. of Anal. Chem*, **74**, 526–529.

Eurachem, 1995. Quantifying Uncertainty in Analytical Measurement.

Ginn, R. E., Packard, V. S. and Fox, T. L., 1986. Enumeration of total bacteria and coliforms in milk by dry rehydratable film methods: a collaborative study, *J. Assoc. Off. Anal. Chem.*, **69**, 527–531.

Jarvis, B., 1989. *Statistical Aspects of the Microbiological Analysis of Foods*, Progress in Industrial Microbiology, Vol. 21, Elsevier, Amsterdam.

Chapter 5.3

Standardization of Water Analysis within the CEN and ISO: The Example of Water Microbiology

R. MAARIT NIEMI AND KIRSTI LAHTI

Hydrological and Limnological Aspects of Lake Monitoring
Edited by Pertti Heinonen, Giuliano Ziglio and André Van der Beken
©2000 John Wiley & Sons, Ltd, ISBN 0 471 89988 7

5.3.1   INTRODUCTION

Why do we need internationally standardized methods for water analysis? Measuring methods are seldom perfect, and they often have limitations to their use because universal methods are rare. Their sensitivity and specificity or selectivity varies, and the detection limit is often strongly affected by the technique applied in analysis. Environmental samples may contain components causing interferences affecting the detection and yield of the measured variable. Differences in costs and feasibility – labour, skill and equipment needed – depend on the technique and method selected. In practice, more than just one method may be needed for measuring the same variable in different samples. On the other hand, research and monitoring programmes spanning wide geographical areas or long periods necessitate compatibility of measurement data. Compatibility of data on the international level is needed, e.g. when studying trans-boundary pollution, in international surveys, in testing compliance with internationally agreed limit values and even for purposes of commerce.

The purpose of standardization is to elaborate methods for international use in a wide range of laboratories in order to enhance monitoring of water quality and compliance with requirements. In many countries, standardization of water analyses has been practised nationally since the early twentieth century (e.g. USA, UK and Germany). Internationally, the ISO (The International Organization for Standardization) has been active in standardizing methods of water analysis since 1971 and the CEN (The European Committee for Standardization) since 1989. The Organization for Economic Cooperation and Development (OECD) is active in harmonizing methods for testing of chemicals, including toxicity tests in the aquatic environment, but this work is not further discussed here.

5.3.2   THE INTERNATIONAL ORGANIZATION FOR STANDARDIZATION (ISO)

The International Organization for Standardization (ISO) is a worldwide federation of national standards bodies presently including 120 members (Figure 5.3.1).

Several technical committees (TCs) are active in the field of environmental standardization (Table 5.3.1). The Technical Committee on Water Quality (ISO/TC 147) has been active since the early 1970s. By 1997, this technical committee had elaborated 113 published ISO standards and 78 proposals were under preparation. The work is carried out in six subcommittees (SCs) (Table 5.3.2).

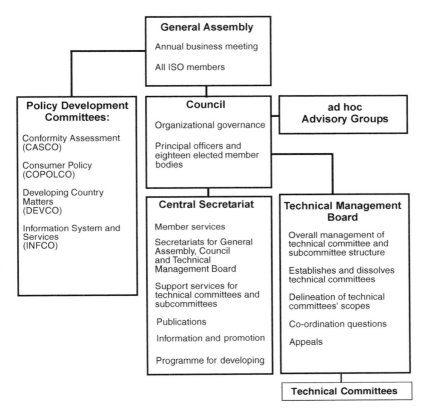

Figure 5.3.1 The organizational structure of The International Organization for Standardization (ISO).

Table 5.3.1 The various technical committees (TCs) on environmental standardization within the ISO and CEN.

ISO committee	Scope	CEN committee	Scope
TC 146	Air quality	TC 164	Water supply
TC 147	Water quality	TC 230	Water analysis
TC 149	Cycles	TC 183	Waste management
TC 190	Soil quality	TC 292	Characterization of waste
TC 200	Solid wastes	TC 308	Sludge classification
TC 207	Environmental management	TC 318	Hydrometry
TC 211	Geographic information/ geomatics		

Table 5.3.2 The subcommittees (SCs) of ISO/TC 147 and working groups (WGs) of CEN/TC 230.

ISO subcommittee	Scope	CEN subcommittee	Scope
SC 1	Terminology	WG1	Physical and chemical
SC 2	Physical, chemical and		methods
	biochemical methods	WG3	Microbiological
SC 4	Microbiological methods		methods
SC 5	Biological methods	WG2	Biological methods
SC 6	Sampling		
SC 7	Precision and		
	accuracy		

The working drafts (WDs) are usually elaborated in specific working groups (WGs) consisting of experts nominated by actively participating (P) member bodies. When consensus on the WD has been achieved within the working group, it is forwarded to the respective subcommittee and circulated to the P member bodies for voting and to the observing (O) members for information as a committee draft (CD). If the voting result is favourable the usually amended version is then circulated as a draft international standard (DIS) for further voting purposes. The amended DIS is then circulated for votes as a final DIS (FDIS). Majority vote is applied and no changes are possible at this stage. Finally, the document is published as an ISO standard in English and French. Every five years, a periodical review is carried out in order to update the standards as needed. ISO standards are recommendations but in national legislation or authorization their use may be mandatory.

Information on ISO activities can be obtained from the regularly published ISO Catalogue, ISO Memento and the Annual Reports of each technical committee via national member bodies.

5.3.3 THE EUROPEAN COMMITTEE FOR STANDARDIZATION (CEN)

The CEN is a co-operative body for the 18 national standards bodies of the European Union (EU) and the European Free Trade Association (EFTA) (Figure 5.3.2). Its importance has increased since the new approach to elaborating EU Directives was adopted. The methodological details are now described in EN standards produced within the CEN and not listed in the directives. The technical committees elaborating environmental standards are listed in Table 5.3.1. The activity in the CEN's Technical Committee on Water Quality (TC 230) started in 1989. Presently, this committee has three working

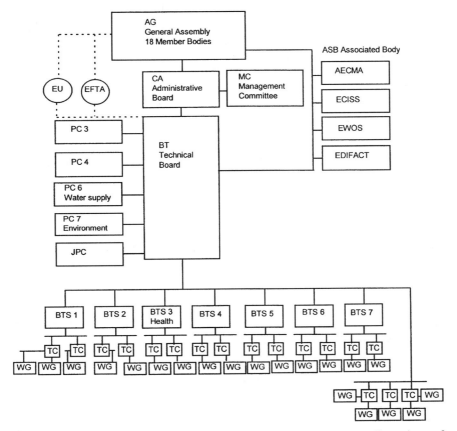

Figure 5.3.2 The organizational structure of The European Committee for Standardization (CEN).

groups (WGs) (Table 5.3.2). In 1997 CEN/TC 230 had elaborated 41 published EN standards and 31 proposals were under preparation.

Each of the working groups has several task groups (TGs) for specific methods. Usually these task groups elaborate the method proposals and when consensus has been achieved they are forwarded to the appropriate technical committee via the working group. They are then circulated as draft European standards (prENs) to the member bodies for voting. The amended version is further circulated for formal voting and finally published as an EN standard in English, French and German. Each member body must publish a national standard based on the EN standard within six months of publishing of the EN standard. Alternatively, a document can be adopted as a European prestandard (ENV), which is in force for a limited period and allows the existence of parallel national standards. The unique acceptance procedure (UAP) is an option to speed up the process when acceptance is highly probable. One difference between the ISO and the CEN is that the voting results in the CEN are calculated from weighted votes (with large nations having more votes than small ones).

Information on CEN activities can be obtained from the annual reports of each technical committee and from the annual report of the CEN via national member bodies.

5.3.4 THE VIENNA AGREEMENT BETWEEN THE CEN AND ISO

The problems associated with two international organizations producing standards for the same purposes have been realized and an agreement between the ISO and CEN was signed in Vienna in 1991. This important Vienna Agreement enables parallel work and parallel voting between the ISO and CEN. This possibility is extensively utilized in the work of CEN/TC 230 and ISO/TC 147.

5.3.5 HOW DOES A NATIONAL BODY PARTICIPATE IN THE WORK OF THE CEN AND ISO?

National member bodies organize the national work by establishing national committees and/or national circulation of proposals for voting. They nominate the experts for ISO working groups and CEN task groups and also the delegates for ISO technical committees and subcommittees and CEN technical committees and working groups. International work is carried out by correspondence and by participation in meetings.

The standardization work is dependent on people participating in different activities. Ideal experts and delegates know the topic thoroughly, have hands-on experience of several alternative methods on a wide variety of samples, and are fluent in the working language. They also ensure information exchange between national and international working groups and committees. Knowledge concerning the characteristics of the natural environment and national legislation affecting the needs for standards is important because it helps the delegates to anticipate the national voting results at the early stages of standardization. The need to demonstrate the requirements for the standard and possible problems associated with the standard proposal as early and as clearly as possible is of major importance. Changes are difficult to incorporate after voting. If the voting result is favourable, the secretariat is not free to make significant changes in accordance with comments from only one member body. It is evident that international consensus usually involves some compromising.

5.3.6 STANDARDIZATION OF WATER MICROBIOLOGICAL METHODS

This work represents the situation in 1997. International standardization of water microbiological methods is carried out within ISO/TC 147/SC 4. This microbiological subcommittee has 24 participating (P) members and 16 observing (O) members. In addition, it has several important liaison members, namely ISO/TC 34/SC 9 (agricultural food products, microbiological methods), the Codex Alimentarius Commission, the European Commission, the Food and Agriculture Organization of the United Nations, the International Association on Water Quality, the World Association of Societies of Pathology, the World Health Organization (WHO) Regional Office for Europe and the World Meteorological Organization. The structure of ISO/TC 147/SC 4 is shown in Table 5.3.3. The standardization of microbiological methods within the CEN was started in 1989 and has intensified recently. The CEN member bodies, i.e. the EU and EFTA countries, are active in the work of CEN/TC 230/WG 3 (Table 5.3.3).

Published ISO standards are available for water analysis by culturing techniques, evaluation of membrane filters and evaluation of culture media. A guidance document for validation of microbiological methods is under preparation. Many of the standards described in the following for specific microorganisms are intended for a limited scope which is not specified here.

The standardization of methods for coliform bacteria and *Escherichia coli* has proved to be difficult due to the importance of these methods, the long but varied traditions in different countries and the need to cope with different requirements for sensitivity, selectivity and different kinds of interference in different water types, and generally insufficient information on validation. The

Table 5.3.3 The working groups (WGs) of ISO/TC 147/SC 4 and the task groups (TGs) of CEN/TC 230/WG 3.

ISO working group	Scope	CEN task group	Scope
WG 1	Cultivation techniques and heterotrophic plate count	TG 1	Heterotrophic colony count
WG 2	Coliforms and *Escherichia coli*	TG 2	*Pseudomonas*
WG 3	*Pseudomonas*	TG 3	Staphylococci
WG 4	Faecal enterococci	TG 4	Enteroviruses
WG 5	Clostridia		
WG 7	*Salmonella*		
WG 10	*Legionella*		
WG 11	Bacteriophages		
WG 12	Analytical quality control		
WG 13	*Cryptosporidium* and *Giardia*		

work was started in 1973 within the ISO and in 1990, two standard methods – a membrane filtration method and a liquid enrichment method – were published. However, they were not accepted as EN standards. These standards are therefore under revision within the ISO and an additional method based on specific substrates and miniaturised technique is under preparation. These methods undergo parallel voting within both the ISO and CEN.

Two ISO standards were published in 1984 for the other widely used indicator bacterial group for faecal contamination, faecal enterococci (earlier faecal streptococci), i.e. one for liquid enrichment and the other for a membrane-filtration technique. These standards were not accepted as EN standards and are under revision within the ISO. Additionally, a miniaturized method is under elaboration. As in the case of the coliform standard proposals, these methods undergo parallel voting within both the ISO and CEN.

Standardization of a detection method for *Salmonella* was started as early as 1973, but the method was only published in 1995 as an ISO standard. It was not directly accepted as an EN standard. During its revision, parallel voting within ISO/TC 147 and CEN/TC 230 is being carried out and co-operation with ISO/TC 34/SC 9 (the Technical Committee on Agricultural Food Products/Subcommittee of Microbiology) is in progress with the aim of producing a common standard for water and food analysis.

Both the liquid-enrichment method and the membrane-filtration method for clostridia, which were elaborated as ISO standards, were also accepted as EN standards. However, they are now regarded as outdated and their revision has been started. One method for heterotrophic plate counting has been published as an ISO standard but this will soon be revised as a result of co-operative work with the CEN. The CEN/TC 230/WG 3 has elaborated a standard proposal for the enumeration of *Pseudomonas aeruginosa*. In addition, one method for the detection of *Legionella* is in its final stages and another is under preparation within the ISO.

Work on *Cryptosporidium* and *Giardia* was recently started within the ISO. Three methods for the enumeration of bacterial viruses are under elaboration, i.e. F-specific RNA phages, somatic coliphages and phages infecting *Bacteroides fragilis*. On the other hand, standardization of a detection method for enteroviruses will soon be started within the CEN. The CEN/TC 230/WG 3 also plans to start the standardization of a method for *Staphylococcus* species.

5.3.7   CHALLENGES IN THE STANDARDIZATION OF MICROBIOLOGICAL METHODS

The scope of microbiological methods may be rather limited. When analysing drinking water samples, the method should be sensitive and be able to detect low numbers, including damaged bacteria. On the other hand, surface waters and wastewater have a dense background flora and the methods applied should be able to inhibit the growth of non-target organisms and specifically detect the target organisms. The technique being applied affects the sensitivity and selectivity. The pour-plate technique may yield lower recoveries than the spread-plate technique due to heat shock, even if the same media and incubation conditions are applied. Membrane filtration enables concentration to be achieved, but may be unsuitable due to simultaneous concentration of inhibitory substances. Liquid enrichment is usually recommended for turbid samples but often necessitates further confirmation steps.

The major obstacle in the standardization of microbiological methods is the lack of knowledge on yield, specificity and applicability to different sample types. Microbiological methods are usually not robust, and the result is easily affected by slight differences in procedures such as incubation temperature, medium composition, background flora of the sample and other sample effects. Identification of presumptive target bacteria is necessary in method comparisons, which further adds to the costs.

Economical restrictions to application are an ever-present reality, especially when global use is intended. This must be considered in international standardization. Standardization is a slow process because processes with voting periods are needed, the experts involved are active for only part of the time and available information is limited. Furthermore, several language versions are needed. Therefore, it is not possible to follow technological development rapidly. In order not to interfere in implementation of technological improvements, it is advisable to be able to use not only standardized methods but also more advanced methods. This, however, necessitates careful validation of the methods used. If validation criteria can be standardized, the comparison of the common 'bench mark', i.e. the international standard method, with the most suitable method for application in each laboratory can be achieved. If the internationally standardized methods are the only permitted methods for important applications (e.g. in testing for compliance with EU Directives), difficulties in securing consensus for the methods are probable. If the international standards are regarded as reference methods, then these problems can be avoided.

The process of standardization brings together experts from different countries. This enhances interchange of information and helps to increase understanding of methodological details and differences between countries, thereby strengthening co-operation.

Part 6
The Management of
Monitoring Results

Chapter 6.1

Methods for Extracting Information from Analytical Measurements

PENTTI MINKKINEN

Hydrological and Limnological Aspects of Lake Monitoring
Edited by Pertti Heinonen, Giuliano Ziglio and André Van der Beken
©2000 John Wiley & Sons, Ltd, ISBN 0 471 89988 7

6.1.1   INTRODUCTION

By using modern instrumental methods, a large number of chemical variables can be determined from a single sample. When these methods are used to monitor the environment or industrial processess, large data sets are produced. These data sets often contain a lot of useful information, but in order to fully exploit their potential, effective (data) analytical methods are needed both to extract and interpret the information that such data sets may contain and also to display the information. Two different chemometric methods for analytical data processing, namely Principal Component Analysis (PCA) and Partial Least-Squares (PLS) Regression, have in recent years found many applications in the interpretation of multivariate data sets and will both be introduced in this present chapter. The general description of these methods and some examples of their applications can be found in Martens and Naes (1989) and Höskuldsson (1996). In aquatic environment, these methods have been used to identify the source of oil spills (Duewer *et al.*, 1975) and to classify aerosol samples (Häsänen *et al.*, 1990). The dispersion and effect of wastewater in the recipient lake has been studied by using these methods for Lake Päijänne and Lake Saimaa, Finland (Yliruokanen *et al.*, 1983; Minkkinen *et al.*, 1988; Sääksjärvi *et al.*, 1989; Mujunen *et al.*, 1996). These methods have also been used for modelling, monitoring and optimizing wastewater treatment processes (Shuetzle *et al.*, 1982; Aarnio and Minkkinen, 1986; Teppola *et al.*, 1997; Mujunen *et al.*, 1998; Teppola *et al.*, 1998).

6.1.2   PRINCIPAL COMPONENT ANALYSIS

PCA is a bilinear projection method where the original n-dimensional measurement matrix $\mathbf{X}$ (n samples, m variables) is projected into a lower, A-dimensional space, by decomposing the $\mathbf{X}$-matrix into a sample score matrix $\mathbf{T}$, a variable loading matrix $\mathbf{P'}$ and residual matrix $\mathbf{E}$, as follows:

$$\mathbf{X} = \mathbf{TP'} + \mathbf{E} \qquad (1)$$

In an ideal case, the useful infromation is extracted into matrices $\mathbf{T}$ and $\mathbf{P'}$ whose product $\mathbf{TP'}$ models the systematic variation in the data set, while the measurement noise is leved into the residual matrix $\mathbf{E}$. By plotting two or three columns of the $\mathbf{T}$ matrix against each other, a two- or three-dimensional projection of the original data set is obtained. Plots of the rows of the $\mathbf{P'}$ matrix show how the variables are correlated and which variables have contributed to each principal component, i.e. which variables discriminate the objects presented by the scores of the $\mathbf{T}$ matrix.

Typically, PCA is used for the following purposes:

- To graphically display multivariate data. Any table of data can be converted into projection pictures by using PCA.
- To filter noise from multivariate data.
- To monitor various kinds of environmental or industrial processes. The score values are not sensitive to random noise but, if some phenomenon is affecting several variables simultaneously in a systematic way and thus changing the correlation structure in the data set, this is immediately seen as shifts in the score values. This makes the score values suitable for detecting trends in multivariate time series.
- To classify samples following multivariate measurements. Figure 6.1.1 illustrates a PCA-based classification method known as Soft Independent Modelling of Class Analogy (SIMCA). This figure also shows the geometric interpretation of PCA in a tri-dimensional space (see also the explanatory text for Figure 6.1.2, below).

6.1.3   PARTIAL LEAST-SQUARES REGRESSION

The PLS method is widely used in multivariate calibration types of problems, where a linear dependence is assumed to exist between two blocks of variables called descriptor variables (matrix $\mathbf{X}$) and response variables (matrix $\mathbf{Y}$). In PLS analysis, both $\mathbf{X}$ and $\mathbf{Y}$ matrices are decomposed into score matrices ($\mathbf{T}$ and $\mathbf{U}$) and loading matrices ($\mathbf{P'}$ and $\mathbf{Q'}$) and to residual matrices ($\mathbf{E}$ and $\mathbf{F}$) as follows:

$$\mathbf{X} = \mathbf{TP'} + \mathbf{E} \tag{2}$$

$$\mathbf{Y} = \mathbf{UQ'} + \mathbf{F} \tag{3}$$

Compared to principal component analysis, the solutions are rotated. In this way, the correlation between the variables (the columns $\mathbf{t}_a$ of $\mathbf{T}$ and $\mathbf{u}_a$ of $\mathbf{U}$) is maximized for each component included in the model. Figure 6.1.2 shows the geometric interpretation of the PLS method.

A typical use of PLS analysis is to calibrate spectroscopic measurements to determine chemical components. It has also found use in modelling various chemical processes. In environmental studies, it can be used to study various exposure–response relationships. A special application of PLS is the so-called discriminant PLS (DPLS), which is used to find those projections of the X-matrix which show best the class separation in linear projection plots. As the first step, the directions in the multivariate space are found by following the PCA application. If these directions are not showing class differences in the projection plots, the optimal class discriminating directions can be found by subsequently using DPLS. In this case, a dummy or class matrix, $\mathbf{Y}$, is created. For each class presented in the calibration set $\mathbf{X}$, the matrix $\mathbf{Y}$ has a column

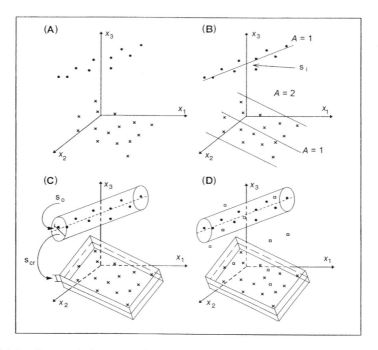

Figure 6.1.1 Geometric interpretation of principal component modeling as it is used for classification in the SIMCA method shown as a three-dimensional example of a multi-dimensional space. (A.) Each object belonging to the training set can be represented by a point in multi-dimensional space. If the measured variables are good classifiers, the objects belonging to a class are close to each other or form a distinct pattern. (B.) Each class of points is approximated by a linear model $X = TP'$ (line, plane or 'hyperplane', according to the dimensions, A, of the model). (C.) Residuals s_i can be used to define the class standard deviations and the confidence intervals for the class models. (D.) New points can be classified either as those belonging to one of the training-set classes or to those being different from all of the training-set classes.

containing ones or zeros. When PLS is run between X and Y, a good class separation in projection plots is usually obtained. Sääksjärvi *et al.* (1989) used this method to study the proportions and mixing of wastewaters from two different sources in the recipient lake.

6.1.4 DATA PREPROCESSING AND OPTIMAL MODEL COMPLEXITY

Both PCA and PLS usually require that the raw data is preprocessed before modelling. The simplest preprocessing method is mean-centring, i.e. the

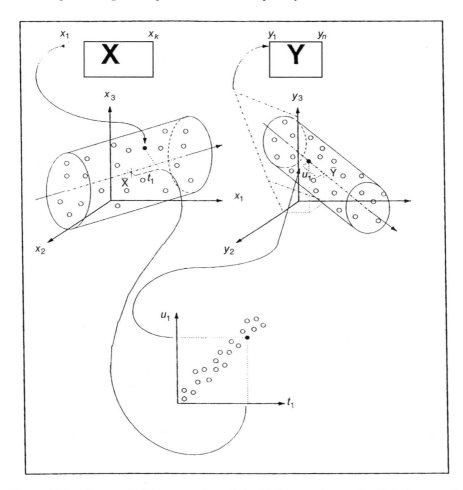

Figure 6.1.2 Geometrical interpretation of the PLS regression method; the spaces of the descriptor variables **X** and **Y** are each modeled separately, both to approximate the training-set points and to maximize the correlation between the projections of the points on the model axes (also called latent variables). The **Y** predictions for new points, whose **X** values are measured, are also obtained by projection. If the new point is within the confidence interval of the model in **X**-space the precision of the predictions is comparable to that of the training set.

variable means of the calibration or modelling set are subtracted from the individual variable values. This is often a sufficient data pretreatment for spectroscopic data. Another commonly used data preprocessing method is autoscaling, where all variables are divided by their standard deviations calculated from the calibration sets. This gives unit variance to all variables of the calibration set. Combined with mean centring, autoscaling is the standard data preprocessing method, especially if the variables are expressed in different units and have numerically different values, or when different types of data are combined. Without preprocessing, variables having numerically high values would also dominate the modelling of the data set. Often in environmental and other related data sets, where the relative ranges of the variables are approximately equal, logarithmic transformation combined with mean centring is an alternative to autoscaling. The basic difference of these two methods is that with autoscaling, those samples having either exceptionally high or low values have high leverage in the modelling, whereas the logarithmic transformation decreases the leverage of high values and increases the leverage of low values.

Both with PCA and PLS, the performance of the models depends on how many components are accepted in the model (complexity of the model). If too few components are used, then all of the useful information is not extracted. If too many components are used, then the models also try to explain the measurement noise, while the predictions (classification or predicted $\mathbf{Y}$ values) on test-set samples that are not used in the calibration become poorer. A cross-validation procedure is commonly used to find the optimal complexity of the models.

Two types of indices of determination are used to describe the 'goodness' of the models: R^2 refers normally to the fit of the calibration samples to the model, and Q^2 refers to the fit of the independent test group or the cross-validation samples. Both values imply how much of the total variance of the measurements is explained by the model.

6.1.5 AN EXAMPLE: EFFECT OF WASTEWATER ON AQUATIC PLANTS

Lake Päijänne is a long lake in central Finland from where the water is presently led through a tunnel to service the needs of Helsinki, the capital of Finland. In August 1972 and 1973, samples of aquatic plants were collected from this area at four different sampling site and analysed for ash content and a number of trace and minor elements. Site 1 was at that time heavily polluted both by municipal wastewater and the effluents from a paper mill. Site 2 was also heavily polluted by effluents from two big pulp and paper mills. Site 3 was at an intermediate zone, while site 4 was practically clean. Since 1978, the municipal wastewaters had no longer been discharged into the lake and better controls on industrial effluents had been applied. In order to speed up the recovery of the lake, the basin had also been aerated. To check whether the recovery of the lake could be tracked in the trace-element patterns of the aquatic plants, the sampling at site 1 was repeated in August 1983. Samples of three species belonging to *Nympheids* (*Potamogeton natans*, *Polygonum amphibium* and *Nuphar luteum*) and one belonging to *Helophytes* (*Phragmites australis*) were collected. The original data, and the results of the data analysis, which was based on autoscaled data (i.e. the variables were mean centred and scaled to unit variance) have been presented in the papers by Yliruokanen *et al.* (1983) and Minkkinen *et al.* (1988). In this example, the data sets obtained for *Nympheids* are reprocessed by using logarithmic scaling and mean centring as the data preprocessing method instead of autoscaling. A constant of 0.05 was added to all values before taking the logarithms, because some not-detected values were coded as zeros (see Table 6.1.1). The scaling affects the numerical results, and the shape of the clusters of data points in plots which are used to display the results. The conclusions that can be drawn from the data analysis are in this case, however, quite similar regardless of the scaling method that is used. The data are given here in Table 6.1.1. Two questions should now be addressed, i.e. (1) are the trace-element patterns of the different species different, and (2) can the effect of the pollution level at the sampling sites be seen in the trace-element patterns of these aquatic plants, because in all of the classes the variables overlap, and no variable alone could discriminate between any of the classes based either on species or on growing site.

6.1.5.1 Discrimination between species

PLS discriminant analysis was carried out on the logarithmically scaled mean centred data. This improved somewhat the class separation of the species classes in comparison to the respective PCA plots. Cross-validation showed that four of the PLS components were significant. Theoretically, a data set presenting three homogenous and separable classes would have three PLS components. Table 6.1.2 gives the variable loading values (P) and the cumulative indices of determination for the fit (R^2) for each **P** component of the **X**-block. In two-dimensional score plots, the classes were not well separated. Figure 6.1.3 shows the scores of the first three PLS components as a three-dimensional projection which explains 75.6% of the total variance of the logarithmically scaled data. Figure 6.1.4 shows the respective variable loadings. As can be seen in in this latter figure, the classes are well separated. Figure 6.1.3 shows that the first component discriminates *Nuphar luteum* (C) from the two other species and its scores have negative values. From Figure 6.1.4 and Table 6.1.2, it can be seen that all variables, except the ash content, plus Rb and Ba, load highly into the first component, having positive loading values. This means that the two other groups tend to have generally higher values for most of the measured variables. The third PLS component is the most important in identifying the two other groups, *Potamogeton natans* (A) and *Polygonum amphibium* (B). By comparing Figures 6.1.3 and 6.1.4, it can be concluded that the most important variables discriminating these groups are Sr

Table 6.1.1 Analytical data obtained for *nympheids* collected in 1971–1972 (upper case letters) and 1983 (lower case letters) from the Lake Päijänne area, Finland: A and a, *Potamogeton natans*; B and b, *Polygonum amphibium*; C and c, *Nuphar luteum*. The sampling locations are indicated by numerals: 1, Jyväsjärvi; 2, Tiirinselkä; 3, Judinsalonselkä; 4, Tehinselkä.

Species/ Location	Ash content (%)	Trace/minor-element content ($\mu g\,g^{-1}$)												
		V	Mn	Fe	Cu	Zn	Rb	Sr	Ba	Y	La	Ce	Pr	Pb
A1	12.3	2.4	1200	2800	24	59	24	120	160	2.0	3.0	6.0	1.0	8.6
A2	12.9	7.7	1600	2200	20	110	28	77	130	1.8	3.8	5.1	0.4	5.1
A2	18.9	30.0	1600	5300	56	140	40	94	190	7.5	13	34	1.5	9.4
A3	9.5	1.0	290	360	9.5	66	19	87	95	1.1	1.5	1.9	0.2	1.3
A3	7.2	0.7	250	280	3.6	57	14	86	36	0.2	0.5	0.7	0	1.4
A4	9.0	0.9	650	1100	5.4	36	16	99	27	0.4	0.6	0.9	0.1	0.9
A4	7.5	3.0	540	970	6.0	22	15	97	75	0.5	3.0	3.3	0.2	3.7
B1	12.4	3.7	740	2300	12	42	32	60	93	2.4	1.2	2.4	0.2	2.4
B1	7.1	2.8	300	2000	14	30	26	36	42	1.4	2.1	2.8	0.2	3.5
B3	6.9	0.7	320	590	3.5	21	22	52	31	1.0	1.4	1.4	0.2	1.4
B3	9.7	5.8	610	1600	2.9	23	33	62	38	1.0	1.0	2.4	0.2	1.9
B4	8.1	0.8	120	340	4.1	21	30	51	36	0.6	0.6	0.4	0.1	2.4
B4	5.5	0.8	209	310	2.7	10	17	40	16	0.2	0.6	0.6	0.05	1.7
C1	13.4	2.7	250	270	20	20	53	19	180	0.3	0.1	0.3	0	21
C2	10.8	2.1	960	520	11	41	38	16	150	0.5	0.6	0.6	0.1	1.0
C2	8.4	1.1	840	400	17	45	50	15	84	0.2	0.3	0.7	0.08	1.7
C3	11.5	0.6	240	140	3.4	65	31	24	180	0.3	0.5	0.9	0.1	2.1
C3	10.1	1	300	110	3.0	40	36	23	120	0.3	0.7	0.8	0.1	0.8
C4	9.0	0.3	93	93	1.9	18	28	65	48	0.1	0.2	0.5	0	0.3
C4	8.5	0.9	140	190	2.6	12	30	29	93	0.2	0.4	0.4	0	1.4
a1	7.91	1.0	1200	240	10	27	20	79	110	0.3	1.2	1.7	0.2	1.0
a1	8.29	0.9	360	250	8.3	21	11	65	41	0.1	1.4	0.9	0.2	8.3
b1	9.17	0.9	1200	440	17	27	29	68	50	0.5	1.3	2.4	0.2	1.1
b1	6.46	1.9	340	430	6.5	9	17	34	29	0.1	1.1	1.4	0.2	1.0
c1	9.24	1.0	1200	160	5.9	19	21	13	130	0.3	1.0	1.1	0.3	1.2
c1	9.27	0.7	650	100	10	18	25	25	140	0.3	0.5	0.5	0.1	1.0
c1	9.02	0.7	330	100	4.7	14	31	12	52	0.3	0.5	0.6	0.1	1.8

Table 6.1.2 Variable loading (*P*) values and cumulative indices of determination (R^2) of the logarithmically scaled and centred X-matrix of the discriminant PLS model with four components, calculated for the analytical data obtained for the *nympheids* collected from the Lake Päijänne area.

Variable	Loading component			
	P1	P2	P3	P4
Ash	0.05	−0.17	0.00	0.02
V	0.35	−0.23	−0.47	0.23
Mn	0.23	−0.31	0.11	−0.22
Fe	0.47	0.11	−0.44	0.26
Cu	0.25	−0.41	0.04	0.21
Zn	0.19	−0.26	0.34	−0.02
Rb	−0.01	−0.24	−0.33	0.00
Sr	0.19	0.33	0.50	0.11
Ba	0.06	−0.59	0.19	−0.06
Y	0.37	−0.15	−0.35	−0.13
La	0.38	0.06	0.17	−0.30
Ce	0.43	−0.04	0.10	−0.27
Pr	0.32	−0.10	0.06	−0.35
Pb	0.21	−0.25	−0.06	0.78
R^2 (%)	49.1	69.5	75.6	81.5

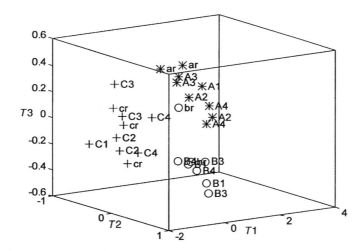

Figure 6.1.3 Three-dimensional score plot obtained from the PLS discriminant analysis showing the grouping of *Nympheids* species: A and a, *Potamogeton natans*; B and b, *Polygonum amphibium*; C and c, *Nuphar luteum*. The capital letters refer to the samples collected during 1972–1973, while the lower case letters refer to those collected in 1983.

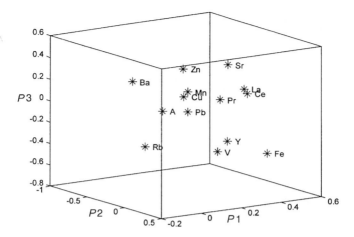

Figure 6.1.4 Three-dimensional variable loadings plot of the PLS discriminant model for the analytical data obtained for the three *Nympheids* species; A represents the ash content in dry matter.

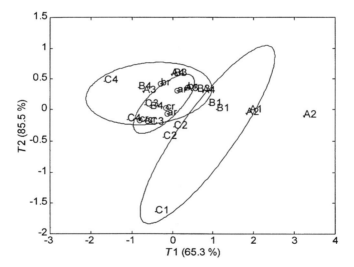

Figure 6.1.5 PCA score plot obtained for the *Nympheid* samples, where the model was calculated by using only those samples which were collected from sites 1 and 4 during 1972–1973; all other sample data were projected onto this model. The samples collected in 1983 from site 1 are indicated by lower case letters, while the values in parentheses given for $T1$ and $T2$ are the indices of determination (R^2) for the fit. The location of the model-group samples (1972–1973) and those samples collected in 1983 from site 1 are shown with ellipsoids.

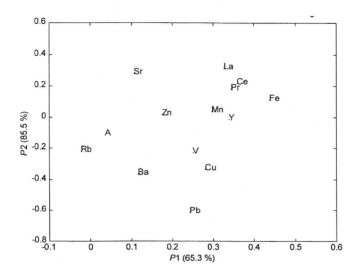

Figure 6.1.6 PCA loading plot of the analytical data obtained for the three *Nympheids* species, where the PCA model was calculated by using only those samples related to Figure 6.1.5; A represents the ash content in dry matter, while the values in parentheses given for $P1$ and $P2$ are the indices of determination (R^2) for the fit.

and Zn, and to a lesser extent, Ba, La, Mn and Ce (high in *Potamogeton natans*), and V, Fe, Rb and Y (high in *Polygonum amphibium*).

6.1.5.2   Effect of pollution level at sampling sites

When the species classes of Figure 6.1.3 are looked at more closely, it seems that the samples taken from the polluted sites (1 and 2) and from the clean site (4) are at the opposite ends of the scale. In order to study the general effect of the pollution level on the measured variable pattern of these plant species, an ordinary PCA model was calculated by using as a modelling group only those samples taken in the years 1972 and 1973 from the clean site (4) and from the contaminated site 1. All other samples were projected on to this model, including those samples taken from site 1 in 1983 after cleaning of the lake had been carried out. Figure 6.1.5 shows the score plot, while Figure 6.1.6 gives the respective loading plot. As can be seen, the samples from the other polluted site (site 2) are close to the samples in polluted site 1. The samples from the transition zone (site 3) partly overlap the clean-site cluster. Samples from site 1 collected in 1983 (denoted by lower case letters) have also shifted towards the clean-site samples, thus indicating the recovery of this lake. The variable-loading plot given in Figure 6.1.6 shows that Fe, Ce, La, Pr, Y, Mn, Cu, V, Pb and Zn are all variables which tend to be high at the polluted site.

REFERENCES

Aarnio, P. and Minkkinen, P., 1986. Application of partial least-squares modelling in the optimization of wastewater treatment plant, *Analy. Chim. Acta*, **191**, 457–460.

Duewer, D. L., Kowalski, B. R. and Schatski, T. F., 1975. Identification of oil spills by pattern recognition analysis of natural elemental composition, *Anal. Chem.*, **47**, 1573–1583.

Häsänen, E., Lipponen, M., Minkkinen, P., Kattainen, R., Markkanen, K. and Brjukhanov, P., 1990. Elemental concentrations of aerosol samples from the Baltic Sea area, *Chemosphere*, **21**, 339–347.

Höskuldsson, A., 1996. *Prediction Methods in Science and Technology*, Vol. 1, *Basic Theory*, Thor Publishing, Copenhagen.

Martens, H. and Naes T., 1989. *Multivariate Calibration*, John Wiley & Sons, New York.

Minkkinen, P., Yliruokanen, I. and Särkkä, J., 1988. Environmental effect on the minor and trace element patterns of some aquatic plants, *Analusis*, **16**, 169–172.

Mujunen, S.-P., Minkkinen, P., Holmbom, B. and Oikari, A, 1996. PCA and PLS methods applied to ecotoxicological data: ecobalance project, *J. Chemom.*, **10**, 411–424

Mujunen S.-P., Minkkinen P., Teppola, P. and Wirkkala, R.-S., 1998. Modelling of activated sludge plants treatment efficiency with PLSR: a process analytical case study, *Chemom. Intell. Lab. Syst.*, **41**, 83–94.

Shuetzle, D., Koskinen, J. R. and Hersfall, F. L., 1982. Chemometric modelling of wastewater treatment processes, *J. Water Pollut. Control Red.*, **54**, 457–465.

Sääksjärvi, E., Khalighi, M. and Minkkinen, P., 1989. Wastewater pollution modelling in the southern area of Lake Saimaa, Finland, by the SIMCA pattern recognition method, *Chemom. Intell. Lab. Syst.*, **7**, 171–180.

Teppola, P., Mujunen, S.-P. and Minkkinen, P., 1997. Partial Least-Squares modelling of activated sludge plant: a case study, *Chemom. Intell. Lab. Syst.*, **38**, 197–208.

Teppola, P., Mujunen, S.-P. and Minkkinen, P., 1998. A combined approach of Partial Least-Squares and fuzzy *c*-means clustering for the monitoring of an activated-sludge wastewater treatment plant, *Chemom. Intell. Lab. Syst.*, **41**, 95–103.

Yliruokanen, I., Minkkinen, P. and Särkkä, J., 1983. Interpretation of trace element data of aquatic plants by using the SIMCA-pattern recognition method, *Kemia-Kemi* **10**, 713–718 (in Finnish).

Chapter 6.2

Water Quality Modelling of Lakes

TOM FRISK

Hydrological and Limnological Aspects of Lake Monitoring
Edited by Pertti Heinonen, Giuliano Ziglio and André Van der Beken
©2000 John Wiley & Sons, Ltd, ISBN 0 471 89988 7

6.2.1   INTRODUCTION

Lakes are complicated ecosystems with very many different factors affecting each other. There are physical and chemical, as well as different biological factors that must be taken into account when considering the functioning of lacustrine ecosystems. In the last few years, special attention has been paid to the biodiversity of ecosystems. Lakes are ecosystems concerning very many different species and thus the concept of biodiversity is very significant when considering the changes of lake ecosystems. However, relatively little is known about the quantitative changes in the biodiversity of lakes.

Because of the complexity of lake ecosystems it is impossible to obtain a complete description of a lake. It is possible to consider only some aspects of a lake at one particular time. However, it must be borne in mind that a lake is a totality, where the different factors affect each other, although the contributions may vary very much.

In studying aquatic ecosystems, it is possible to apply water quality models. These can be regarded as follows:

- simplified mathematical descriptions of a lake;
- tools that can be used in making water quality assessments;
- methods in scientific limnological research.

A water quality model may be a computer program, but this is not necessarily the case. It can also be a collection of equations or just a theory on the basis of which the calculations are made.

One problem in studying lake ecosystems is actually the complexity mentioned above. We can see very many factors but it is difficult to see the essential ones. In this situation, water quality models can be used for reducing the amount of details that must be considered. A Finnish limnologist, K.M. Lappalainen, has compared a water quality model to a *macroscope*, i.e. the 'opposite' of a microscope. If the objects to be studied are very small, a microscope is useful for achieving this. If all of the details of the lake prevent us from seeing the general patterns, we may use a water quality model as a macroscope (Lappalainen, 1978).

Water quality models have been developed for lakes, rivers and their catchments, as well as coastal and open sea areas. In this present chapter, the main emphasis is on lake modelling, even though most of the general principles are also valid for other water quality modelling studies.

6.2.2   THE STRUCTURE OF WATER QUALITY MODELS

There are many different types of water quality models and therefore it is useful to have some classifications for them. The models can be classified in several ways depending on the criteria involved. They can, for example, be classified on the basis of time and thus we speak then about dynamic models, steady-state models and transient models. In dynamic models, time is an explicit variable, whereas in steady-state models time is not considered at all. Transient models may be formally similar to dynamic models, but they do not describe the real dynamics of the system but instead the transition between successive steady states.

Another way of classifying the models is to look at their mathematical structure. They may be, e.g. statistical models, mechanistic models or neural-network models. Statistical models are based on data material and they do not contain information on causal relationships between the different affecting factors. Mechanistic models describe the physical, chemical and biological processes of the system. Neural-network models imitate the object systems in a way similar to the nervous system of human beings. In a way, neural-network models are forms of statistical models.

The models can also be mathematically divided into deterministic, stochastic or probabilistic models. In deterministic models, only one output is produced, whereas stochastic or probabilistic models produce different output values in different simulations. If there is a probability distribution, then the model is probabilistic; otherwise it is stochastic. However, probabilistic models are also often called stochastic. In describing the details of the models, deterministic mechanistic models, which that are the most common and traditional water quality modes, will only be dealt with here.

The models can also be classified into transport-oriented models and ecology-oriented models. If the spatial distribution of different substances is considered to be important, a transport-oriented model is most probably applied. If the structure of the ecosystem of the lake is considered to be important in the assessment, an ecology-oriented model is applicable. The models can also be classified on the basis of the dimension. There are three-dimensional (3D), two-dimensional (2D), one-dimensional (1D) and zero-dimensional (0D) models. Transport-oriented models are usually 3D or 2D, whereas ecology-oriented models are usually 1D or 0D. However, 0D and 1D models do not necessarily contain a complicated description of the ecosystem. A zero-dimensional model describes the lake as homogenous with no spatial variability.

Mechanistic water quality models are based on balance considerations, the kinetic principle and the stoichiometric principle. There are three important balance principles that are taken into account:

- mass balance;
- water balance;
- energy balance.

Mass and water balances must be considered in all water quality models, while the energy balance needs to be considered in the models in which temperature is simulated as a state variable. A water quality model consists of two compartments, namely an hydraulic compartment and a water quality compartment. The balance principles are taken into account in the hydraulic compartment, and the principles of kinetics and stoichiometry in the water quality compartment.

In water quality models, the lake system is characterized with *state variables*, e.g. temperature, different concentrations and different biomasses. The system is affected by *forcing functions* (*driving variables*). Examples of forcing functions are incoming radiation, incoming discharges and inputs of the different substances.

The interactions between the different state variables and factors affecting them are *processes* of the model. In the equations describing the processes, there are coefficients and other constants which are called *parameters*. It is important to see the difference between state variables and parameters: a state variable is a variable simulated by the modes, while a parameter is a constant in an equation of the model.

6.2.3 HYDRAULICS OF WATER QUALITY MODELS

In water quality models, the lake is normally described in a simple way. Most model makers prefer a way which is as simple as possible, but there are also modellers who always prefer sophisticated hydraulic descriptions. Advection is simply the phenomenon in which (dissolved or suspended) substances flow along with the water. It can be mathematically described as the product of discharge and concentration. Because discharge is defined as the product of the cross-sectional area and (mean) velocity, advection can be calculated as follows:

$$M_{adv} = Qc = A_x uc \tag{1}$$

where M_{adv} is the advective mass flow (MT^{-1}), Q is the discharge ($L^3 T^{-1}$), c is the concentration (ML^{-3}), A_x is the cross-sectional area (L^2), u is the (mean) velocity of flow (LT^{-1}), and x is the distance coordinate (L). The general dimension symbols used in parentheses are M (mass), T (time), and L (length).

Dispersion results from advective and molecular diffusion, as well as differences in the velocities of flow. Dispersion is usually described according to Fick's diffusion law as follows:

$$M_{disp} = -D_x A_x \frac{\partial c}{\partial x} \tag{2}$$

where M_{disp} is the dispersive mass flow (MT^{-1}), D_x is the longitudinal dispersion coefficient ($L^2 T^{-1}$), and x is the distance coordinate (L).

Dispersive processes tend to smooth the distributions of concentrations. For this reason, there is a minus sign on the right-hand side of equation (2), indicating a backward mass flow in the case where the concentration is increasing along with the x-axis (i.e. the concentration gradient, $\partial c/\partial x$, is positive).

The total mass flow can be calculated as the sum of the advective and dispersive mass flows, as follows:

$$M = M_{\text{adv}} + M_{\text{disp}} \tag{3}$$

where M is the total mass flow ($M\ T^{-1}$).

The most common hydraulic descriptions in water quality models are the following:

- continuously stirred tank reactor (CSTR);
- plug flow reactor (PFR);
- advection–dispersion reactor (ADR).

CSTR is a zero-dimensional description of the lake. The basic idea is that the lake is considered as being completely mixed and the mixing of the incoming substances is assumed to take place instantaneously. The concentration at the outflow is thus assumed to be equal to the concentration in the lake. Accordingly, no horizontal or vertical differences of concentrations are assumed.

CSTR models can be applied to shallow lakes that are not thermally stratified. They can also be applied to holomictic lakes that have periods of stratification if long-term average values of concentrations are considered. In these cases, the CSTR description is applied in steady-state models. This kind of an approach has been common in phosphorus-balance models (e.g. Dillon and Rigler, 1974; Lappalainen, 1974).

PFR is in a way the opposite of CSTR. In the PFR description, no longitudinal dispersion is assumed and so the only process affecting the distribution of concentration is advection. Thus, it could be also called advection hydraulics. PFR can be applied to rivers (e.g. Streeter and Phelps, 1925), and elongated lakes with short detention times (the so-called river-run lakes) (Chapra and Reckhow, 1983). However, they can also be applied to elongated lakes with long detention times, if differences of concentration in the longitudinal directions are important and long-term average values are considered (e.g. Frisk, 1989).

In ADR hydraulics both advection and dispersion are included. ADR can be one-, two- or three-dimensional in nature. Horizontal 1D models can be applied to rivers (e.g. Norton *et al.*, 1974) or elongated lakes. Vertical 1D models can be applied to stratified lakes in which horizontal differences of water quality are not great (e.g. Chen and Orlob, 1972; Gaume and Duke,

1975). The 2D ADR models are usually horizontal and can be applied to shallow, unstratified lakes or coastal areas when areal differences of water quality are important. The 3D models (e.g. Virtanen *et al.*, 1986) are the most general, and can in principle be applied to all cases. However, at the present time there are very few specialists on the practical application of these kinds of models.

The partial-dispersion situation described by ADR can also be achieved by forming a linear combination of PFR and CSTR (e.g. Frisk, 1989). In this model, there is a coefficient indicating to what extent the system has a PFR behaviour, as follows:

$$c = Dc_p + (1 - D)c_c \qquad (4)$$

where c is the concentration calculated by using the total model ($M\,L^{-3}$), c_p is the concentration calculated by using the PFR model (ML^{-3}) c_c is the concentration calculated by using the CSTR model (ML^{-3}), D is the plug flow coefficient.

6.2.4 DESCRIPTION OF THE INTERNAL PHYSICAL, CHEMICAL AND BIOLOGICAL PROCESSES

The non-hydraulic processes can be described as being internal. For lakes, the following can be regarded as examples of these kinds of processes:

- sedimentation;
- reaeration;
- oxidation;
- reduction;
- growth;
- decomposition;
- uptake of substances;
- release of substances;
- sorption (absorption, adsorption and desorption);
- volatilization.

According to the kinetic principle mentioned above, the rate of change of a concentration is calculated as a product of a concentration or as a function of a concentration (or several concentrations) and a reaction-rate coefficient. The order of the reaction depends on the power to which the concentration is raised, e.g. first-order reaction kinetics can be expressed as follows:

$$\frac{dc}{dt} = k_1 c \qquad (5)$$

where k_1 is the first-order kinetic coefficient (T^{-1}); and c is the concentration $(M \, L^{-3})$.

The derivative on the left-hand side of equation (5) represents the rate of change of the concentration c. The direction of the reaction is dependent on the sign of k_1, i.e. when $k_1 > 0$, the concentration is increasing, if $k_1 < 0$, it is decreasing.

Second-order kinetics can be expressed as follows:

$$\frac{dc}{dt} = k_2 c^2 \tag{6}$$

where k_2 is the second-order reaction rate coefficient $(M^{-1} \, L^3 \, T^{-1})$.

However, there is also another possibility for second-order kinetics in which the reaction rate is dependent on two different concentrations, c_1 and c_2:

$$\frac{dc}{dt} = k_2 c_1 c_2 \tag{7}$$

The order of the reaction can also be zero, which means that the reaction rate is constant:

$$\frac{dc}{dt} = k_0 \tag{8}$$

where k_0 is the zero-order reaction rate coefficient $(M \, L^{-3} \, T^{-1})$.

In addition to the various types of 'order reactions' described above, there are also other kinds of kinetic descriptions. One of the most common of these are the Michaelis–Menten kinetics, as follows:

$$\frac{dc}{dt} = k_m \frac{c}{k_c + c} \tag{9}$$

where: k_m is the maximum value of the reaction rate (at $c = \infty$) $(M \, L^{-3})$, and k_c is the half-saturation constant for concentration c $(M \, L^{-3})$.

The half-saturation constant is equal to the concentration at which the reaction rate is half that of the maximum value $(= 0.5 \, k_m)$.

The other main principle of mechanistic water quality models, i.e. the principle of stoichiometry, means in practice that the reactions of different substances take place in constant ratios that are described with stoichiometric coefficients. This can be explained by means of an example which describes the oxidation processes of ammonium nitrogen. In the first step, i.e. nitritation, ammonia is oxidized to nitrite, according to the following:

$$\frac{dN_1}{dt} = -\beta_1 N_1 \tag{10}$$

where N_1 is the concentration of ammonium nitrogen ($M\ L^{-3}$), and β_1 is the nitritation coefficient (T^{-1}).

Nitritation consumes oxygen, and the rate of consumption can be expressed by the following equation:

$$\frac{dO_2}{d_t} = -\alpha_1\beta_1N_1 \tag{11}$$

where O_2 is the concentration of dissolved oxygen ($M\ L^{-3}$), and α_1 is the stoichiometric-oxygen-consumption coefficient of nitritation ($\approx 3.4\,\mathrm{mg\,l^{-1}}$ dissolved oxygen per 1 $\mathrm{mg\,l^{-1}}$ ammonium nitrogen).

The concentration of nitrite nitrogen is dependent on nitritation, while the second step, i.e. nitratation, depends on the oxidation of nitrite to nitrate, as follows:

$$\frac{dN_2}{dt} = \beta_1N_1 = -\beta_2N_2 \tag{12}$$

where N_2 is the concentration of nitrite nitrogen ($M\ L^{-3}$), and β_2 is the nitratation coefficient (T^{-1}).

Oxygen consumption in nitratation can be expressed in a similar way to equation (11), as follows:

$$\frac{dO_2}{dt} = -\alpha_2\beta_2N_2 \tag{13}$$

where α_2 is the stoichiometric-oxygen-consumption coefficient of nitratation ($\approx 1.1\,\mathrm{mg\,l^{-1}}$ dissolved oxygen per 1 $\mathrm{mg\,l^{-1}}$ ammonium nitrogen).

Finally, the rate of change of nitrate nitrogen concentration can be expressed as follows:

$$\frac{dN_3}{dt} = \beta_2N_2 \tag{14}$$

where N_3 is the nitrate nitrogen concentration ($M\ L^{-3}$).

In equations (10), (12) and (14), other processes affecting the concentrations of the different forms of nitrogen (such as uptake by organisms and release from the sediment, or denitrification) have not been included.

The reaction-rate coefficients of the different processes have very different values, e.g. the value of the nitratation coefficient is much greater than that of nitritation coefficient, and therefore the concentration of nitrite in lakes is usually very small.

The reaction-rate coefficients are often described as dependent on temperature, where the latter can be either simulated by the model or given as a driving variable in the input data. There are several ways of describing the temperature dependence. The most widely used is the monotonic function of

Streeter and Phelps (1925), in which the reaction rate always grows along with rising temperature, as follows:

$$k(T) = k(T_s)\theta^{T-T_s} \tag{15}$$

where $k(T)$ is the reaction-rate coefficient at temperature T, T is the temperature, T_s is the standard temperature (usually 20 °C), and θ is an empirical coefficient.

The empirical coefficients θ can have different values for different processes, even if the value of 1.047 given by Streeter and Phelps (1925) has been adopted for very many different processes, without any theoretical or empirical arguments being considered. Equation (15) cannot be applied to describe the optimum temperatures of biological reactions. For this purpose, more sophisticated temperature-dependence functions (e.g. Lassiter and Kearns, 1973; Frisk and Nyholm, 1980) should be used.

6.2.5 HOW TO USE WATER QUALITY MODELS

There are many different purposes for which models are applied. They are often used within a planning project when there is a need to know in advance the effects of the planned measures on water quality. Models are often applied to make assessments on how future development will affect the water quality of a lake under different scenarios. Models are also needed for calculating the contributions of the different factors affecting the detected change of water quality which is impossible only on the basis of observations. Models can also be used as tools in scientific research, where they can, for example, help us to explain different phenomena.

In principle, all kinds of water quality problems can be dealt with by using models. The prerequisites are that there are either sufficient data available or the behaviour of the state variables in the lake is otherwise known. The oldest water quality model (Streeter and Phelps, 1925) was developed to describe the effect of biological oxygen demand (BOD) decay on dissolved oxygen concentration in rivers. Today, primary BOD is usually no significant problem in lakes, with the oxygen depletion being largely caused by decomposition of autochtonous biomass. Dissolved oxygen concentration in lakes can be calculated either with statistical models (e.g. Lappalainen, 1974), or ecological-simulation models (e.g. Chen and Orlob, 1972).

Another phenomenon which is often calculated by using models is eutrophication. The simplest way is to apply steady-state, phosphorus-balance models (e.g. Vollenweider, 1969; Lappalainen, 1974; Dillon and Rigler, 1974; Frisk *et al.*, 1981). The only state variable in these models is the average total phosphorus concentration, which is assumed to be correlated with the trophic status of the lake. In more advanced eutrophication models, phytoplankton is also simulated. Phytoplankton can be simulated as one community, or it can be divided into different groups. In principle, it is possible to simulate even the different species, but in practice this is a very inconvenient process.

Water quality models can also be used for calculating the acidification of lakes. However, acidification of lakes is often calculated with a catchment model in which a lake sub-model is included. Models are often applied in describing the behaviour of toxic substances in aquatic ecosystems. These may be inorganic substances, mercury, or organic compounds. In models, often bulk variables, such as adsorbable organically bound halogens (AOX) are used, even though these are not necessarily correlated with the toxic effect. If the toxic effects are to be modelled, then the results obtained from toxicity tests must be used.

Models can be applied to study the effects of different factors. They have very often been used to calculate the effects of wastewaters, which still remains important in many countries. In countries in which the level of wastewater treatment is high, more attention has been paid to diffuse loading in recent years. If the amount of diffuse loading cannot be measured, it can then be calculated by using a model which is based on watercourse observations. Models have also been used in calculating the effects of atmospheric inputs, as well as the anticipated climate change on lakes (Frisk *et al.*, 1997).

The selection of the state variables of the models to be used is mainly dependent on the problem that will be studied and the factors affecting the lake. It also depends on the data available and the possibilities to make complete measurements. In selecting the state variables, knowledge of the kinetics and other behaviours of the substances are also taken into account. The resources available in general are decisive when the state variables are being selected.

The resources considered in a modelling effort are time, availability of experts, data material, possibilities of obtaining new data, and computers. If there is only a very short time (e.g. one or two days), then only simple models can be used. However, in a planning or research project lasting several years, a more sophisticated modelling approach is usually preferred. Experts on modelling are needed for applying sophisticated water quality models whereas simple mass-balance models can in principle be applied by any limnologist. In earlier years, the availability of computers had set restrictions on the usage of models, but today ordinary microcomputers are generally sufficient for most calculations of water quality models.

In the modelling procedure, calibration, validation and sensitivity analysis are important steps. Calibration is a process in which the parameters of the model are tuned so that the output of the model corresponds well enough to the observed data. There are methods for automatic calibration, but visual calibration is still more common because the automatic calibration approach does not necessarily produce the best result. Validation is a process in which the calibrated model, with no changes in the parameter values, is applied by using an independent data set. If the agreement between the observed and the calculated results is good, then the model is considered to be validated. Validation is sometimes also called verification. In sensitivity analysis the values of the parameters and the driving variables of the model are changed in order to see how much these changes affect the calculated results. Sensitivity analysis is very important in identifying the behaviour of the system. Unfortunately, too little attention has been paid to sensitivity analysis, in water quality modelling. The following data are needed for model applications: in calibration, the values of the state variables and driving variables representing the calibration period, for validation, a similar data set is needed but from another period, and for predictions, the initial conditions, i.e. the values of the state variables in the beginning of the simulation, as well as the values of the driving variables (in different scenarios), are required.

One problem in the general application of water quality models is that there tend to be too many alternatives to be studied, i.e. when the different loading alternatives and hydrological situations are combined. Special attention should be paid to selection of the relevant scenarios.

Water quality models may have very different-time scales. In predictions of the possible effects of climate change, the simulation period may even cover several centuries. In these cases, the problem may be an accumulation of the error in the calculated results. The effects of wastewaters are often studied by using a research period of only one summer. In these kinds of short simulations, the problem is that the initial conditions may have a profound effect on the final results. The dynamics simulated with the model may be caused by the fact that the initial conditions do not represent the equilibrium towards which the model is tending. Therefore, the effect of the forcing functions cannot be seen in the correct way.

6.2.6 CONCLUDING REMARKS

Mechanistic water quality modelling is an effective tool in making water quality assessments. It also helps us in limnological research. However, this approach does have its restrictions. It is much easier to describe chemical variables than it is biological variables, and the exact description of the whole ecosystem is very difficult. Therefore, an angle of vision must be chosen in building a water quality model. The development of mathematical models has become easier and easier thanks to the availability of modern computers. However, limnological expertise is always needed in water quality modelling, at least in the interpretation of the results. Models are not universal, and they must be calibrated. The reliability of the model is dependent on the data material being used in the calibration and validation of the model. Finally, very much attention should be paid for the selection of the alternatives for the driving variables (e.g. loading), the effects of which are studied by using the models.

REFERENCES

Chapra, S. C. and Reckhow, K. H., 1983. *Engineering Approaches for Lake Management*, Vol. 2, *Mechanistic Modeling*, Butterworth Publishers, Boston, MA, USA.

Chen, C. W. and Orlob, G. T., 1972. Ecologic Simulation for Aquatic Environments, Final report for the Office of Water Resources Research, US Department of the Interior, Water Resources Engineers, Walnut Creek, CA, USA.

Dillon, P. J. and Rigler, F. H., 1974. A test of a simple nutrient budget model predicting the phosphorus concentration in lake water, *J. Fish. Res. Board Can.*, **31**, 1771–1778.

Frisk, T., 1989. Development of Mass Balance Models for Lakes, Publications of the Water and Environment Research Institute Report No. 5, National Board of Waters and the Environment, Helsinki, Finland.

Frisk, T. and Nyholm, B., 1980. Lämpötilan vaikutuksesta reaktionopeuskertoimiin vedenlaatumalleissa (The effect of temperature on reaction rate coefficients in water quality models), *Vesitalous*, **21**(5), 24–27 (in Finnish, with an English summary).

Frisk, T., Niemi, J. S. and Kinnunen, K. A. I., 1981. Comparison of statistical phosphorus-retention models, *Ecol. Modelling*, **12**, 11–27.

Frisk, T., Bilaletdin, Ä., Kallio, K. and Saura, M., 1997. Modelling the effects of climate change on lake eutrophication, *Boreal Environ. Res.*, **2**, 53–67.

Gaume, A. M. and Duke Jr, J. H., 1975. Computer Program Documentation for the Reservoir Ecologic Model EPAECO, prepared for the US Environmental Protection Agency, Planning Assistance Branch, Water Resources Engineers, Walnut Creek, CA, USA.

Lappalainen, K. M., 1974. Kehitysarviot eri kuormitusvaihtoehdoilla – Kallaveden reitti ja Haukivesi (Water quality predictions with different loading alternatives – the watercourse of Lake Kallavesi and Lake Haukivesi), Report No. 59, National Board of Waters, Helsinki, Finland, 1–84 (in Finnish).

Lappalainen, K. M., 1978. Vesistöjen ravinnemallit ja niiden yhteys happimalleihin (The nutrient models of watercourses and their connection with oxygen models), in Nyroos, H. (Ed.), *Vesistömallit ja niiden soveltaminen käytäntöön*, Vesi-ja kalatalousmiehet ry, Helsinki, Finland, 53–60 (in Finnish).

Lassiter, R. R. and Kearns, D. K., 1973. Phytoplanktonpopulation changes and nutrient fluctuations in a simple aquatic ecosystem model, in Middlebrook, E. J., Falkenborg, D. H. and Maloney, Th.E. (eds), *Modeling the Eutrophication Process*, pp. 131–138.

Norton, W. R., Roesner, L. A., Evenson, D. E. and Monser, J. R., 1974. Computer Program Documentation for the Stream Water Quality Model QUAL-II, prepared for the Environmental Protection Agency, Systems Development Branch, Water Resources Engineers, Walnut Creek, CA, USA.

Streeter, H. W. and Phelps, E. B., 1925. A study of the pollution and natural purification of the Ohio river, Public Health Bulletin No. 146, United States Public Health Service, Treasury Department, Washington, DC, USA.

Virtanen, M., Koponen, J., Dahlbo, K. and Sarkkula, J., 1986. Three-dimensional water-quality-transport model compared with field observations, *Ecol. Modelling*, **31**, 185–199.

Vollenweider, R. A., 1969. Möglichkeiten und Grenzen elementarer Modelle der Stoffbilanz von Seen (Possibilities and limits of elementary models concerning the budget of substances in lakes), *Arch. Hydrobiol.*, **66**, 1–36 (in German, with an English abstract).

Chapter 6.3

The Water Quality Classification Systems in Sweden

TORGNY WIEDERHOLM

Hydrological and Limnological Aspects of Lake Monitoring
Edited by Pertti Heinonen, Giuliano Ziglio and André Van der Beken
©2000 John Wiley & Sons, Ltd, ISBN 0 471 89988 7

6.3.1   INTRODUCTION

Swedish "Environmental Quality Criteria" have been developed for the forest landscape, the agricultural landscape, groundwater, coasts and seas, contaminated sites and lakes and watercourses by the Swedish Environmental Protection Agency (SEPA). They were developed to provide an easy way for local and regional authorities and others to assess the state of the environment, based on monitoring and surveillance data.

This chapter describes the EQC for lakes and watercourses. The classification system is described in full in SEPA (1999) and SEPA (2000). A somewhat abbreviated version may be seen at http://www.environ.se. Background data and a more complete description of how the guidelines were derived occur in Wiederholm (1999a, 1999b).

Another classification system, primarily intended for the assessment of conservation values in Swedish lakes and watercourses, is also described briefly in the present paper.

6.3.2   PRINCIPLE AND PARAMETERS[1]

The EQC for lakes and watercourses are based on a previous system (SEPA, 1991), which was similar in structure and purpose, but more limited in scope. Basically, the criteria enable the user to classify a water body as belonging to one of, usually, five classes with respect to a number of physico-chemical and biological parameters. The assessment involves two aspects. One is the status as such, i.e. whether the value of a specific parameter is high or low or intermediate in perspective of the range that occurs in Swedish water bodies. For example, the concentration of total phosphorus in a lake may be classified as being low, moderately high, high, very high or extremely high. The other aspect involves the extent to which the state deviates from a reference value or "comparative value", which ideally represents an estimate of a "natural" state. For example, the phosphorus concentration may also be classified as showing no or insignificant deviation from reference conditions or significant, large, very large or extreme deviation from reference conditions, respectively. Deviation from reference value is usually expressed as the quotient between recorded value and reference value.

The classification system involves parameters that are generally considered to be important indicators of water quality in a wide sense, and for which methods and a reasonable amount of data from a variety of geographical

---

[1]Parameter is used as synonymous with variable in SEPA's publication and the former wording is retained here.

settings were available (Table 6.3.1). The chemical parameters indicate the status of a water body with respect to nutrients/eutrophication, oxygen conditions, transparency, buffering capacity/acidification and the occurrence of metals. The biological parameters reflect both plant and animal communities. In general the biological parameters do not reflect specific environmental threats, but do rather provide an integrated measure of the environmental situation and the impact to which a water body may be exposed. For various reasons, it has not been possible to include some parameters, which do in fact represent important aspects of water quality. Hydrological and morphological conditions are two aspects that are not covered by the system. Nor does the system include the assessment of changes in water quality over time.

In general, assessments are to be based on data established through sampling and laboratory analyses as outlined in SEPA's guidelines for environmental monitoring.

6.3.3   CLASSIFICATION AND CLASS DELINEATION

The first type of scale – the state scale – was developed with the ambition to set boundaries at "ecological break points". For chemical parameters this might be levels at which clear biological effects occur; for biological parameters it might be levels where changes are perceived as being particularly evident. Where levels representing biological effects or other changes could not be identified, boundaries were decided statistically on the basis of the most representative data possible, or arbitrarily on the basis of on overall judgement of what can be considered reasonable.

In general, the state scale cannot be interpreted as representing "good" or "bad" environmental quality. Instead, the parameters must be evaluated individually in the light of the quality aspects that they are intended to reflect. This may be illustrated with some examples. Metals and alkalinity are fairly closely related to water quality in the sense that increasing concentrations (decreasing for alkalinity) reflect a growing risk of negative effect on aquatic organisms or use of water. Here the lowest class may be rather easily thought of as representing the "best" conditions. Increasing concentrations of phosphorus reflect a growing risk of increasing quantities of planktonic algae. From an aesthetic viewpoint, for bathing and for water supply this is generally undesirable, but in terms of production and as a basis for biomass of fish it is not. Since phosphorus is primarily intended to indicate conditions for the presence of planktonic algae and associated adverse effects, the scale has the same direction as that for algae with the lowest class representing "good" conditions, and hence reflecting the lower risk of high biomasses of algae. The same rationale underlies the assessment scales for other parameters, i.e. where

Table 6.3.1 Parameters of the Swedish Environmental Quality Criteria for lakes and watercourses (SEPA, 1999, 2000); + and − denotes occurrence or absence of criteria for a specific parameter.

Parameter	Lakes/water-courses	State	Deviation from reference conditions
Nutrients/eutrophication			
Total phosphorus	l	+	+
Total nitrogen	l	+	−
N/P-quotient	l	+	−
Area specific loss of total nitrogen	w	+	+
Area specific loss of total phosphorus	w	+	+
Oxygen and oxygen consuming substances			
Oxygen content	l/w	+	−
TOC	l/w	+	−
COD_{Mn}	l/w	+	−
Transparency			
Absorbance	l/w	+	−
Colour	l/w	+	−
Turbidity	l/w	+	−
Secchi depth	l	+	−
Acidity/acidification			
Alkalinity	l/w	+	+
pH	l/w	+	+
Metals			
Metals in water, sediment, water moss and fish	l/w*	+	+
Planktonic algae			
Total volume	l	+	+
Chlorophyll	l	+	−
Diatoms	l	+	+
Water blooming cyanobacteria	l	+	+
Potentially toxic cyanobacteria	l	+	+
Biomass of *Gonyostomum semen*	l	+	+
Macrophytes			
Submersed and floating-leaved plants, number of species and indicator value**	l	+	+
Periphyton—diatoms			
IPS-index	w	+	−
IDG-index	w	+	−
Macroinvertebrates			
Shannon's diversity index***	l/w	+	+
Danish fauna index***	l/w	+	+
ASPT-index***	l/w	+	+
Acidity index***	l/w	+	+
BQI-index****	l	+	+
O/C-quotient****	l	+	+

(*continued*)

Table 6.3.1 (*continued*)

Parameter	Lakes/ water- courses	State	Deviation from reference conditions
Fish			
Number of naturally occurring species	l/w	+	+
Diversity of naturally occurring species	l	+	+
Biomass of naturally occurring species	l/w	+	+
Abundance of naturally occurring species	l/w	+	+
Relative abundance of cyprinids	l	+	+
Relative abundance of piscivorous percids	l	+	+
Relative abundance of salmonids	w	+	+
Reproduction of salmonids	w	+	+
Acid-sensitive species and life stages	l/w	+	+
Species tolerant to low oxygen levels	l	+	+
Relative abundance of exotic species	l/w	+	+
Index based on (part of) above indicators	l/w	+	+

* sediment in lakes; ** indicator value used for calculation of deviation from reference; *** in lakes: littoral zone; **** profundal zone of lakes.
Acronyms in table (for complete reference see SEPA, 1999, 2000)
IPS = "Indice de polluo-sensibilité" (CEMAGREF, 1982); EDG = "Indice diatomique génerique" (Rumeau and Coste, 1988; ASPT = Average score per taxon (Armitage *et al.*, 1983); O/C = Oligochaeta/Chironomidae (Wiederholm, 1980).

contradictory interpretation or evaluations are possible, the direction of the scales has been decided by the environmental quality aspect each parameter is primarily intended to indicate and what is deemed essentially "good" or "bad" in this respect.

Assessing deviation from reference values is generally less of a problem than assessing state. Increasingly pronounced deviation from the reference values, i.e. from natural conditions, is usually regarded as negative. Class 1 therefore represents the most favourable conditions and class 5 the least favourable. It should be kept in mind, however, that the assessment applies to the quality aspect that the parameter is primarily intended to reflect. From other perspectives, growing deviations from reference values may be favourable.

6.3.4 REFERENCE VALUES

When assessing deviation from reference values, it is of course important that these values are representative of the water body that is to be assessed. The guidelines suggest different ways to achieve this. Reference values may be derived from historical data, palaeolimnological records, by inference from other, unaffected water bodies or through modelling. For some parameters the guidelines give reference values for geographic regions or types of water bodies

Table 6.3.2 EQC for nutrients/eutrophication. Upper class limits for classification of state and deviation from reference values, respectively, and algorithms for estimation of site-specific reference values (cf. text) (SEPA, 1999, 2000).

Parameter	Class				
	1	2	3	4	5
State					
Total P μg/l, May–Oct.	12.5	25	50	100	>100
Total P μg/l, Aug.	12.5	23	45	96	>96
Total N mg/l, May–Oct.	0.300	0.625	1.25	5.0	>5.0
Total N/Total P, June–Sept.	>30	30	15	10	5
Area specific loss of total P, kg/ha y	0.04	0.08	0.16	0.32	>0.32
Area specific loss of total N, kg/ha y	1.0	2.0	4.0	16	>16
Deviation from reference					
Total P*	1.5	2.0	3.0	6.0	>6.0
Area specific loss of total P*	1.5	3	6	12	>12
Area specific loss of total N*	2.5	5	20	60	>60

Algorithm for reference values for lakes

Total P_{ref} (μg P/l) = $\quad 5 + 48 \cdot abs\ f_{420/5}.$

Algorithms for reference values for watercourses

Total P_{ref} (kg P/ha y)

$$0.002 \cdot x_1 + 0.015$$
$$0.10 \cdot x_2 + 1.2/(5 \cdot x_2 + 12)$$
$$0.91 \cdot x_3 \cdot 10^{-3} + 0.02$$
$$2.45 \cdot x_4 \cdot 10^{-3} + 0.024$$
$$3.15 \cdot x_1 \cdot 10^{-4} \cdot (5 + 60 \cdot x_5)$$

Total N_{ref} (kg N/ha y) =

$$0.018 \cdot x_1 + 0.85$$
$$-0.023 \cdot x_2 + 1.25$$
$$0.008 \cdot x_3 + 0.85$$
$$0.03 \cdot x_4 + 0.90$$
$$3.15 \cdot x_1 \cdot 10^{-4} \cdot (125 + 500 \cdot x_5)$$

where
x_1 = specific flow (l/km^2 sec)
x_2 = lake percentage in drainage area
x_3 = area specific loss of COD_{Mn} (kg/ha y)
x_4 = area specific loss of silica (kg/ha y)
x_5 = flow weighted absorbance at 420 nm

*measured/reference.

that may be used in the absence of site-specific values, with due cautiousness. For other parameters models are suggested, by which reference values may be calculated. In both cases, data from the national monitoring network or other studies with broad coverage have been used as a basis. Stations considered to be affected by man have usually been excluded. However, for fish calculations have been made using national or supra-regional data bases in their entirety, since it was not possible to sort out unaffected waters. Thus the reference values for fish represent the mean situation in Swedish lakes and watercourses, differentiated with respect to certain morphological characteristics.

6.3.5 NUTRIENTS AND EUTROPHICATION – AN EXAMPLE

Nutrient status and eutrophication may be assessed in several ways (Table 6.3.2). Total phosphorus, total nitrogen and phosphorus/nitrogen-quotient may be used for lakes. Total phosphorus has been selected as a key parameter due to its analytical simplicity and general use in monitoring programmes. Total nitrogen was kept from the previous guidelines due to a widespread wish to have a scale by which to separate lakes with respect to their nitrogen content. The N/P-quotient indicates the potential for nitrogen fixation and mass development of cyanobacteria. Area specific loss of nitrogen and phosphorus, respectively, is used for the assessment of watercourses. The parameters are common in monitoring programmes and used to calculate loadings of nutrients on lakes and coastal areas. They also give an indirect measure of the production potential for plant and animal communities in watercourses.

Each parameter requires a certain amount of data. For example, the assessment of total phosphorus in lakes should be based on monthly samples (May–Oct.) from the epilimnion or surface water (0.5 m) during one year. The assessment may be made based on late season samples, but only at low concentrations and then by use of data from three consecutive years.

The class boundaries each have their specific basis and significance. Thus the total phosphorus classes are commonly recognized as representing oligotrophic (1), mesotrophic (2), eutrophic (3 + 4) and hypertrophic conditions (5) in lakes. The area specific losses of total phosphorus to watercourses are typical of forested land unaffected by man (1), common forest land in Sweden (2), clear cut areas, bogs and agricultural land with little erosion (3), field cultivated agricultural land (4) and agricultural land with a tendency to erosion (5).

Algorithms are suggested by which reference values may be estimated in the absence of site-specific measured values (Table 6.3.2). They express the relationship between EQC parameters and physico-chemical parameter which are more conservative and less influenced by human activities. The algorithms have been derived from national monitoring data and are considered to represent, as much as possible, relationships characteristic of "natural" conditions. All algorithms are expected to give low estimates, and the highest value obtained by any of the equations should be used as a reference (some are not to be used at all under certain circumstances, which are described in the guidelines).

Table 6.3.3   Criteria and parameters in System Aqua (cf. text).

I. Catchment area

Structural diversity
   Number of lakes and lake area
   Diversity of vegetation types/land use
   Topographic relief

Naturalness
   Physical manipulations
   Degree of perturbation
   Water quality

II. Object: Lake or watercourse

Structural diversity
   Riparian land use categories
   Bottom substrate
   Shore line developmental/fluvial features
   Structure of aquatic vegetation

Naturalness
   Long-lasting physical encroachment
   Flood control
   River corridor vegetation
   Biotic changes
   Water quality

Rarity
   Endangered species
   Vulnerable species
   Rare species
   Care-demanding species
   Regionally threatened species

Species richness
   Macrophytes
   Phytoplankton
   Benthic macroinvertebrates
   Fish
   Nesting birds

Representatives
   Macrophytes
   Phytoplankton
   Benthic macroinvertebrates
   Fish
   Nesting birds

6.3.6 SYSTEM AQUA

The need for a system for identification of water bodies with particular nature conservation value, but also for impact assessment and evaluation of restoration requirements, made SEPA fund the development of what became named System Aqua (Willén *et al.*, 1996, 1997). The system was developed and modelled after SERCON (System for the Evaluation of Rivers for Conservation) (Boon *et al.*, 1997). It has both general and specific features in common with SERCON, but also several characteristics that are unique.

System Aqua was developed for use on lakes and watercourses. Both the water bodies as such and their catchment areas are evaluated. The evaluation is based on habitat characteristics and biological measures. The catchment area is evaluated by two criteria and the water body by five criteria, each one characterized by a number of variables (Table 6.3.3). Each variable is scored from 0 to 5. After a simple weighing procedure the results may be presented in a graph describing the specific criteria profile of an object and its catchment area. Background information on each object and its catchment is given as supplementary information.

An important aspect is that the system can be used in a step-wise manner. In the first step, a preliminary or limited assessment can be made by use of information available from maps, statistics on land use and/or the analysis of aerial photos. This may be followed by a broader assessment, involving the remaining variables or criteria. The methods that are to be used for field inventories are the same ones that are being prescribed by SEPA for national and regional environmental monitoring.

REFERENCES

Boon, P. J., Holmes, N. T. H., Maitland, P. S., Rowell, T. A. and Davies, J., 1997. A system for Evaluating Rivers for Conservation (SERCON): Development, Structure and Function. In: Boon, P. J. and Howell, D. L. (Eds.). *Freshwater Quality: Defining the Indefinable*. Scottish Natural Heritage. Edinburgh: The Stationery Office.

SEPA, 1991. Quality Criteria for Lakes and Watercourses. A System for Classification of Water Chemistry and Sediment and Organism Metal Concentrations. Swedish Environmental Protection Agency.

SEPA, 1999. Bedömningsgrunder för miljökvalitet – Sjöar och vattendrag. Naturvårdsverket Rapport 4913. In Swedish with English summary.

SEPA, 2000. Environmental Quality Criteria – Lakes and Watercourses. Swedish Environmental Protection Agency Report 5050.

Wiederholm T. (Ed.), 1999a. Bedömningsgrunder för miljökvalitet – Sjöar och vattendrag. Bakgrundsrapport 1. Kemiska och fysikaliska parametrar. Naturvårdsverket Rapport 4920. In Swedish with English summary.

Wiederholm T. (Ed.), 1999b. Bedömningsgrunder för miljökvalitet – Sjöar och vattendrag. Bakgrundsrapport 2. Biologiska parametrar. Naturvårdsverket Rapport 4921. In Swedish with English summary.

Willén, E., Andersson, B. and Söderbäck, B., 1996. System Aqua. Underlag för karakterisering av sjöar och vattendrag. Naturvårdsverket Rapport 4553. In Swedish.

Willén, E., Andersson, B. and Söderbäck, B., 1997. System Aqua: A Biological assessment Tool for Swedish Lakes and Watercourses. In: Boon, P. J. and Howell, D. L. (Eds.). *Freshwater Quality: Defining the Indefinable.* Scottish Natural Heritage. Edinburgh: The Stationery Office.

Chapter 6.4

Classification of the Environmental Quality of Freshwater in Norway

JON LASSE BRATLI

Hydrological and Limnological Aspects of Lake Monitoring
Edited by Pertti Heinonen, Giuliano Ziglio and André Van der Beken
©2000 John Wiley & Sons, Ltd, ISBN 0 471 89988 7

6.4.1 INTRODUCTION

A classification system for the environmental quality of freshwater in Norway has been developed by the Norwegian Institute for Water Research (NIVA), under contract from the Norwegian Pollution Control Authoritiy (SFT). The Norwegian health authorities have agreed to the criteria proposed for drinking and bathing water.

The main purpose of the classification system is to give different people in the central, regional and local administrations, consulting engineers and scientific researchers a uniform and objective tool for the evaluation of environmental quality status and trends in Norwegian watercourses.

This system will be of help in the development of goals for environmental quality, and 'translates' environmental observations obtained from biological and chemical variables and concentrations into concepts which are useful for decision makers and are of interest to the general public.

Earlier versions of the classification system were published in 1989 and in 1992. This paper presents a shortened version of the revised guideline published by the Norwegian Pollution Control Authority in 1997 (SFT, 1997). The revised edition includes some adjustments of a practical and technical character, due to the availability of new data and information over the past few years. A new chapter with detailed information regarding the use and constraints of the system will hopefully eliminiate earlier misuse of the system. New national regulations and the adoption of various EU Directives have made such a revision necessary.

6.4.2 SYSTEM STRUCTURE AND LIMITATIONS

This present paper contains tables and verbal descriptions for the classification of environmental *quality status* and *suitability* related to adequate usage of watercourses; details are shown in Table 6.4.1.

The **classification of quality status** is based on measured values which have two components, i.e. a natural component which stems from natural processes in the catchment area, and a component which stems from human influence, i.e. acid rain, effluents from industry and sewage, and agricultural runoff. The latter is defined as pollution. This classification is illustrated in Figure 6.4.1.

The human influence on water quality can vary considerably, and it is important to estimate the natural water quality when the goals for the water quality have been set. As an example, Table 6.4.2 shows the expected natural water quality and the observed quality status for a shallow lake in the south-eastern part of Norway.

Table 6.4.1 Concepts used in the classification system.

Basis	Quality status		Suitability
	Measured values		Adequate usage associated with a given water quality
Classes	Nutrients, organic matter etc.	Micro pollutants	Four classes
	I = Very good	I = Slightly polluted	1 = Well suitable
	II = Good	II = Moderately polluted	2 = Suitable
	III = Fair	III = Markedly polluted	3 = Less suitable
	IV = Bad	IV = Severely polluted	4 = Unsuitable
	V = Very bad	V = Extremely polluted	

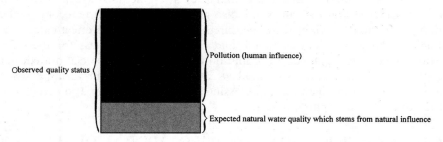

Figure 6.4.1 A measured quality status can be divided into an expected natural water quality and various contributions from human activities.

Table 6.4.2 Illustration of expected natural water quality and observed quality status for a typical shallow lake in the south-eastern part of Norway with most of its catchment in marine clay.

Effect category	Quality class				
	I	II	III	IV	V
Nutrients		▨		▆	
Organic matter	▨			▆	
Acidifying components					
Micro-pollutants					
Particles			▨		▆
Faecal bacteria	▨			▆	

▨ Expected natural water quality; ▆ Observed quality status, when it is not identical to the expected natural water quality.

The difference between the observed quality and the expected natural quality represents the pollution, and a goal for future quality should lie between these two. A class-II goal for particles in this lake is therefore meaningless in this respect.

The **classification of suitability** is based on the evaluation of the pollution control and health authorities that are appropriate for the environmental quality related to different usage of the water, i.e. for drinking water, bathing, fishing and irrigation.

6.4.3   METHODS AND DATA REQUIREMENTS

As shown in Table 6.4.3, there are six different effect categories or pollution types in the system. Each of these effect categories has a number of variables to describe the pollution types. The variables shown in italics are the so-called key variables. The sampling frequencies and calculation methods which are used to obtain the classification values, are also given in the table. It is recommended that two or more variables are measured within each of the effect categories. Each of the variables is classified, and merged into one class for the effect category concerned. A general pollution class should not be elaborated, and each of the effect categories should be treated separately. Some variables which are commonly examined but not classified in this system, are also included in the table (shown in parentheses).

6.4.4   CLASSIFICATION OF ENVIRONMENTAL QUALITY STATUS

The basis for the division of variables into quality classes is a combination of statistical information about the distribution of the substances in Norwegian watercourses, and a knowledge of the effects of the substances on the ecology in the water environment.

Tables 6.4.4 and 6.4.5 show the classification of the *quality status*, where the key variables are shown in italics.

For the micro-pollutants, a so-called 'high diffuse background level', based on a large statistical material, has been established. This background level is somewhat higher than the mean value, being the value of 75–90%, and represents the limit between classes I and II. Classes II–V are a result of an upscaling of the background level after an assessment of the different substances according to the following:

- how hazardous each of them are;
- if they are observed in low or high concentrations in the watercourses;
- how large a change in concentration a given effluent will entail;
- the health risk (only for mercury in fish).

Table 6.4.3 Requirements for classification of each of the effect categories.

Effect category	Ecosystem type	Variable[a,b]	Sampling frequency	Calculation method
Nutrients	Lakes	*Total phosphorus* *Chlorophyll* a *Secchi depth* Primary production Total nitrogen (Orthophosphate)[c] (Phytoplankton) (Zooplankton)	At least monthly; mixed sample (May–October); deep-profile (3–5 samples), late-summer and late-winter	Arithmetic mean
	Rivers	*Total phosphorus* Total nitrogen (Periphyton) (benthic fauna)	At least monthly	Arithmetic or time-weighted mean
Organic matter	Lakes	*TOC* *Colour* *Oxygen* *Secchi depth* COD_{Mn} Fe Mn	Deep-profile (3–5 samples) in spring, late-summer, fall and late-winter	Arithmetic mean; Oxygen, lowest value; Fe and Mn, highest value
	Rivers	*TOC* COD_{Mn} (Periphyton) (Benthic fauna)	At least monthly[d]	Arithmetic or time-weighted mean
Acidifying components	Lakes and rivers	*Alkalinity* *pH* (Benthic fauna)	Spring, summer, fall and winter in lakes; monthly in rivers	Lowest value
Micro pollutants (heavy metals)	Lakes and rivers	Dependent on problematic component(s)	Spring, summer, fall and winter in lakes; monthly in rivers	Highest value
Particles	Lakes and rivers	*Turbidity* *Suspended matter* *Secchi depth (in lakes)*	At least monthly	Arithmetic or time-weighted mean
Faecal bacteria	Lakes and rivers	*Thermotolerant coliform bacteria*	At least monthly[e] deep-profile (3–5 samples)	Highest 90%

[a]Key variables are shown in italics.
[b]Variables which are commonly examined, but not classified in this sytem, are shown in parentheses.
[c]Measured in smaller rivers and in deep-profile in lakes.
[d]More frequent sampling in small rivers.
[e]If drinking or bathing interests (bathing season) prevail, weekly sampling may be necessary (SFT, 1997).

Table 6.4.4 Classification of the quality status for nutrients, organic matter, acidifying components, particles and faecal bacteria.

Effect category	Variable[a]	Quality class I 'very good'	II 'good'	III 'fair'	IV 'bad'	V 'very bad'
Nutrients	*Total phosphorus,* ($\mu g\,P\,l^{-1}$)	<7	7–11	11–20	20–50	>50
	Chlorophyll a ($\mu g\,l^{-1}$)	<2	2–4	4–8	8–20	>20
	Secchi depth (m)	>6	4–6	2–4	1–2	<1
	Primary production (g C m^{-2} y^{-1})	<25	25–50	50–90	90–150	>150
	Total nitrogen ($\mu g\,l^{-1}$)	<300	300–400	400–600	600–1200	>1200
Organic matter	*TOC* (mg C l^{-1})	<2.5	2.5–3.5	3.5–6.5	6.5–15	>15
	Colour (mg Pt l^{-1})	<15	15–25	25–40	40–80	>80
	Oxygen (mg $O_2\,l^{-1}$)	>9	6.4–9	4–6.4	2–4	<2
	Oxygen (%)	>80	50–80	30–50	15–30	<15
	Secchi depth (m)	>6	4–6	2–4	1–2	<1
	COD_{Mn} (mg O l^{-1})	<2.5	2.5–3.5	3.5–6.5	6.5–15	>15
	Iron (μg Fe l^{-1})	<50	50–100	100–300	300–600	>600
	Manganese (μg Mn l^{-1})	<20	20–50	50–100	100–150	>150
Acidifying components	*Alkalinity* ($mmol\,l^{-1}$)	>0.2	0.05–0.2	0.01–0.05	<0.01	0.00
	pH	>6.5	6.0–6.5	5.5–6.0	5.0–5.5	<5.0
Particles	*Turbidity* (FTU)	<0.5	0.5–1	1–2	2–5	>5
	Suspended matter ($mg\,l^{-1}$)	<1.5	1.5–3	3–5	5–10	>10
	Secchi depth (m)	>6	4–6	2–4	1–2	<1
Faecal bacteria	Thermotolerant coliform bacteria (no. per 100 ml)	<5	5–50	50–200	200–1000	>1000

[a]Key variables are shown in italics.

In classes IV and V, there are usually known effects from the substances on one or several of the elements in the ecosystem. These will, however, vary a great deal because of variable bioavailability, where the latter varies according to the content of organic and particulate matter, conductivity and pH.

6.4.4.1 Biodiversity

Freshwater species have different tolerances to different environmental impacts. For instance, different species of benthic fauna will react differently to acidification. Some species will have great problems with reproduction even

Table 6.4.5 Classification of the quality status for micro-pollutants in water, sediment and fish.

Effects of micro-pollutants (heavy metals)	Variable	Quality class				
		I 'slightly polluted'	II 'moderately polluted'	III 'markedly polluted'	IV 'severely polluted'	V 'extremely polluted'
In water[a]	Copper	< 0.6	0.6–1.5	1.5–3	3–6	> 6
	Zinc	< 5	5–20	20–50	50–100	> 100
	Cadmium	< 0.04	0.04–0.1	0.1–0.2	0.2–0.4	> 0.4
	Lead	< 0.5	0.5–1.2	1.2–2.5	2.5–5	> 5
	Nickel	< 0.5	0.5–2.5	2.5–5	5–10	> 10
	Chromium	< 0.2	0.2–2.5	2.5–10	10–50	> 50
	Mercury	< 0.002	0.002–0.005	0.005–0.01	0.01–0.02	> 0.02
In sediment[b]	Copper	< 30	30–150	150–600	600–1800	> 1800
	Zinc	< 150	150–750	750–3000	3000–9000	> 9000
	Cadmium	< 0.5	0.5–2.5	2.5–10	10–20	> 20
	Lead	< 50	50–250	250–1000	1000–3000	> 3000
	Nickel	< 50	50–250	250–1000	1000–3000	> 3000
	Arsene	< 5	5–25	25–100	100–200	> 200
	Mercury	< 0.15	0.15–0.6	0.6–1.5	1.5–3	> 3
In fish[c]	Mercury	< 0.2	0.2–0.5	0.5–1	1–2	> 2

[a]Content in water is measured as μg substance per litre.
[b]Content in sediment is measured mg substance per kg sediment (dry weight).
[c]Mercury content in fish measured as mg Hg per kg muscle (wet weight).

if the degree of acidification is not very high. The general picture is that the number of species will drop when the environmental stress is enhanced. Table 6.4.6 shows how a number of different indicator species display different tolerances to acidification.

6.4.4.2 Suitability for drinking water

The suitability of raw water for drinking purposes is linked to the treatment it has undertaken before delivery to the consumer. A relatively bad raw water quality can, if the appropriate technology is applied, result in a very good tap water quality. In Norway, unlike continental Europe, most of the drinking water plants operate with low degrees of treatment. The raw water quality is therefore associated with simple treatment procedures (fine screening, desinfection and possible pH adjustment).

Table 6.4.6 Tolerance to acidification displayed by a number of common benthic (indicator) species in Norwegian freshwater fauna.

Species		V < 4.5	IV 4.5–5.0	III 5.0–5.5	II–I 5.5–6.0	 > 6.0
Leuctra hippopus	Stonefly	•	•	•	•	•
Amphinemura sulcicollis	Stonefly	•	•	•	•	•
Rhyacophila nubila	Caddisfly	•	•	•	•	•
Polycentropus	Caddisfly	•	•	•	•	•
Leptophlebia spp.	Mayfly	•	•	•	•	•
Isoperla spp.	Stonefly		•	•	•	•
Brachyptera risi	Stonefly		•	•	•	•
Hydropsyche siltalai	Caddisfly		•	•	•	•
Pisidium spp.	Mollusc		•	•	•	•
Ameletus inopinatus	Mayfly			•	•	•
Heptagenia sulphurea	Mayfly			•	•	•
Diura nanseni	Stonefly			•	•	•
Capnia atra/pygmea	Stonefly			•	•	•
Lymnea peregra	Snail				•	•
Gyraulus acronicus	Snail				•	•
Gammarus lacustris	Crustacea				•	•
Baetis spp.	Mayfly				•	•
Ephemeralla aurivillii	Mayfly				•	•
Caenis spp.	Mayfly				•	•

In Table 6.4.7 the most important variables describing the quality of raw water have been classified. The Norwegian Health Authority issued a drinking water provision in 1995, based on a number of EU Directives. In this provision, a number of other variables were included, e.g. micro-pollutants. In most watercourses in Norway, these requirements are easy to meet, and potential water sources with, e.g. heavy metals, can thus be avoided.

Water in the suitability class 4 is unsuited for simple treatment, but it can provide good quality tap-water if extensive physical and chemical treatment is provided.

In addition to the key variables, some eutrophication variables should be included, mainly because of the possible presence of algal blooms in lakes, which can induce taste/odour problems and possible toxin production. These support variables are shown in Table 6.4.8.

6.4.4.3 Suitability for bathing and recreation

Recreation includes various water-related activities in direct contact with water.

Table 6.4.7 Key variables used in the evaluation of raw water quality.[a]

Effects of	Variable	Suitability class			
		1 Well suitable	2 Suitable	3 Less suitable	4 Un- suitable
Faecal bacteria	TCB[b] (no. per 100 ml)	0[c]	0[d]	–	> 0[e]
Organic matter	Colour (mg Pt l^{-1})	< 20	< 20	–	> 20
	Iron (μg Fe l^{-1})	< 50	50–200	–	> 200
	Manganese (μg Mn l^{-1})	< 20	20–50	–	> 50
	Oxygen (%)	> 70	< 70	–	–
Physico-chemical variable	pH	7.5–8.5	6.5–8.5	< 6.5/ > 8.5	–
	turbidity	< 0.4	0.4–14	–	> 4

[a]The absence of data indicates that no meaningful values are available.
[b]TCB, thermotolerant coliform bacteria.
[c]90% of the samples must meet the requirements, while the remainder must be in the range 0–10 TCB per 100 ml.
[d]For waterworks supplying > 10 000 people, a minimum of 70% of the samples must meet the table value; corresponding values are minima of 60 and 50% for waterworks supplying > 1000 and > 100 people, respectively; remainder of the samples must be in the range 0–10 TCB per 100 ml.
[e]Less than 50% of the samples must meet the table value, or single sample values are higher than 10 TCB per 100 ml.

Table 6.4.8 Support variables used in the evaluation of raw water quality.

Effect of	Variable	Suitability class			
		1 well suitable	2 suitable	3 less suitable	4 un- suitable
Nutrients	Total phosphorus (μg l^{-1})	< 7	7–11	11–20	> 20
	Chlorophyll *a* (μg l^{-1})	< 2	2–4	4–8	> 8

The health authorities issued guidance standards for bathing water quality In Norway in 1994. In addition to the health aspects, the published criteria included conditions associated with the peoples' well-being and aesthetic aspects (Tables 6.4.9 and 6.4.10). The water samples must be taken at the place where people regularly bathe, at at least 1 m depth, and at a mininum distance of 2–3 m from the shore. The support variables given in Table 6.4.10 are included for the same reason as for drinking water, i.e. the possibility of

Table 6.4.9 Key variables used in the evaluation of water quailty for bathing and recreation.

Effect of	Variable	Suitability class			
		1 Well suitable	2 Suitable	3 Less suitable	4 Unsuitable
Faecal bacteria	Thermotolerant coliform bacteria (no. per 100 ml)	<100	<100	100–1000	>1000
	Faecal streptococci (no. per 100 ml)	<30	<30	30–300	<300
Physico-chemical variables	pH	5.0–9.0	<5.0 />9.0	–	–
	Turbidity, FTU	<1	1–2	2–5	>5

Table 6.4.10 Support variables used in the evaluation of water quality for bathing and recreation.

Effect of	Variable	Suitability class			
		1 Well suitable	2 Suitable	3 Less suitable	4 Unsuitable
Nutrients	Total phosphorus (μg P l^{-1})	<7	7–11	11–20	>20
	Chlorophyll a (μg l^{-1})	<2	2–4	4–8	>8
	Secchi depth (m)	>4	2–4	1–2	<1
Organic matter	Colour (mg Pt l^{-1})	<25	>25	–	–

algal blooms, with sampling usually being made at the deepest point of the lake.

Other conditions of relevance here could include the following:

- water temperature;
- sun and wind conditions, and currents;
- shore and bottom conditions;
- floating objects and garbage;
- algae and parasites (*Gonyostomum* and cercarie-larva);
- macrovegetation, such as water lilies.

Table 6.4.11 Variables used in the evaluation of water quality for recreational fishing.

		Suitability class			
		1 Well suitable	2 Suitable	3 Less suitable	4 Unsuitable
Effect of	Variable				
Organic matter	Oxygen (surface) (%)	80–110	110–130	130–160	>160
	Oxygen (deep water) (%)	>70	30–70	15–30	<15
Acidifying components	pH^a	>6.0–8.5	5.5–6.0	5.0–5.5	<5.0
	Alkalinity, mmol l^{-1}	>0.05	0.05–0.01	<0.01	0
Micro-pollutants (heavy metals)	Mercury in fish (mg kg^{-1}) (fillet, freshweight)	<0.2	0.2–0.5	0.5–1.0	>1.0
Nutrients^b	Total phosphorus, (μg P l^{-1})	<11	11–20	20–50	>50
	Chlorophyll a (μg l^{-1})	<4	4–8	8–20	>20
	Secchi depth (m)	>4	2–4	1–2	<1

^{a}The effect is dependent upon the concentration of Ca, TOC and labile Al.
^{b}Applies for where redfish are spawning.

6.4.4.4 Suitability for fishing (recreational)

The basis for this classification is the environmental requirement for reproduction of Salmonoid fish (redfish) and the animals that they feed upon. A notable exception in this context is the presence of mercury in fish, where the health aspects are all prevailing (Table 6.4.11).

Class 1: Well suitable

The water quality creates no problem for the fish or related organisms in the water, and presents no health risks on eating the fish. In addition, the quality of the fish is good.

Class 2: Suitable

The water quality can create some problems for the most important animals that Salmonoid fish feed upon, e.g. *Gammarus lacustris*. The fish fauna itself is often not affected, apart from the overall quality of the fish, which can be somewhat reduced.

Class 3: Less suitable

The water quality can be a significant stress factor, especially for Salmonoid fish. Among other things, the reproduction and breeding areas can suffer. Odour, a poorer taste and an elevated content of micro-pollutants can appear in the fish fillet.

Class 4: Unsuitable

Salmonoid fish find it difficult to live and breed in this quality of water, and the latter is also critical for other fish species, such as white fish. In highly eutrophicated waters, many of the fish species are unsuitable for human consumption. The odour and taste are affected and an elevated content of micro-pollutants is common in the fish fillet. Parasites and diseases can also occur.

6.4.4.5   Suitability for irrigation

This section is based on the water quality criteria for irrigation which have been proposed by a working group under the agricultural authorities. In addition to the proposed criteria concerned with bacteria, some eutrophication variables are also included (Table 6.4.12).

Table 6.4.12   Variables used in the evaluation of water quality for irrigation.

		Suitability class			
Effect of	Variable	1 Well suitable	2 Suitable	3 Less suitable	4 Unsuitable
Nutrients	Total phosphorous (μg P l^{-1})	< 11	11–20	20–50	> 50
	Chlorophyll *a* (μg l^{-1})	< 4	4–8	8–20	> 20
Faecal bacteria	Thermotolerant colioform bacteria (no. per 100 ml)	< 2	2–20	20–100[a]	> 100[a]
	Coliform bacteria (no. per 100 ml)	< 20	20–200	200–1000[a]	> 1000[a]

[a]For plants in category III (see text) values up to 150 thermotolerant coliform bacteria and 1500 coliform bacteria are tolerated.

The plants are divided into three categories as follows:

(I) fruit, berries, lettuce, Chinese cabbage, cauliflower, broccoli, carrot and other types of vegetables which are eaten raw and without peeling;
(II) plants that are peeled or heat-treated before eating, e.g. potato, common cabbage, onion and fodderplants, which are not dried or ensilaged;
(III) cereal or leguminous plants and fodderplants which are dried or ensilaged, as well as plants in sports and park installations.

Class 1: Well suitable

The water can be used for all types of plants until the day of harvesting.

Class 2: Suitable

The water can be used for the plants in category (I) until two weeks before harvesting or until harvesting, if drip-watering is applied. It can be used without any restrictions on other types of plants.

Class 3: Less suitable

The water should not be used for plants in category (I), although it can be used for plants in category (II) until two weeks before harvesting. It can be used for plants in category (III) without any restrictions (for these plants, values of up to 150 thermotolerant coliform bacteria and 1500 coliform bacteria, both per 100 ml volume, are acceptable).

Class 4: Unsuitable

The water should not be used for any type of plants.

REFERENCE

SFT, 1997. Klassifisering av miljøkvalitet i ferskvann. SFT Veiledning 97:04. TA-nr 1468/1997. 31 s.

Chapter 6.5
Water Quality Classification in Finland

PERTTI HEINONEN

Hydrological and Limnological Aspects of Lake Monitoring
Edited by Pertti Heinonen, Giuliano Ziglio and André Van der Beken
©2000 John Wiley & Sons, Ltd, ISBN 0 471 89988 7

6.5.1   INTRODUCTION

The first official water quality monitoring network for the main rivers and the biggest lakes in Finland was established by the water authority in 1962 and 1965, respectively. The assessment of the monitoring data, and general watercourse classification on the basis of the physical, chemical, bacteriological and biological variables, measured from the national freshwater network, had already started at the end of the 1960s. The first version of the classification system used by the Finnish water authority was very similar to the Swedish guidelines for practical assessment of water quality of inland waters. In the 1980s, the Finnish National Board of Waters started to develop a new classification system for inland waters. These new classification guidelines were accepted for official use in 1987 (Heinonen and Herve, 1987; National Board of Waters, 1988).

6.5.2   THE STRUCTURE OF THE FINNISH CLASSIFICATION SYSTEM

There are three different utilization-specific watercourse classifications, namely a quality-based classification of watercourses for recreational purposes, a raw water (for domestic use) classification and a quality classification of watercourses for fishing. However, the most common application in the practical assessment of watercourses is the general water quality classification system, which is in a way a summary of these three specific classifications. The quality criteria for water are divided into five groups, so that the first two classes represent natural waters with no or very slight anthropogenic impact (see Appendix 1). The third class, i.e. 'satisfactory', consists of two subclasses, namely one of natural eutrophic or polyhumic waters and the other of watercourses with clear human impact. The last two mentioned classes (poor and bad) are reserved for waters that are polluted as a result of different human activities.

The most important variables in estimating the general water quality classification are closely connected to eutrophication processes, i.e. total phosphorus and chlorophyll *a*. The other variables are water colour and Secchi depth, turbidity, oxygen concentration during stagnation periods, especially in the hypolimnion, faecal bacteria and certain harmful or dangerous substances, such as As, Hg, Cd, Cr, Pb and CN. There are no limit values for any of the biological characteristics in the classification system. In addition, there is only a general description of each class to clarify the classification with the aid of non-measurable factors.

Table 6.5.1 The usability of Finnish lakes estimated for three periods during the 1980s and 1990s (assessment carried out in the Finnish Environment Institute).

Class	Distribution among class of lake (%)		
	In the middle of the 1980s (87% coverage of lakes)	1990–1993 (72% coverage of lakes)	1994–1996 (79% coverage of lakes)
I Excellent	36.5	37.7	38
II Good	40.8	42.9	42
III Satisfactory	18.5	15.7	16
IV Poor	3.7	3.3	4
V Bad	0.5	0.3	0.3
	(140 km^2)	(75 km^2)	(67 km^2)

6.5.3 THE USABILITY OF FINNISH LAKES IN THE 1980s AND 1990s

The usability of Finnish inland waters has been estimated on three occasions during the 1980s and 1990s by using the general water quality classification system. The results obtained are shown in Table 6.5.1.

The total lake area in Finland is some $32\,000 \text{ km}^2$, and nowadays only 67 km^2 of lakes belong to Class V (bad). The trend of the improvement has been very positive during the last two decades, especially in the receiving water bodies of the pulp and paper industry, where the latter had earlier been the worst polluter of Finnish inland waters.

6.5.4 CONCLUSIONS

The present water quality classification system has been used in practice now for some 15—20 years, and has proved to be a very easy way to demonstrate the monitoring results in a very concrete form. However, there is an evident need to revise the classification system in the very near future, on account of the following reasons:

- there are new variables to be considered, which have already been monitored long enough for reliable classification (periphytic growth, species composition of phytoplankton, bioaccumulation of harmful substances (Herve *et al.*, 1988), etc.);
- the classification system should clearly be divided into a lake classification and a river classification;
- pelagic variables are not adequate for lake classification, and profundal characteristics are also needed.

In Finland, the water quality classification system is based on the experiences which have been gained mainly from the 1970s and the 1980s. A very practical reason for the revision arises from the new proposed EU Water Framework Directive (Commission Proposal, 1998) which is already scheduled to come into force (concerning the monitoring programmes, Article 10) at the end of the year 2000, and which will require the establishment of a specific monitoring and quality assessment programme for all of the watercourses in the EU countries.

REFERENCES

Heinonen, P. and Herve, S., 1987. Water quality classification of inland waters in Finland, *Aqua Fenn.*, **17**, 147–156.

Herve, S., Paasivirta, J. and Heinonen, P., 1988. Use of mussels (*Anodonta piscinalis*) in the monitoring of organic chlorine compounds, *Water Sci. Technol.*, **20**, 163.

National Board of Waters, 1988. Vesistöjen laadullisen käyttökelpoisuuden luokittaminen (Water quality classification for different uses), Vesi ja Ympäristöhallinnon Julkaisuja 20, Finnish National Board of Waters, Helsinki, Finland (in Finnish only).

APPENDIX 1: GENERAL WATERCOURSE CLASSIFICATION CRITERIA

Class	Watercourse description	Variables and their threshold values
I Excellent	The watercourse is in a natural state and usually oligotrophic, clearwatered or with only low levels of humus. The water is well suited for various different modes of utilization	Colour value, < 50 Visible depth, $> 2.5\,\text{m}$ Turbidity, < 1.5 FTU Faecal coliforms or fecal streptococci, < 10 CFU per 100 ml Mean chlorophyll a in the growing season, $< 3\,\mu\text{g}\,\text{l}^{-1}$ Total phosphorus, $< 12\,\mu\text{g}\,\text{l}^{-1}$
II Good	The watercourse is in a near-natural state or only slightly eutrophied. The water and watercourse are still both well suited to various different modes of utilization	pO_2 in epiliminion, 80–100%; no oxygen depletion in hypolimnion Colour value, 50–100 (except in naturally humic waters, < 200) Secchi depth, 1–2.5 m Faecal coliforms or fecal streptococci, < 50 CFU per 100 ml Mean chloropyll a in the growing season, $< 10\,\mu\text{g}\,\text{l}^{-1}$ Total phosphorus, $< 30\,\mu\text{g}\,\text{l}^{-1}$
III Satisfactory	The watercourse is slightly affected by effluents, non-point loading or other activities, or is naturally remarkably eutrophic. The water and the watercourse are usually satisfactory for most modes of utilization	pO_2 in epilimnion, 70–120%, oxygen depletion may occur in hypolimnion Colour value, < 150 Faecal coliforms or fecal streptococci, < 100 CFU per 100 ml Mean cholorphyll a in the growing season, $< 20\,\mu\text{g}\,\text{l}^{-1}$ Total phosphorus, $< 50\,\mu\text{g}\,\text{l}^{-1}$ In areas affected by pulping effluents, NaLS 2–5 mg l^{-1}
IV Poor	The water is strongly polluted by effluents, non-point loading or other activities. The watercourse is suited only to those modes of utilization with the least stringent quality requirements.	pO_2 in epilimnion, 40–150%, oxygen depletion in hypolimnion Faecal coliforms or fecal streptococci, < 1000 CFU per 100 ml Mean chlorophyll a in the growing season 20–50 $\mu\text{g}\,\text{l}^{-1}$ Algal blooms frequently recorded Total phosphorus, 50–100 $\mu\text{g}\,\text{l}^{-1}$ In areas affected by pulping effluents, NaLS 5–10 mg l^{-1}

Continued

Appendix 1 *Continued.*

Class	Watercourse description	Variables and their threshold values
IV Poor (*Cont.*)		The following threshold values apply for elements with deleterious health effects: As, $< 50\,\mu\mathrm{g\,l^{-1}}$; Hg, $< 2\,\mu\mathrm{g\,l^{-1}}$; Cd, $< 5\,\mu\mathrm{g\,l^{-1}}$; Cr, $< 50\,\mu\mathrm{g\,l^{-1}}$; Pb, $< 50\,\mu\mathrm{g\,l^{-1}}$; total cyanide, $< 50\,\mu\mathrm{g\,l^{-1}}$ Off-flavours frequently observed in fish
V Bad	The watercourse is completely polluted by effluents, non-point loading or other activities and is poorly suited to any mode of utilization	Major disturbances of oxygen economy, where the oxygen saturation level in the epilimnion may exceed 150% in summer, while on the other hand, total oxygen depletion may also be observed in the epilimnion, and at the end of the stratification season the hypolimnion is usually completely deoxygenated Mean chlorophyll *a* in the growing season, $> 5\,\mu\mathrm{g\,l^{-1}}$ Total phosphorus, $> 100\,\mu\mathrm{g\,l^{-1}}$ In areas affected by pulping effluents, NaLS $> 10\,\mathrm{mg\,l^{-1}}$ One or more of the following elements exceeds the maximum threshold values set for Class IV: As, Hg, Cd, Cr, Pb or total cyanide Mercury concentrations in carnivorous fish species, $> 1\,\mathrm{mg\,kg^{-1}}$ Oil film often observed on the water surface

Chapter 6.6

Use and Impact of Monitoring Results for Water Protection Management

PERTTI HEINONEN, GIULIANO ZIGLIO AND GUIDO PREMAZZI

Hydrological and Limnological Aspects of Lake Monitoring
Edited by Pertti Heinonen, Giuliano Ziglio and André Van der Beken
©2000 John Wiley & Sons, Ltd, ISBN 0 471 89988 7

6.6.1   INTRODUCTION

Monitoring programmes for the states of the watercourses and especially the important lakes are nowadays rather extensive in most European countries, but at the same time these operations are relatively expensive. Therefore, it is quite natural that society every now and then questions the necessity of this monitoring. The most common statement is, 'you are collecting enormous amount of data, but are these data of any use to our society?'. Usually, people are in many cases even ready to cut down the amount of monitoring in areas where no changes seem to have taken place over a period of some years. On the other hand, people tend to require far more extensive sampling and research, when, for example, hazardous algae already bloom by the shores.

The well organized research and monitoring of water resources is, however, an unquestionable starting point for all sustainable decision making in water management and in the implementation of different measures used for improving the quality of waters and the ecological status of watercourses. In Finland, water resources monitoring has been based for a long time on versatile national monitoring systems of surface waters, ground waters, and pressures directed towards these resources. This long-term and scientifically high-grade monitoring material has been used in compiling reliable assessments and summaries of the state of the Finnish inland and coastal waters. It has even revealed positive long-term trends in water quality, particularly in the waters where wastewater loading has decreased.

6.6.2   MONITORING NETWORKS IN FINLAND

Hydrological monitoring in Finland started during the last decades of the 1800s, with this first being carried out in the biggest rivers and lakes. In water level monitoring, a complete national coverage was achieved in 1911. At present, the national water level network is based on some 300 stations. The other valuable national hydrological network is that of discharge observations in rivers, which presently consists of some 350 stations.

Water quality was not a particular problem at the beginning of this century, although the first signs of pollution, especially in the neighbourhood of some pulp and paper mills, were visible. However, the quality monitoring of surface waters was not started earlier than the beginning of the 1960s, simultaneously with the new water legislation. At that time, particular attention was paid to different sources of loading, e.g. to the wastewaters discharged from the biggest communities and factories to inland and coastal waters. Their effects on water quality were monitored. In addition, estimations of the non-point loading, especially that from agricultural areas, were occasionally carried out.

The first national water quality monitoring network was established in 1962 by the Finnish Water Authority. At this time, the network consisted of some 150 river stations situated all over the country. The national monitoring network of lakes, including some 150 stations situated in the biggest and most important lakes, was started in 1965. Mainly physical and chemical methods were first used, but the proportion of biological methods soon increased. In addition, there are special monitoring programmes, e.g. for harmful substances and acidification of lakes. However, the largest water quality monitoring network consists of several statutory monitoring programmes including about 4500 sites, which are carried out by the 'polluters' and supervised by the Environment Authorities.

Wastewater loading from the biggest communities and factories has been monitored continuously since the first years of the 1970s. The results of this monitoring have shown that the total loading has significantly decreased and the water quality of inland waters has simultaneously improved. The maximum input of total wastewater loading were measured in 1973. After this, the turning point was reached and the loading started to decrease. Presently, the purification efficiency in wastewater treatment plants is about 90–95%. Municipal sewage is treated most effectively, but the efficiency in the purification of industrial wastewaters has improved as well, e.g. at the moment all Finnish pulp and paper mills have biological wastewater treatment plants.

6.6.3   THE RISKS IN CHANGING THE MONITORING PROGRAMMES

The monitoring of surface waters based on unified programmes has been going on in Finland since the beginning of the 1960s. These monitoring programmes have been critically assessed several times during this period, and they have been developed considerably, both qualitatively and quantitatively. We now have enough result material from many watercourses, even for studying possible change trends in the water quality. This applies above all to water areas in the vicinity of industry and population centres that used to receive an extensive amount of effluents, in which the efficiency of wastewater treatment has been improved considerably and where the water quality has, as a result of this, improved drastically.

The reliable estimation of less remarkable and slower changes and their development trends, is, however, still uncertain due to the relatively short time-scale of the monitoring period. Above all, in smaller watercourses where the water changes quickly, variations due to meteorological and hydrological reasons are also naturally extensive and often sudden. It is, therefore, not possible to estimate the qualitative trends or possible cycles of these water bodies, which often suffer from non-point pollution and minor wastewater load, which are statistically fully reliable with the presently applied frequencies of taking samples.

The hydrological measurements and monitoring, which have been carried out for a clearly longer time than qualitative monitoring, have, in addition to certain development trends, shown that the variations in water amounts are at times very extensive. As the amount of water that flows in nature at a certain time is a significant factor influencing water quality, one can assume that significant variations also take place in the quality of the water without any direct impact from human activities. It is, for example, known that the natural leaching of nutrients varies significantly according to the amount of water. The variation of water quantities is also a central factor with regard to the extent of all non-point pollution. The larger the water amount that flows in nature, then the more leaching from nature, e.g. agricultural land, there will be. The more precise evaluation of the development trends in the quality of water would therefore require the use of water quality data from which the effects of water quantity variations have been excluded.

Another factor that has a significant impact on water quality and the state of water bodies is the temperature and its annual variations. Water temperature has a central impact on, for example, the size of primary production and the algal blooms in waters, the extent of hygienic nuisances and (indirectly) on, e.g. the development of the oxygen situation in deep lakes during the winter. A reliable estimation of the development trend of water bodies would also require the exclusion of variations caused by temperature or the length of the growth period from water quality data.

6.6.4   HOW DO WE SET THE GOALS FOR WATER PROTECTION?

Water protection activities have been carried out in Finland for about 40 years. When more extensive operations started at the beginning of the 1960s, there were hardly any research data available on most of the watercourses, not to mention any kind of organized monitoring of their state. The gathering of data on watercourses and the load on them started right at the beginning of the 1960s, but the actual water protection decisions were based, on one hand, on a general knowledge of the harmful effects of wastewater on water quality and, on the other hand, on reports by users of the watercourses, above all professional fishermen, about negative changes that had taken place both in the watercourses and in the fish stock.

The establishment of durable water protection solutions always requires that these solution have their starting point in the watercourses. The main features of the process of carrying out water protection measures are presented in Figure 6.6.1. The operating unit of water protection is always an entire

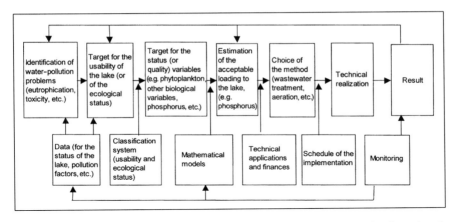

Figure 6.6.1 Schematic representation of the main steps in water-production planning and the implementation of water-protection measures.

watercourse area, with the absolute pre-condition of the operations being that both those who design the measures and those who carry them out constantly have at their disposal a sufficient amount of relevant information about the water resources of the area, the function and possible disturbances in the ecosystems of the watercourses and, on the other hand, information about the total load on the watercourse and other activities that change it.

The identification of the main problems must always be based on material that has specifically been gathered in order to study the quality and quantity of problems that are already known, e.g. wastewater that is discharged to a lake. When problems are identified, all of the factors that may influence the water quality should, however, be considered as thoroughly as possible in advance. Different checking lists are suitable for this purpose. For example, factors that increase the nutrient load on a watercourse can be considered by making use of the checklist presented in Table 6.6.1.

The target state of a watercourse can be expressed by using usability targets with different classification systems. For example, the quality of a strongly eutrophic lake that is now in a poor state must be made satisfactory. This target can be further expressed as various quality targets for a watercourse, e.g. a phosphorus content that is $80 \, \mu g \, L^{-1}$ when the measurements are made must be cut down to $35 \, \mu g \, L^{-1}$. Different models can be used for achieving this, e.g. the largest possible phosphorus load on the watercourse can be estimated with the help of such a model. When calculating, for example, the maximum allowed load from a single source of effluents, the nutrient load on the entire planning area must, however, be taken into consideration.

The protection of waters must be regarded as a set of continuous operations, where the solutions that have been carried out and the results achieved with

Table 6.6.1   The factors increasing eutrophy in watercourses.

A. Pollution from point sources
Domestic wastewater
Industrial wastewater
Load caused by fish farming
Load caused by peat-production areas
Load caused by fur farming
Direct agricultural effluents
Waste tips
Artificial lakes

B. Pollution from non-point sources
Scattered settlement
Leaching caused by agriculture
Leaching caused by livestock farming
Leaching caused by forestry
Leaching caused by road traffic
Pollution through the air

C. Other changing activities
Changes made in the water level
Changes made in the flow
Water constructions

D. Internal load
Dissolving of nutrients caused by the lack of oxygen near the bottom during the
   stagnation periods
Accelerated circulation of nutrients due to too dense fish stock

them play a central role (Nyroos, 1994). For example, wastewater treatment
plants must be monitored frequently enough in order to see that the fixed
targets will be achieved and above all, in order to obtain precise information
about the actual load on the watercourse. The monitoring of the watercourse
gives, in its turn, important information about the present state of the loaded
watercourse and shows whether the measures taken have already improved the
state of the watercourse. The monitoring also reveals possible new problems
that call for remedial measures. The comprehensive evaluation of monitoring
data obtained for different purposes (water quality, pollutant loads, discharges,
pressure factors, etc.) is, therefore, the most essential part of the management
and decision process.

6.6.5   CASE STUDY FROM A LAKE LOADED BY AGRICULTURE

There are many lakes in Finland where the ecological status and usability of
the watercourse has relatively quickly improved due to effective wastewater

treatment. In these cases, the change in the wastewater loading has usually been significant, and therefore it is also easy to understand the rapid improvement in the state of the lakes concerned. The situation is quite different for the lakes, which are affected by non-point loading such as agriculture and forestry. Most of the agricultural areas in Finland are situated in the southern part of the country, where the lake percentage is very low. However, there are some smaller lakes, which are situated right in the areas of intensive agricultural activity. These have already probably been originally slightly eutrophic, because of the clayey ground. However, many of these lakes have eutrophied during this century as a result of agriculture (Kauppi *et al.*, 1993; Rekolainen *et al.*, 1995). Different methods to decrease the nutrient loading from agricultural areas have been used since the last years of the 1980s and especially in the 1990s. The changes in the recipient lakes are, however, very slow (Figure 6.6.2).

The case study presented here is for Lake Pyhäjärvi in the south-eastern part of Finland. The problem of Lake Pyhäjärvi is the increasing eutrophication due to the non-point loading originating from agriculture. This lake has been monitored at least twice a year, during the stagnation periods, since 1965. The trends of the nutrient concentrations during the period 1966–1998 have been presented. The increase in the phosphorus and nitrogen concentrations has been evident up to the beginning of the 1990s. Since 1990, the concentrations have, however, decreased, but very slowly. This may be due to the changes in the tillage practices and to other means suggested in the Finnish Agri-Environmental Programme (Valpasvuo-Jaatinen *et al.*, 1997).

6.6.6 MONITORING AND MANAGEMENT OF LAKE QUALITY IN ITALY: THE EXPERIENCE OF THE LOMBARDY REGION

The most important Italian lake district is located in Lombardy (northern Italy). It includes the deep insubrian lakes (Maggiore, Como, Iseo and Garda) and some small-medium prealpine lakes (e.g. Varese and Briantei). The Lombardy authorities have shown particular concern for water resources management and safeguarding in its environmental policies over the last two decades. These actions emerge clearly from both a legislative and technical standpoint. Beginning with the first regional law regarding sewage disposal, the safeguarding and restoration programmes have been extended in several directions.

In order to comply with the requirements of the National Law No. 319/1976, the regional authorities set up and implemented the Water Clean-up Plan in 1984, plus some important eutrophication control measures, such as the adoption of a limit value of $0.5 \, mg \, L^{-1}$ phosphorus for all effluents entering a lake or within 10 km from the lake shore line. In order to obtain this phosphorus level in urban wastewaters, about 95% of the phosphorus-

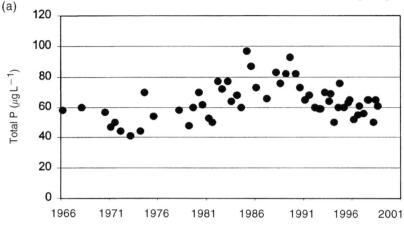

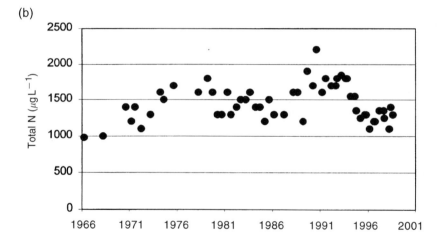

Figure 6.6.2 Total phosphorus (a) and total nitrogen (b) concentrations taken at a depth of 35 m from Lake Pyhäjärvi (Artjärvi municipality, southern Finland) in spring (March, during winter stratification) and autumn (August, during summer stratification), measured over the period 1966–1998 (redrawn from Kauppi et al., 1993, by adding new data obtained from the 1990s).

generated loads must be removed. A full compliance with the law is expected to improve the quality status in the majority of the aquatic environments leading to a condition of oligotrophy, and to reduce drastically the cases of hypereutrophy and eutrophy throughout the region.

The primary objectives of the surface water quality monitoring programme carried out in the Lombardy region are as follows:

- to characterize the quality of the water resources and identify problem waters;
- to support the development of water quality management priorities and plans;
- to evaluate the effectiveness of pollution-control actions.

The sampling frequency is twice a year for natural lakes and assimilated water bodies; i.e. one at the beginning of spring (February–March), during the complete overturn of the water column, and the other from September to October at the end of the summer stratification.

In addition to the emission limit value approach, the regional authorities introduced the concept of receptive capacity of the water body in their Regional Water Clean-up Plan. This gives the possibility of fixing more or less stringent limits, according to the natural lake characteristics and to the characteristics of the contaminants (PRRA, 1992). This policy anticipated, in fact, the so-called 'combined approach' of the European Union (i.e. the integration of quality objectives for water with emission standards for water contaminants), as indicated in the forthcoming Water Framework Directive (1997).

Environmental quality objectives (both ecological and management) have been established after evaluating the natural phosphorus levels for each lacustrine environment. The ecological objectives were quantified as an increase of 25% in the natural phosphorus concentration for oligo-mesotrophic waters, with a 50% increase in phosphorus level for meso-eutrophic lakes. The intermediate (managerial) objectives were identified as an increase in the natural concentration of 50% for oligo-mesotrophic lakes and a 100% increase for eutrophic waters.

In order to achieve the established environmental quality objectives, considerable technical, administrative and economical efforts have been made. During the last decade, several agreements among the Ministry of the Environment, the Lombardy Region and the European Commission were signed, with these aiming to assess the environmental benefits of planned restoration programmes. In addition, intensive research activities have taken place and various projects have been developed (Chiaudani and Premazzi, 1990, 1993; Premazzi *et al.*, 1995; Premazzi *et al.*, 1998).

The survey programme aimed to collect physical, chemical, biological and hydrological data in order to allow the verification of the effectiveness of the pollution-control measures. Samples taken systematically have created the basis for the description of the trophic evolution of the lakes by means of the main water quality indicators (e.g. algal nutrients, dissolved oxygen and chlorophyll).

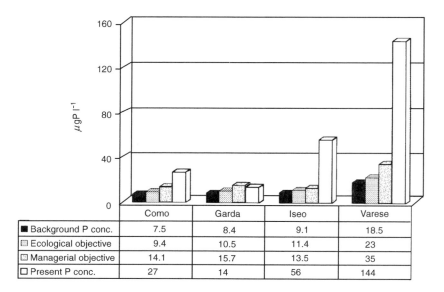

	Como	Garda	Iseo	Varese
■ Background P conc.	7.5	8.4	9.1	18.5
▨ Ecological objective	9.4	10.5	11.4	23
▨ Managerial objective	14.1	15.7	13.5	35
□ Present P conc.	27	14	56	144

Figure 6.6.3 Present conditions, water-quality objectives and natural background concentrations of phosphorus shown for the four case studies of Lakes Como, Garda, Iseo and Varese (Lombardy region of northern Italy).

The reductions of phosphorus input due to the implementation of the measures mentioned above have in several cases reduced the lake phosphorus concentration and, in turn, resulted in favourable shifts in the phytoplankton communities and in the chlorophyll-*a* concentrations in the majority of the more productive lakes. A survey of the case studies, for Lakes Garda, Como, Iseo and Varese, where point source inputs of phosphorus into the lakes have been reduced, demonstrated a general improvement in the conditions of all of the lakes. Figures 6.6.3, 6.6.4 and 6.6.5 illustrate the present trophic levels of these lakes and their evolution with respect to the natural background phosphorus levels and the water quality objectives established by the authorities.

6.6.7 CONCLUSIONS

Monitoring the state of lakes is a demanding task and requires, before all else, great patience. Useful results can not be expected until after many years, and maybe even decades. For example, the monitoring of the results for phosphorus contents or for algal biomass and species composition of phytoplankton, obtained from a period of only a few years, cannot show

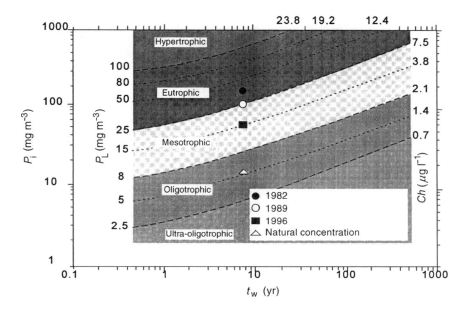

Figure 6.6.4 The OECD diagram for Lake Como which correlates the concentration of incoming total phosphorus (P_i) with both the concentrations of total phosphorus (P_L) and chlorophyll *a* (*Ch*) in the lake as a function of water-residence time (t_w).

whether the eutrophication process is still advancing or if it has actually stopped, because the variations in the concentrations due to the different hydrological and meteorological situations may be significant. The possibility of making false assessments after too short a monitoring time may be even greater. The monitoring programmes must, consequently, have a long duration. For example, in Finland the national monitoring programme has remained essentially the same during the whole of the 1990s. We now follow a procedure according to which the national monitoring programme is approved and adjusted at intervals of every three years.

However, the monitoring of data is already the most crucial part in the planning of water protection measures for lakes, and especially in the assessment of the effectiveness of water protection policy. If we keep the monitoring programmes as constant as possible (sampling sites, sampling times, sampling depths, measurements and methodology, etc.) from one year to another, and we use appropriate statistical tools for interpreting the data obtained, the gathered data themselves will improve every year and thus become 'real' information. As a consequence, the usability of the data for different purposes will increase (e.g. in trend calculations, in model defining,

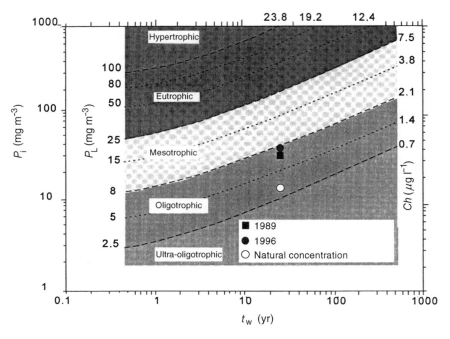

Figure 6.6.5   The OECD diagram for Lake Garda which correlates the concentration of incoming total phosphorus (P_i) with both the concentrations of total phosphorus (P_L) and chlorophyll *a* (*Ch*) in the lake as a function of water-residence time (t_w).

etc.) and the complex decisions that need to be made will thus be based on scientifically sound and validated data and information.

REFERENCES

Chiaudani, G. and Premazzi, G., 1990. Il Lago di Garda – Evoluzione Trofica e Condizioni Attuali. EUR 12925 IT, European Commission, 196 p.

Chiaudani, G. and Premazzi, G., 1993. Il Lago di Como – Condizioni Ambientali Attuali e Modello di Previsione dell'Evoluzione della Qualità delle Acque, EUR 15267 IT, European Commission, 270 p.

Ekholm, P., 1998. Algal-available Phosphorus Originating from Agriculture and Municipalities, Monographs of the Boreal Environment Research, No. 11, Finnish Environment Institute, 60 p.

European Commission, 1997. Proposal for a Council Directive establishing a Framework for a Community Action in the Field of Water Policy, COM(97)49 Def., OJ L137, European Commission.

Kauppi, L., Pietiläinen, O.-P. and Knuuttila, S., 1993. Impacts of agricultural nutrient loading on Finnish watercourses, *Water Sci. Technol.* **28**, 461–471.

Nyroos, H., 1994. Water Quality Assessment in Water Protection Planning. Publications of the Water and Environment Research Institute No. 14, National Board of Waters and the Environment, Helsinki, Finland.

Premazzi G. and Chiaudani, G., 1995. Il Lago di Varese – Condizioni Ambientali e Soluzioni per il Risanamento, EUR16233 IT, European Commission, 116 p.

Premazzi G. and Chiaudani, G., 1998. Il Lago d'Iseo – Condizioni Ambientali e Prospettive di Risanamento, EUR 17720 IT, European Commission, 154 p.

PRRA, 1992. Criteri di Pianificazione in Rapporto alla Gestione delle Risorse Idriche Lombarde, Piano Regionale di Risanamento delle Acque, 190 p.

Rekolainen, S., Pitkänen, H., Bleeker, A. and Felix, S., 1995. Nitrogen and phosphorus fluxes from Finnish agricultural areas to the Baltic Sea, *Nordic Hydrol.*, **26**, 55–72.

Valpasvuo-Jaatinen, P., Rekolainen, S. and Latostenmaa, H., 1997. Finnish agriculture and its sustainability: Environmental impacts, *Ambio*, **26**, 448–455.

Index